Bridging the Seas

**Transformations: Studies in the History of Science and Technology**

Jed Z. Buchwald, general editor

Dolores L. Augustine, *Red Prometheus: Engineering and Dictatorship in East Germany, 1945–1990*

Lawrence Badash, *A Nuclear Winter's Tale: Science and Politics in the 1980s*

Lino Camprubí, *Engineers and the Making of the Francoist Regime*

Mordechai Feingold, editor, *Jesuit Science and the Republic of Letters*

Larrie D. Ferreiro, *Ships and Science: The Birth of Naval Architecture in the Scientific Revolution, 1600–1800*

Gabriel Finkelstein, *Emil du Bois-Reymond: Neuroscience, Self, and Society in Nineteenth-Century Germany*

Kostas Gavroglu and Ana Isabel da Silva Araújo Simões, *Neither Physics nor Chemistry: A History of Quantum Chemistry*

Sander Gliboff, *H.G. Bronn, Ernst Haeckel, and the Origins of German Darwinism: A Study in Translation and Transformation*

Niccolò Guicciardini, *Isaac Newton on Mathematical Certainty and Method*

Kristine Harper, *Weather by the Numbers: The Genesis of Modern Meteorology*

Sungook Hong, *Wireless: From Marconi's Black-Box to the Audion*

Jeff Horn, *The Path Not Taken: French Industrialization in the Age of Revolution, 1750–1830*

Alexandra Hui, *The Psychophysical Ear: Musical Experiments, Experimental Sounds, 1840–1910*

Myles W. Jackson, *The Genealogy of a Gene: Patents, HIV/AIDS, and Race*

Myles W. Jackson, *Harmonious Triads: Physicists, Musicians, and Instrument Makers in Nineteenth-Century Germany*

Myles W. Jackson, *Spectrum of Belief: Joseph von Fraunhofer and the Craft of Precision Optics*

Paul R. Josephson, *Lenin's Laureate: Zhores Alferov's Life in Communist Science*

Mi Gyung Kim, *Affinity, That Elusive Dream: A Genealogy of the Chemical Revolution*

Ursula Klein and Wolfgang Lefèvre, *Materials in Eighteenth-Century Science: A Historical Ontology*

John Krige, *American Hegemony and the Postwar Reconstruction of Science in Europe*

John Krige, *Sharing Knowledge, Shaping Europe: U.S. Technological Collaboration and Nonproliferation, 1955–1970*

Janis Langins, *Conserving the Enlightenment: French Military Engineering from Vauban to the Revolution*

Wolfgang Lefèvre, editor, *Picturing Machines 1400–1700*

Staffan Müller-Wille and Hans-Jörg Rheinberger, editors, *Heredity Produced: At the Crossroads of Biology, Politics, and Culture, 1500–1870*

Staffan Müller-Wille and Christina Brandt, editors, *Heredity Explored: Between Public Domain and Experimental Science, 1850–1930*

William R. Newman and Anthony Grafton, editors, *Secrets of Nature: Astrology and Alchemy in Early Modern Europe*

Naomi Oreskes and John Krige, editors, *Science and Technology in the Global Cold War*

Gianna Pomata and Nancy G. Siraisi, editors, *Historia: Empiricism and Erudition in Early Modern Europe*

Alan J. Rocke, *Nationalizing Science: Adolphe Wurtz and the Battle for French Chemistry*

George Saliba, *Islamic Science and the Making of the European Renaissance*

Suman Seth, *Crafting the Quantum: Arnold Sommerfeld and the Practice of Theory, 1890–1926.*

William Thomas, *Rational Action: The Sciences of Policy in Britain and America, 1940–1960*

Leslie Tomory, *Progressive Enlightenment: The Origins of the Gaslight Industry 1780–1820*

Nicolás Wey Gómez, *The Tropics of Empire: Why Columbus Sailed South to the Indies*

Roland Wittje, *The Age of Electroacoustics: Transforming Science and Sound, 1863–1939*

# Bridging the Seas

The Rise of Naval Architecture in the Industrial Age, 1800–2000

Larrie D. Ferreiro

The MIT Press
Cambridge, Massachusetts
London, England

© 2020 Massachusetts Institute of Technology

All rights reserved. No part of this book may be reproduced in any form by any electronic or mechanical means (including photocopying, recording, or information storage and retrieval) without permission in writing from the publisher.

This book was set in Stone Serif and Stone Sans by Westchester Publishing Services. Printed and bound in the United States of America.

Library of Congress Cataloging-in-Publication Data

Names: Ferreiro, Larrie D., author.
Title: Bridging the seas : the rise of naval architecture in the industrial age, 1800–2000 / Larrie D. Ferreiro.
Other titles: Rise of naval architecture in the industrial age, 1800–2000
Description: Cambridge, MA : MIT Press, [2019] | Series: Transformations: studies in the history of science and technology | Includes bibliographical references and index.
Identifiers: LCCN 2019015423 | ISBN 9780262538077 (pbk. : alk. paper)
Subjects: LCSH: Naval architecture--History--19th century. | Naval architecture--History--20th century.
Classification: LCC VM19 .F47 2019 | DDC 623.8/109034--dc23
LC record available at https://lccn.loc.gov/2019015423

10 9 8 7 6 5 4 3 2 1

# Contents

Preface   ix
Terms, Symbols, Units of Measure, and Money   xiii
Acknowledgments   xvii

**Prologue: A Bridge Too Far—Brunel and *Great Eastern*   1**

1   Improving Naval Architecture   23

2   Steam, Iron, and Steel   57

3   The Quest for Accuracy   95

4   The Demand for Standardization   147

5   The Need for Professionalization   191

6   Laboratory Life   233

7   The Ghost in the Machine   277

**Epilogue: From Metacenter to Metasystem   303**

Notes   311
Selected Bibliography   345
Index   381

# Preface

**Rationale for This Work**

As a boy I was transfixed by the television series *The Undersea World of Jacques Cousteau*. Not the sharks or the whales or the dolphins, but by his ship *Calypso* and its diving saucer (*soucoupe*) *Denise*. I drew them incessantly, then other ships and other submarines, even when I was supposed to be learning in class. When I was about 12 years old I picked up a copy of the book *Your Future in Naval Architecture* by Harry Benford. I hadn't known that drawing ships could actually be a profession, and I was hooked. Harry would become one of my professors at the University of Michigan, and I was lucky to meet and work with many of the people profiled in his book (and, by great good fortune, I met Jacques Cousteau himself). I subsequently led a career in naval architecture that spanned almost 40 years, working and studying at military and civilian institutions in three nations, which allowed me to know professionals in the field from all over the world. Those experiences and their stories have deeply enriched this book.[1]

This book charts the rise of naval architecture in the industrial age, from 1800 to 2000. It follows on directly from my previous book *Ships and Science: The Birth of Naval Architecture in the Scientific Revolution, 1600–1800*. In that work I explain how I came to this project of writing the history of naval architecture, detail the prior studies of the profession, and explain its evolution during the period of the age of sail that directly overlapped the Scientific Revolution. I will, in this preface, repeat only the highlights and expand on a few topics not covered in the previous book. The term "naval architecture" arose in the early 1600s, originally meaning an "architecture of the sea," similar to terrestrial architecture (a view that is once again gaining some traction, thanks to more recent work on maritime aesthetics[2]). By the 1700s it described the application of geometry to ship design, and by the mid-1800s it had become a branch of engineering, applying ship theory to ship design and construction. My working definitions for that book and for this one are as follows:

*Ship design and construction:* The technology of creating the ship, from conception to fabrication.

*Ship theory:* The science explaining the physical behavior of a ship, through the use of fundamental mathematics or empirically derived data.

*Naval architecture:* The branch of engineering concerned with the application of ship theory within the design and construction process, with the purpose of predicting the characteristics and performance of the ship before it is built.

The thesis of *Ships and Science* is that naval architecture was developed and used by navies, starting in the 1600s, in response to a bureaucratic need by naval administrations for greater control over their constructors and for standardization of the ship design process. The systematic use of ship theory made sense only within the bureaucratic organization established for naval construction, which came to include a strong central control of design and a system of professional formation that enabled constructors to learn and carry out the calculations. The thesis of this book, *Bridging the Seas*, is much the same: naval architecture was adopted and practiced by commercial shipyards, starting in the 1800s, in response to a bureaucratic need by shipowners who demanded greater control over the ships they were buying, including tighter scheduling, improved performance, and the safe delivery of cargoes. This in turn required shipbuilders to develop more sophisticated and cost-effective means of controlling the design and construction process, notably vertical integration and the development well-defined standards and design practices. The systematic use of ship theory made sense only within this strong centralized control framework, which in turn demanded a system of professional formation that enabled naval architects to learn and carry out the calculations.

This story takes place in the industrial age, which almost exactly coincides with the nineteenth and twentieth centuries. This period was marked by widespread use of steam for power and iron for construction, not just at sea but also on land. These radically changed the global economy and connectedness of populations, creating an explosive growth in the flow of people and goods. Much of the story told in this book focuses on Britain, for one reason: it was the world's economic, military, naval, shipbuilding, and shipping powerhouse for the majority of this period, roughly 1800 until 1920. The shipping, shipbuilding, and economic figures bear this out. Britain's access to global capital and to important natural resources like coal and iron, both at home and in its expanding empire, gave it enormous leverage in shipping and shipbuilding over competitor nations. Even after the repeal of the Navigation Acts in 1849 (cabotage laws that prohibited non-British hulls to be British registered), shipping companies still flocked to British shipbuilders in lieu of cheaper competition overseas.

By 1890, British shipyards were building 75 percent of the world's commercial tunnage. They were also outproducing every other nation's military; fully one-third of the world's naval fleet was British, and its naval expenditures were larger than the next three navies, France, Russia, and the United States, combined. But even before World War I, competitor nations like the United States were already wrestling for power. The postwar period from the 1920s saw the decisive shift in naval power to the United States, which for a short time also outproduced every other nation in terms of merchant shipbuilding. The end of World War II ushered in the historically unusual period we are in today, in which the dominant naval power (the United States) is not the same as the dominant merchant maritime powers (Japan, South Korea, and China). This story ends with the dawn of the information age, marked by the widespread availability of computers and the advent of a global network, which are even now radically changing the global economy and connectedness of populations, creating an explosive growth in the flow of knowledge and ideas.[3]

**Structure of This Work**

The prologue sets the stage for this book, beginning with the launch of the iron steamship *Great Eastern*. Its creator, Isambard Kingdom Brunel, is the central figure, for he drew into this project the very engineers—John Scott Russell, William Froude, William Fairbairn—who would lay the foundations for modern naval architecture.

Chapter 1 explains that iron steamships like *Great Eastern* coincided with the last gasp of wooden sailing ships, which appeared to be fighting back against the new technologies with improvements to structures and hull shape that led, for example, to clipper ships. In fact, these improvements arose even before iron and steam came of age and were applied equally to the new iron steamships.

Chapter 2 describes the technologies that gave shipowners the attribute they most desired—*predictability*. Steam allowed independence from wind and tide; iron and steel reliably replaced dwindling timber supplies. Predictable ocean transportation meant higher profits for shipowners but played havoc with wood shipbuilders, most of whom went out of business while new iron shipyards sprang up on greenfield sites or alongside locomotive works.

The introduction of steam, iron, and steel meant that the knowledge about wooden sailing ships accumulated over centuries was now almost useless and that new rules for designing and building ships had to be developed and verified. Chapter 3 explores these theoretical developments in naval architecture—ship motions, speed and power, propellers, maneuvering, and structural design—that gave engineers and scientists an accurate understanding of their behavior and performance.

Predictability was the motivation for adopting the technologies of steam, iron, and steel in shipbuilding; theory and experimentation were the mechanisms to achieve accuracy in predicting their performance; and standardization was the reason that practicing constructors and shipbuilders integrated naval architecture theory and experimentation into new, vertically integrated shipyards and industry-wide rules and regulations. Chapter 4 describes how shipbuilders integrated naval architecture theory and experimentation into ship design and construction. Standardization established minimum levels of safety and quality, allowed economies of scale, and reduced risk for shipowner and shipbuilder alike.

The creation of industry-wide rules and vertically integrated shipyards required scientifically trained engineers and naval architects to administer the scale and complexity of these new enterprises and to meet increasingly stringent demands of the customer. Chapter 5 shows that the professionalization of naval architects during the nineteenth and twentieth centuries followed the same pattern as that of other engineers: first, creating specialized careers within commercial shipyards; next, the establishment of formal training and education and the formation of professional associations; and finally, the modest but important role of women in the profession during that period.

As both ships and shipbuilding grew more complex, so too did research into the various aspects of naval architecture. Chapter 6 shows how a consortium of navies, other government-supported institutions, and academia began developing and maintaining the laboratory infrastructure needed to carry out fundamental research, including testing tanks, maneuvering basins, and structural laboratories. Chapter 7 describes the tools that naval architects traditionally used in their profession, from drawing curves to mechanical integrators, and how the rise of computers replaced these machines with software. The epilogue forecasts how the transition of naval architecture to the information age will demand a shift from the well-worn design spiral to systems engineering and systems-of-systems engineering methods. Of necessity, I have provided only a cursory overview of modern computer-based techniques and processes, such as numerical hull design and probabilistic analyses for reliability and stability; these are still very much works in progress, and moreover, they are inextricably enmeshed with rapid developments in other fields like aerospace engineering. It is simply too early in their evolution to examine them adequately, so we must wait for future historians to take a clear-eyed view of these advances.

Which brings me to the statement I made in *Ships and Science*, which I stand behind for this work: I fully expect others to carry on researching the history of naval architecture and even overturn some of what I say here. This work will have succeeded if it becomes the standard reference in 5 years' time. It will have failed if, a generation from now, it continues to be the standard reference 35 years hence.

# Terms, Symbols, Units of Measure, and Money

I generally avoid using archaic technical terms and symbols, although I define them where used, and provide modern symbols from the 1988 edition of *Principles of Naval Architecture*.[1] These correspond to the widely accepted nomenclature adopted by the International Towing Tank Conference (ITTC). I have also converted most measures of length, weight, and so on, into modern SI (metric) values, with the exception of sea distances in nautical miles and speed in knots, which is standard maritime practice today. Below are some specific definitions and conversions that I use throughout this work.

| | |
|---|---|
| Length ($L$): | Distance of ship from bow to stern |
| Beam ($B$): | Width of ship |
| Depth ($D$): | Height of ship from keel to open deck |
| Draft ($T$): | Immersion from waterline to bottom of keel |
| Freeboard ($f$): | Clearance from waterline to open deck |
| Deadrise: | Angle of the ship's bottom measured transversally from the keel |
| Displacement ($\Delta$): | Weight of ship—actually, the weight of its displaced water |
| $\overline{GM}$ Metacentric height | Measure of initial stability |
| $\overline{GZ}$ Righting arm | Measure of ability to recover from heeling moment |
| (First) moment of inertia: | A body's resistance to angular acceleration |
| Second moment of inertia: | A body's resistance to bending |
| Neutral axis: | The location in a body where no tension or compression occurs in bending |
| Stress and strain: | Respectively, for a material, force per unit area and displacement per unit area |
| Ships' plans: | Waterline (horizontal cuts fore and aft), profile or side (buttock) lines (horizontal cuts vertically), body section lines (vertical cuts transversally) |

IHP: Indicated horsepower (theoretical engine power)
BHP: Brake horsepower (measured engine power)
SHP: Shaft horsepower (actual power delivered to the propeller)
kW: Kilowatts

Tests are conducted at model scale. Trials are conducted at full scale.

"Model basin," "testing tank," and "towing tank" are used interchangeably to indicate ship model testing facilities.

Many navies, not just Britain's, were royal, so I refer to them by nation.

In *Ships and Science*, I differentiated eighteenth-century shipyards from dockyards by the latter having greater industrial capacity. By the nineteenth century that difference had largely disappeared. Moreover, the term "dockyards" came to denote naval shipbuilding and repair complexes, while "shipyards" generally denoted commercial facilities, a usage still current. This is how I use the terms in this work.

"Naval architect" and "constructor" are sometimes used interchangeably, though the latter term generally connotes a person who designs military ships. I explain the evolution of the term "naval architect" in chapter 5.

Almost all the people in this book are male, so I sadly but unflinchingly use the terms "he," "him," and "men" throughout. I discuss women in the profession in chapter 5.

I use Britain, Great Britain, and the United Kingdom interchangeably. I use Russia and Russian throughout for consistency, even during the Soviet regime (1922–1991).

Motions are described in the table below.

The term "ton" can describe both weight and volume, so to avoid confusion, I use *tun* and *tunnage* for volumetric measures (admeasurement) and *ton (tonne)* and *tonnage* for weights and displacements. I discuss this more thoroughly in *Ships and Science*.

I convert the currency in 1800–2000 to U.S. dollars in 2002 (to keep in line with *Ships and Science*). I have used two broad comparators depending upon what is being

|  | Static motion | Dynamic motion |
| --- | --- | --- |
| **Vertical** | Sinkage | Heave |
| **Side to side** |  | Sway |
| **Side-to-side rotation** | Heel | Roll |
| **Fore-aft vertical rotation** | Trim | Pitch |
| **Fore-aft horizontal rotation** |  | Yaw |

Terms, Symbols, Units of Measure, and Money

measured. When I am discussing the price of individual goods, outlays, and salaries, I use the real price comparator that measures consumer goods and services and is based on the consumer price index (CPI). When I am discussing large national outlays like government budgets and shipbuilding projects, I use the economy cost comparator that is based upon the national gross domestic product (GDP) deflator. These give very different modern values, as GDP growth has been much faster than CPI growth over the last 200 years.[2] At 25-year intervals, these conversions are as follows:

| Year | Based on CPI (e.g., salaries) UK (£) | Based on CPI (e.g., salaries) US ($) | Based on GDP (big projects) UK (£) | Based on GDP (big projects) US ($) |
| --- | --- | --- | --- | --- |
| 1800 | $ 80 | $15 | $4,800 | $23,000 |
| 1825 | $ 80 | $19 | $3,500 | $13,400 |
| 1850 | $100 | $18 | $3,200 | $ 4,300 |
| 1875 | $ 90 | $16 | $1,400 | $ 1,300 |
| 1900 | $100 | $19 | $ 940 | $ 530 |
| 1925 | $ 55 | $ 9 | $ 400 | $ 120 |
| 1950 | $ 33 | $ 6 | $ 140 | $ 40 |
| 1975 | $ 8 | $ 3 | $ 16 | $ 7 |
| 2000 | $ 2 | $ 1 | $ 2 | $ 1 |

# Acknowledgments

First and foremost, from my wife Mirna and me to our sons Marcel and Gabriel, who are our true legacies.

I especially acknowledge the late David K. Brown (1928–2008), Philip J. Sims (1949–2014), and Francisco Fernández González (1939–2018). They were national treasures in, respectively, Britain, the United States and Spain, who combined hands-on experience in ship design with a deep knowledge of the history of naval architecture. David was always able to sum up a lifetime's experience in a pithy phrase—my favorite, concerning the process of ship design, from early-stage preliminary design to construction and launch, was that "conception is more fun than birth." Phil was always able to connect ship design with the wider world and lucidly explain how it fit within social and economic history. Francisco (Paco) was the epitome of a courtly Renaissance man who brought a thoroughly modern analysis to understanding how the great sailing ships of yesteryear were designed and built. I counted all as friends and mentors who firmly set me on the path to writing the history of our profession of naval architecture.

Although many people assisted me in this work, I list here (alphabetically and by nation) those whose help was absolutely indispensable to its completion. Some have already passed, and I make no distinction here:

Britain: David J. Andrews, Trevor Blakeley, Louise Bloomfield, Peter Froude, Robert Gardiner, Barbara Jones, Andrew Lambert, Stuart A. McKenna, Michael J. Pryce, Louis Rydill, Jo Stanley, Fred Walker

France: Régis Beaugrand, Alain Bovis, Alexandre Sheldon-Duplaix

Finland: Jouni Arjava, Armas Rahola

Germany: Dirk Böndel, Michael Eckert, Jobst Lessenich, Horst Nowacki

Japan: Takao Inui, Hitoshi Narita

Netherlands: Bart Boon, Jan M. Dirkzwager, Alan Lemmers, Gerbrand Moeyes, Jeroen ter Brugge, Bruno Tideman

Portugal: Sandro Mendonça, Rogério D'Oliveira

South Africa: William (Bill) H. Rice

Spain: José María de Juan-García Aguado, José María Sánchez Carrión

United States of America: Harry Benford, Susan Caccavale, Milka Duno, Peter Gale, Lawrence L. Goldberg, Robert S. Johnson, Carl Kriegeskotte, Thomas Messenger, James L. Mills Jr., Alex Pollara, Sidney Reed, John Rosborough, Bruce Rosenblatt, Theodore H. Sarchin, Andrew B. Summers, David Walden, Robert Wasalaski, Dana Wegner, Arthur Welch, Paul Wlodkowski

All translations are my own, as are any errors in them, as are any errors in facts or analysis in this work.

# Prologue: A Bridge Too Far—Brunel and *Great Eastern*

The launch of *Great Eastern* was an abject failure, but it launched naval architecture into the modern age. The ship was the brainchild of the British civil engineer Isambard Kingdom Brunel, and in that magnificent but ill-fated vessel, he wove together all the threads of Victorian engineering into a single, unified vision of modern naval architecture. Conceived to steam around the globe without ever taking on coal, its structure was carefully designed, following the lessons of the most advanced engineering of the day, to resist the worst actions of waves, weather, and grounding. The shape of its hull was based upon an equally advanced wave-line theory, which promised to give the least resistance for its state-of-the-art steam engines that drove both paddle wheels and a newfangled screw propeller. Even its stability and motions had been analyzed to ensure the enormous vessel would first enter the water safely and ride securely across the wide oceans. *Great Eastern* was, in sum, the first ship ever designed that synthesized all the elements of theoretical naval architecture—strength, stability, hydrodynamics, and seakeeping—into an integrated whole.

Brunel himself stood at the center of Victorian engineering, in an age when Britain was the foremost industrial power in the world. British engineers held a global monopoly on expertise in design and construction of steamships, railways, bridges, and tunnels, and Brunel was the acknowledged master of them all. In each of these fields he brought his vast, interdisciplinary knowledge to bear, utterly transforming every one of them. Along the way he brought into his circle all the great engineers of his day, who were themselves transformed by Brunel's vision and accomplishments.

Brunel and his *Great Eastern* marked the turning point in the development of modern naval architecture. Brunel brought together, in that one ship, the concepts that would revolutionize the practice of ship design and construction and the very engineers who would carry on that revolution. Brunel himself, however, would not live to see this revolution in the shipbuilding industry come to fruition.

## Isambard Kingdom Brunel and the Engineers of the Victorian Era

The Industrial Revolution marked the transition of national economies away from agriculture and crafts and toward machine-based manufacture. It began in Britain in the mid-eighteenth century, and from the beginning it was dominated by British inventors, scientists, and engineers. Iron slowly replaced wood as the material of choice for building major structures, gaining notoriety with Abraham Darby's construction of the Iron Bridge in Shropshire in 1778. The steam engine, which had seen limited utility in pumping out mines, became a widely used engine for manufacturing and motive power after 1780, with the innovations of James Watt and Matthew Boulton (notably the use of a steam jacket and separate condenser) greatly improving its efficiency.

The use of iron and steam revolutionized the way industry operated, and from the bottom up it completely changed Britain's physical and social landscape. When Brunel was a boy in the early 1800s, British people were generally living in towns and villages and tilling the pastoral countryside, which was dotted with small mills alongside streams that furnished waterpower for their simple machines. When Brunel died in 1859, much of the population was living in urban squalor, employed in the large coal-fired, steam-driven factories that belched smoke and soot across vastly enlarged cities. But it was trade, more than coal, that powered these factories. Raw materials like cotton and iron ore were imported from abroad and were transformed into the textiles, rails, and locomotives that were exported to colonies and allies. In the space of two generations, the Industrial Revolution transformed Great Britain from a nation of shopkeepers into the workshop of the world.[1]

The Industrial Revolution was just getting underway in Britain when, across the English Channel, the French Revolution broke out in 1789. That Revolution overturned the French social order from the top down, replacing a centuries-old monarchy with a series of increasingly radical and ruthless governments that destroyed the ancien régime, closed the influential French Academy of Sciences, and laid waste to France's intellectual elite, causing many of them to flee the country. One such émigré was Marc Brunel, the son of a Normandy farmer and who showed particular aptitude for mechanics while serving as a naval cadet. His skill did not extend to politics, for his very public sympathy toward the monarchy forced him to flee France for the United States, where he was employed for several years as an engineer and architect. In 1798 he moved to Britain, where he worked with the machine toolmaker Henry Maudslay and the naval inspector general Samuel Bentham to develop innovative steam-powered machinery that would mass produce ships' pulley blocks for the British navy, which were soon used in battles against the French navy that Marc Brunel had once served in.

While in Britain, Marc Brunel married Sophia Kingdom, and together they had three children, of whom the youngest was Isambard Kingdom Brunel (figure P.1), born in Portsmouth on 9 April 1806. His father taught him basic engineering at an early age, after which he was sent to France for two years to continue his mathematical and scientific education and to serve a short apprenticeship with the clockmaker Abraham-Louis Breguet, in which he honed his drafting and mechanical skills. After his return to Britain in 1822, then age 16, he worked in Maudslay's shops and as an engineer for his father. By his late teens he was supervising workers on the construction of the Thames Tunnel, his father's most important project, although he was almost killed when the unfinished tunnel flooded in 1828. Within a few years the young Brunel, only in his

**Figure P.1**
Isambard Kingdom Brunel standing before the launching chains of the *Great Eastern*.
Credit: Metropolitan Museum, Gilman Collection, Purchase, Harriette and Noel Levine Gift, 2005.

20s, struck out on his own and won major contracts, such as the Clifton Suspension Bridge in Bristol, and became chief engineer for important projects, most notably the Great Western Railway that ran between London and Bristol. By age 30 he was responsible for over £5 million worth of development (about $1.5 billion in today's money) and had established his offices and new family in one of the most exclusive parts of London, near the government offices of Whitehall and the Parliament.[2]

Brunel came into his own just as Queen Victoria came to the throne in 1837, ushering in what was soon dubbed the Victorian era. It was a period of relative peace in Europe and its colonies, referred to as Pax Britannica for its roots in Britain's domination of maritime trade routes and unchallenged naval supremacy around the globe. It was also marked by unparalleled prosperity at both the national and the family level; personal income, birth rates, and lifespan all increased, and investments in infrastructure projects such as canals, railways, and shipping lines returned enormous profits to government and industry coffers.

The engineers of the Victorian era, responsible in large measure for creating this prosperity, were the acknowledged heroes of the age. The most celebrated engineers built the railway networks that spanned Britain, fanned out to continental Europe, and spread throughout the British Empire. George Stephenson, working with his son Robert Stephenson, developed the first practical railway lines in the 1820s to carry coal and passengers. Within a decade, rail lines had sprung up around Britain. Robert Stephenson and Brunel were the most closely associated with this railway expansion and soon became household names; when in 1843 Brunel suffered an accident that threatened his life, the *Times* regularly reported on his recovery.[3]

Victorian engineers were at once fierce business rivals and close personal friends. This was nowhere more evident than in the relationship between Brunel and Robert Stephenson. As the two men were working on expanding Britain's rail network, they became the central figures in the notorious gauge wars that pitted their rival industrial consortia against each other. In Britain's northern and central regions, railways followed the standard-gauge width of track (4 feet, 81/2 inches, or 1,435 millimeters), which had been pioneered by the Stephenson family. Brunel, convinced that a wider track would allow greater stability and comfort, settled on a broad gauge (7 feet, 1/4 inch, or 2,140 millimeters) for his Great Western Railway. Robert Stephenson was opposed to the differing gauges, on the grounds that it prevented interoperability across all railways. Brunel argued that the greater efficiency of his broad gauge would more than make up for any inconvenience. To show this, in 1838 he had his friend Charles Babbage (best known for developing the difference and analytic engines that are the direct ancestors to the modern computer) to run a series of carefully instrumented trials on both types of train, which demonstrated that the broad-gauge trains were more efficient and had far fewer vibrations than the standard-gauge ones.[4] Yet despite this business rivalry, Brunel and Stephenson struck up a lifelong friendship and often sought each other's help with difficult technical problems. For example, when Brunel ran into trouble with his first locomotives, he turned to Stephenson to provide him with two high-performance locomotives, which Brunel had converted to broad gauge, thus ensuring the railway's early profitability.

The expansion of railways meant spanning rivers and waterways. Bridge engineering quickly evolved as both the strengths and the limitations of structural iron were discovered through experimental trial and error. One of the most challenging projects began in 1845 with the construction of a 500-meter-long Britannia railroad bridge across the Menai Strait in the north of Wales. Robert Stephenson took on this task with the help of his friend William Fairbairn (figure P.2), a well-established civil engineer and co-owner (with David Napier) of the Thames shipbuilding firm Millwall Iron Works. Stephenson and Fairbairn decided against spanning the Menai with either an arched or suspension bridge (both of which supported loads with additional structure at the top or bottom), instead opting to build a long, wrought iron tubular bridge that supported its load with no additional structure. Since no one had ever built a tubular bridge of this size, Fairbairn, aided by his colleague Eaton Hodgkinson, used his shipyard facilities to carry out a series of scale-model experiments from 1845 to 1846 on tubular beams of various configurations, based upon structural beam theory that had recently been popularized by mathematicians such as Henry Moseley. After determining that the primary constraint for tubular bridges was buckling (bending and folding in compression), Stephenson and Fairbairn came up with a design having a rectangular cross section with smaller rectangular cells at the top and bottom to resist buckling (see figure P.5). Brunel was on hand to assist Stephenson in 1849 when the wrought iron tubes were raised into position.[5]

As the railways spread across the nation, engineers had to deal with the speed-killing hills and valleys that dotted the landscape. On the South Devon Railway, which ran along the undulating south coast of Britain, Brunel used the novel atmospheric railway system. Large pneumatic pipes, connected at intervals to steam-powered pumping stations, ran between the tracks in the hilly sections. Internal pistons connected to the trains through a leather-covered slot in the pipe; as air was pumped out from the front of the pistons, the trains were dragged up the hills. The leather seals quickly degraded, however, causing loss of pressure and efficiency. In 1847 Brunel asked one of his brightest young engineers, William Froude, to study the problems associated with friction and loss of power. Although Froude improved the design, the atmospheric railway proved untenable, and it was soon abandoned.[6] At the same time, the British

**Figure P.2**
William Fairbairn.
Credit: Creative Commons.

Parliament required that all new railroads be built to Stephenson's standard gauge, effectively ending the broad gauge, Brunel's other great railroad experiment.

**From Railways to Steamships**

Throughout his career, Brunel drove himself, and his fellow workers, at a relentless pace. He was up before dawn and worked well after midnight, rarely sleeping more than four or five hours a night, stoking his nervous energy with the cigars that were perpetually in his mouth. He was loath to delegate work, often carrying out the survey, design, and calculations for a project by himself. Even when reluctantly delegating work to subordinates and contractors, he carefully checked it and browbeat them mercilessly when they failed to meet his exacting standards. When one project became overwhelming, he compensated by taking on additional, even more difficult projects.

In October 1835, Brunel was just starting engineering work on the Great Western Railway and had not even settled on the track design, when the railroad's board of directors met to review the plans. One of the directors complained about the length of the proposed line, which stretched more than 100 miles from London to Bristol. Brunel bristled and countered with the remark, "Why not make it longer, and have a steamboat to go from Bristol to New York, and call it the *Great Western*?"[7] Most of the directors took it as a joke. Sailing packets had been making regular, monthly crossings between New York and Liverpool since 1818, but these vessels were still subject to the vagaries of the wind. The idea of steaming across the Atlantic had been in the air for years; two sailing ships fitted with small steam-powered paddle wheels had previously made the journey—the American *Savannah* in 1819, and the Canadian *Royal William* in 1833—but no ship had made the journey entirely under steam power alone.[8]

One of the directors present did not take Brunel's remark lightly. Thomas R. Guppy was an engineer who had turned his talents to running a sugar refinery and was keen to see his home port of Bristol supplant Liverpool as the shipping capital of Britain. That night he and Brunel—it is not clear who initiated the conversation—discussed the very real possibility of starting a transatlantic steam service from Bristol to New York. They formed a committee with other members of the railway board to explore the idea. The first step was for Guppy and another prominent Bristol figure, Christopher Claxton (a retired naval officer and harbormaster of the port) to visit shipbuilders around Britain to determine the feasibility of building such a ship. Encouraged by their findings, they created the Great Western Steamship Company (GWSS), which raised money to finance construction of a wooden-hulled 1,340 tun paddle wheel ship, powered by a 420 IHP (indicated horsepower) steam engine and auxiliary sails. The vessel, retaining Brunel's appellation *Great Western*, would be designed and built by the Bristol shipbuilder William Patterson, although under the direct supervision of the GWSS.[9]

The port of Liverpool would not lightly give up its predominant position. Within a few months of the GWSS being formed, plans for a transatlantic steamship service from that city were being debated in public. At one lecture given in December 1835 at the Liverpool Mechanics' Institute, one of many technical schools for workingmen, a vociferous critic named Dionysius Lardner heaped scorn on the idea, claiming that "the project … of making the [steam] voyage directly from New York to Liverpool was … perfectly chimerical; and they might as well talk of making a voyage from New York or Liverpool to the moon."[10]

Lardner was a popularizer of engineering and technology, well respected for his many books and articles on steam power and railways. Nevertheless, he frequently got his physics wrong. During one run-in with Brunel over his proposed Great Western Railway, Lardner claimed that a runaway train on a downhill gradient in one of the tunnels would accelerate to almost 200 kilometers per hour, which would suffocate the passengers. Brunel pointed out that Lardner had neglected rail friction and air resistance in his calculations, so that a train would not attain even half that speed. Lardner's objections to the 3,000-mile voyage between New York and Liverpool, the most profitable centers of commerce, were even more erroneous. He wrongly assumed that the resistance of a ship increased in direct proportion to its size, so no matter how much coal was loaded, a steamship could never travel more than 2,000 miles; the best one could hope for was a voyage between southwest Ireland and Newfoundland, which were seen as remote, unprofitable outposts of the British Empire. Brunel correctly saw the flaw in Lardner's logic; resistance increases as the square of the vessel size, but coal capacity increases as the cube. Thus, a sufficiently large vessel could carry enough coal to take it across the ocean, and Brunel made certain this fact was inserted in the report that detailed the future steamship *Great Western*.[11]

Though Brunel was occupied full time with his railway and had never built a ship in his life, he became immediately and intimately involved with the design and construction of *Great Western*. As a major shareholder in the GWSS, he had both a personal and a financial interest in the project. Availing himself of his close ties to the engine-building firm Maudslay, Sons, and Field of London (run by direct descendants of the shops where he apprenticed), Brunel selected the propulsion machinery and oversaw its fabrication. On his weekly visits to Bristol he meddled in the design and construction of the wooden hull, though his naval colleagues Claxton and Patterson would continually remind him of the difference between building a ship and building a bridge.

The ship was launched in July 1837 and traveled under sail to London to have its engines installed. In March 1838, as the ship was preparing for its maiden voyage under steam to Bristol, a fire broke out onboard, almost killing Brunel. While he was laid up recovering, the ship coaled at Bristol for its transatlantic journey, which the owners were certain would be the very first made entirely under steam. Departing April 8, *Great Western* made the voyage in just over two weeks (a sailing packet typically took six weeks), arriving in New York harbor on 23 April 1838. There they found that another British vessel, *Sirius*, had anchored just a few hours earlier after completing its own voyage from Cork, Ireland. That voyage was a one-off stunt, financed by rival shipowner Junius Smith, who was loath to see his first-across-the-Atlantic crown go to someone else. He had hired *Sirius*, a small Channel steamer, and overloaded it with coal to compete with

*Great Western*, although it could not make the transatlantic journey on a regular basis. For the time being, at least, Brunel's steamship had the Atlantic to itself.[12]

Within two years that situation changed, as other companies competed for the transatlantic market. The most aggressive was the passenger service established in 1840 by a Nova Scotia businessman named Samuel Cunard, who operated (with significant government-mail subsidies) four wooden paddle steamers between Liverpool and Halifax.[13] The GWSS had initially considered meeting this competition by building a running mate, a wooden paddle steamer similar in size to the eminently successful *Great Western*. Brunel, always intent on advancing the state of the art, quickly pushed for an iron-hulled ship after a chance visit to Bristol by the iron Channel steamer *Rainbow*. Once again, he enlisted his colleagues Guppy, Claxton, and Patterson to make extensive trips examining the merits of iron versus wood, and under Brunel's influence they soon reported their preference for building in iron.

The iron ship that rose in Patterson's new building dock was originally nicknamed *Mammoth*, for at 3,400 tuns (and 3,675 tonnes displacement) it was three times the size of *Great Western* and almost six times larger than any iron vessel then afloat. The shaft of the paddle wheel had to be so large that it could not be forged with existing tools, which led James Nasmyth, one of the principal machinery contractors, to invent the powerful steam hammer that would go on to revolutionize large-scale metalworking. This steam hammer became moot in May 1840, when Brunel noted the arrival in Bristol of a screw-propelled ship named *Archimedes*. The vessel, named for the purported Greek inventor of the screw, had been constructed to demonstrate the effectiveness of a screw propeller built by a farmer and inventor named Francis Pettit Smith, one of several competing propeller types being developed at that time. Brunel had been looking for ways to improve the paddle wheel, which was generally seen as cumbersome and inefficient. The screw propeller fit perfectly with Brunel's predilection to push the state of the art, so once again he changed direction and ordered a halt to the construction of the half-built *Mammoth* so that he could hire *Archimedes* to test the newfangled propeller.

For three months Brunel tested eight different configurations of the screw propeller on *Archimedes*, which demonstrated its superiority over paddle wheels. By December 1840 the GWSS voted to fit the ship with a propeller. The original engines were scrapped, new machinery built, and the hull refitted to accommodate the screw, almost doubling the cost over the initial estimates. The ship was finally floated out of its building dock in July 1843 under the name *Great Britain*. Brunel, already noting the similarities between his bridges and the ship's hull girder, ensured that it was fitted with heavy longitudinal stiffeners for strength and divided into five watertight compartments in case of damage. It was the most seaworthy ship then afloat.[14]

After a further two years of fitting out and benefiting from additional propeller trials, *Great Britain* entered service in July 1845, making several passages between Liverpool and New York. The following year its captain lost his way in heavy seas and grounded the ship on a beach in northeast Ireland. The enormous strength that Brunel had designed into *Great Britain* now proved its worth, for the ship was refloated intact after a year of being pounded by the surf; according to the surveyors, a lesser ship would have been utterly destroyed. The state of the GWSS at that time (1847) was far more fragile. Having recently sold *Great Western* to finance the salvage operations of *Great Britain*, it had no source of revenue and could not afford to repair its remaining iron ship. The company folded in 1851 after selling *Great Britain* for a tenth of its original cost, after which the ship sailed between Britain and Australia for another 26 years. The vessel has survived to the present day; after an extensive restoration it is now a museum ship in Bristol, displayed in the very building dock where it was born.

**The Ship as a Bridge: The Design and Construction of *Great Eastern***

By the time *Great Britain* had been sold, Brunel was already thinking about his next ship project. The nature of that project had been foretold by *Great Britain*'s new life in the Britain–Australia trade. In 1851 the discovery of large deposits of gold near Melbourne sparked an emigration boom that rivaled the recent California gold rush—within the space of three years, over 100,000 people had left Britain to settle in Australia. To meet this demand for passenger traffic, the Australian Royal Mail Steam Navigation company (whose ships carried both mail and passengers) engaged Brunel to advise them on the purchase of two new iron steamships. He examined the issue of worldwide steaming and concluded that coal supplies were the limiting factor in long-distance trade. Each time a ship had to make port to recoal—with their inefficient engines, early steamships might require up to five coaling stops on the around-Africa voyage—the voyage lost time and money. Furthermore, the company had to pay three to four times the domestic price per ton to deliver coal to those distant ports.[15] Brunel's initial recommendation for a pair of 5,000–6,000-tun ships, each twice as large as any ship then plying the oceans, was predicated on the ships carrying enough coal to require only one refueling stop each way, at the Cape of Good Hope. The company directors rejected this idea, instead settling on a more conventional pair of 2,000-tun steamers. Brunel unhesitatingly recommended John Scott Russell, a well-respected and scientifically minded shipbuilder, to bid for the ships, which were subsequently built under the names *Adelaide* and *Victoria* in Russell's Millwall shipyard on the Thames.

John Scott Russell (figure P.3), two years younger than Brunel, was born near Glasgow on 9 May 1808. His mother, Agnes Scott, died soon after his birth. His father, David Russell, an itinerant minister, moved around Scotland with his new wife and growing family. At the young (but not unusual) age of 13, Russell began attending Glasgow University, taking his mother's maiden name as his own middle name. His classical studies did not include engineering, for the university did not add such a chair until 1840. Graduating in 1825 at age 17, he taught mathematics and science around Edinburgh, while independently learning the mechanical trades to build and test steam carriages. This combination of a theoretical education and solid practical training was

**Figure P.3**
John Scott Russell.
Credit: Royal Institution of Naval Architects.

almost unmatched in Britain, save by Brunel himself. It was this shared background that would later draw the two men together and ultimately drive them apart.[16]

During the 1820s and 1830s, steam was also powering canal barges, and engineers like William Fairbairn were trying novel concepts to gain a competitive advantage. One of the canal owners, impressed with Russell's careful work on steam carriages, commissioned him to carry out experiments aimed at verifying some of the claims that, under certain conditions, canal boats could be propelled along a surface wave with surprisingly little resistance (which could translate to smaller engines). The discovery came in 1829 when, as Russell later described it, "a spirited horse in the boat of William Houston [a canal owner] took fright and ran off, dragging the boat with it, and it was then observed ... that the vessel was carried on through water comparatively smooth, with a resistance greatly diminished." Russell's research into this phenomenon from 1834–1835 involved towing canal boats connected by pulleys to a heavy weight dropped from a tall trestle and measuring the resistance with a spring dynamometer. During these trials, Russell observed the phenomenon of solitary waves (solitons) moving long distances through canals with no decrease in speed, and he suspected that wavemaking resistance was the key to Houston's discovery. Russell confirmed that boats accelerating in shallow, confined waters have increased wavemaking resistance until reaching what is today called critical velocity (in that case, about eight knots), above which the boat climbs over its own bow wave with correspondingly diminished resistance.[17]

In 1835, Russell used his initial observations to develop a ship form of least resistance. Russell believed that conventional hulls pushed a bow wave of water in front of them, thus retarding their speed. His solution was to shape the bow in a way that avoided this buildup of water. Using a complex mathematical formulation based on the geometry of waves, he determined that the bow waterlines should be hollow (concave) to form a set of curves like the waves it generates. These lines, he claimed, would uniformly move the water to the sides of the vessel without creating a bow wave. He initially tested the idea with a boat named *Wave*, built with parabolic waterlines. This led to his so-called wave-line theory of hull design, described in chapter 1.

Russell's mathematical pursuits did not land him his hoped-for academic position in Edinburgh, so in 1838 he accepted a managerial position at the Caird shipyard in Greenock on the Clyde. Over several years he gained practical engineering experience overseeing the construction of numerous iron-and-steam vessels while continuing his scientific pursuits of the wave-line theory. For Russell, this was merely a stepping-stone to greater achievements; in 1844 he moved his growing family to London to edit a railway newspaper and joined the Institution of Civil Engineers, where he frequently met with luminaries such as Stephenson, Fairbairn, and Brunel. Unlike most of his fellow engineers, he also became a member of the scientifically oriented Royal Society of London. In 1847 he joined a shipbuilding firm that took over William Fairbairn's old shipyard in Millwall on the Thames, where he soon became sole owner. His ships sported his now-famous wave-line hull form, including the yacht *Titania*, which unsuccessfully sailed against the U.S. yacht *America* after it had won the Hundred Sovereigns Cup (later known as the America's Cup).

Russell, now a fixture in the worlds of engineering and the sciences, was equally at home in the rarefied worlds of politics and arts (Arthur Sullivan, of Gilbert and Sullivan fame, was a frequent guest at his home). As secretary of the Royal Society of Arts, he rubbed shoulders with Queen Victoria's husband, Prince Albert. In 1849 at the instigation of Prince Albert, Russell began working with Henry Cole, a fellow member of the society and an up-and-coming civil servant, to showcase Britain's engineering might as part of the first-ever world's fair in Hyde Park and South Kensington, London. The 1851 Great Exhibition, as it became known, would house 13,000 exhibits from almost 80 colonies and countries, many in the massive iron-and-glass Crystal Palace. Russell, who knew Brunel from the 1835 debates with Lardner over transatlantic steamships, invited the famous engineer to be president of the Building Committee that oversaw the construction of the Crystal Palace. The exhibition was a resounding success, and Brunel soon returned the favor by ensuring that Russell received the commission to build the steamers *Adelaide* and *Victoria* for the Australia trade.

In March of that same year, while Brunel was working on plans for the new Paddington Station in London, he began sketching an "East India Steamship," scribbling beneath it, "Say 600 ft x 65 ft x 30 ft." With a projected displacement of roughly 20,000 tonnes—six times larger than any ship then afloat or under construction—it was his first inkling of "the Great Ship," as he referred to it, which became known as *Great Eastern* (figure P.4). As with his project for the Australian Royal Mail company, Brunel was deeply concerned by the problems with coaling en route and envisioned a ship that could carry its own coals the entire route, from Britain to Australia and back (coal had not yet been discovered in that distant colony). He carefully calculated the power needed to steam at 14–15 knots, using formulas derived from experiments on ship resistance carried out in 1793 by Mark Beaufoy. From there he derived the required coal capacity and iterated his calculations in a remarkably modern fashion to balance weight and displacement (the two must be equal for a ship to float), using "parent hull" data from existing ships to estimate the hull weights on the basis of ship dimensions.[18]

Within a few weeks Brunel's calculations were sufficiently advanced to show to Russell, who confirmed his initial estimates. Brunel and Russell together submitted the scheme to the newly formed Eastern Steam Navigation Company (ESN), which eagerly

**Figure P.4**
*Great Eastern.*
Credit: Author's collection.

latched on to the concept as a way to outflank its competition. From that point the project moved swiftly. By July, Brunel was appointed engineer to the company, which set about raising shares for the project as the new ship slowly took form in Brunel's and Russell's notebooks.[19]

Through 1852 and 1853 Russell and Brunel refined the plans for the ship, which grew to 680 feet (207 meters) and 27,300 tonnes displacement, to carry 4,000 passengers and 12,000 tonnes of coal. The ship would require about 8,500 IHP (6,400 kilowatts) to steam at 14 knots; this was too great to be carried via a single screw propeller, so they agreed on a combined screw-and-paddle-wheel arrangement with auxiliary sails. Russell used his wave-line theory to design the hull, although he considerably modified the hull lines away from his own theory to provide the necessary buoyancy. The two men also modeled the hull structure on the basis of Fairbairn's Britannia tubular bridge, with cellular structures top and bottom for unprecedented strength (figure P.5).[20]

The ESN went through the motions of inviting bids for construction, but it was a foregone conclusion that they would give Russell the contract. This was undoubtedly helped by his suspiciously low bid of £373,200 ($40 million today) to build the ship and engines; at just £20 per tun, it was about half the cost per tun spent to build

## A Bridge Too Far—Brunel and *Great Eastern*

*Great Britain*. This would not be Russell's biggest problem; it was, instead, the contracts he signed with ESN that gave Brunel "full control and supervision over every part of the work ... with very large powers of interpretation."[21] In the contracts, Brunel was referred to as "the Engineer," Russell as "the Contractor." Although Russell was sole owner of his shipyard and the technical equal of Brunel, he would no longer be the unquestioned master of the project but rather Brunel's employee in all but name, subject to the same remorseless bullying as the lowliest apprentice draftsman.

The ship began to rise from the ways in February 1854, with scheduled completion two years later. Russell had originally intended to construct a dry dock in which to build the enormous ship, but Brunel rejected it on cost grounds. The vessel was far too long to build and launch on a traditional inclined slipway perpendicular to the shore; the bow would have stood 30 meters above the ground, too high to reach with cranes.[22] It could be built and launched only sideways (parallel to the river), so Russell bought part of David Napier's adjacent shipyard to accommodate the ship. While Russell was producing detailed drawings just ahead of each phase of construction, Brunel became an ever-present fixture in the shipyard, cajoling and browbeating Russell with incessant demands and "suggestions" for changes that were really direct orders. In one instance, Brunel brutally chastised Russell for using an "unnecessary" structural element that added 40 tonnes to the weight, a vanishingly minor element of the ship's 27,000 tonnes displacement.[23] Russell gamely replied to each request, usually closing his letters with "Faithfully Yours" or "Your Obedient Servant," in an attempt to placate the increasingly anxious and belligerent "Engineer."

Brunel's anxiety was due in large part to a rise in shipyard wages and costs of iron, which had not been accounted for in the original budget and were now threatening to derail the whole project. Some of the inflation was associated with the need for warships to fight in the Crimean War, which had broken out in 1853; Russell, for example, was constructing an ironclad battery (floating gun platform) and a series of gunboats simultaneously with *Great Eastern*. The demand for iron passenger ships also pushed up prices; for example, Cunard had ordered the massive (5,400 tonne) paddle wheeler RMS *Persia* from the Glasgow shipbuilder Robert Napier, the same year *Great Eastern* began construction. Russell struggled to hold costs under control while keeping his shipyard going, but a fire in May 1855 forced the underinsured Russell to absorb the equivalent of $5 million in losses. Brunel began to suspect (wrongly) that Russell was diverting some of the iron to other purposes and intentionally understating the weights:

> How the devil can you say you satisfied yourself of the weight of the ship when the figures your clerk gave you are 1000T less than I make it or than you made it a few months ago—*for shame*—I wish you *were* my obedient servant, I should begin by a little flogging.[24]

**Figure P.5**
Britannia tubular bridge (this page) and *Great Eastern* midship section (opposite page).
Credit: Author's collection.

# A Bridge Too Far—Brunel and *Great Eastern*

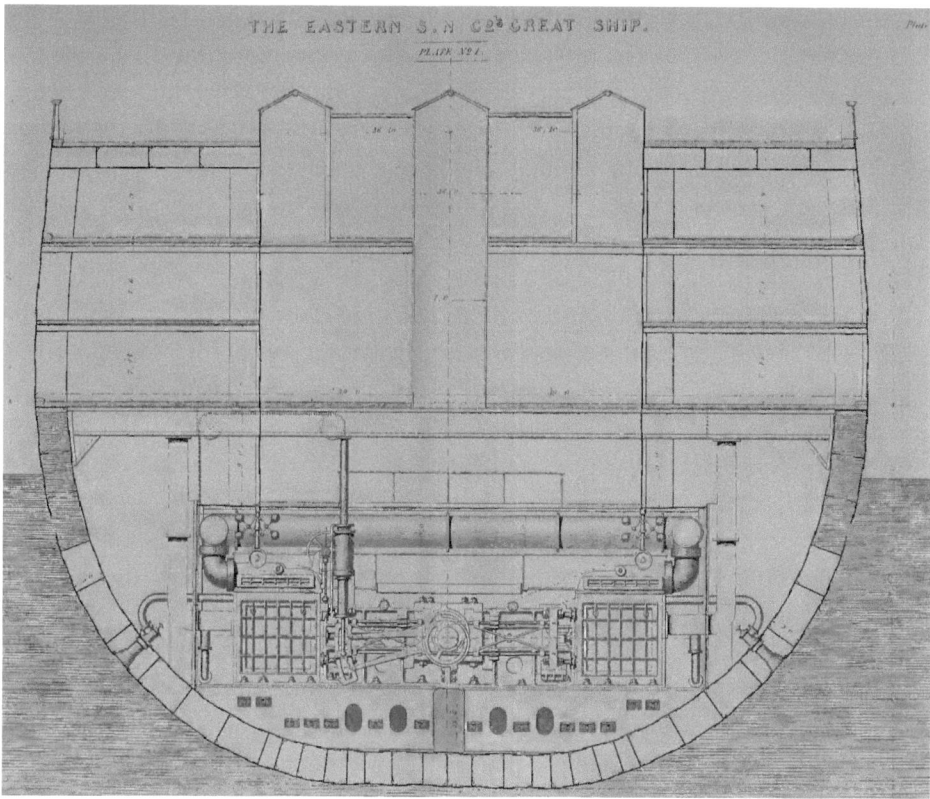

By 1856 Brunel was withholding payments, forcing Russell to liquidate the company and put the construction of *Great Eastern* on hold. The ship languished four months on the ways until Russell secured new credit and restarted work. By now the relations between the "Engineer" and the "Contractor" had been strained to the breaking point. Brunel took total control of the project, appointing his own managers as overseers and completely sidelining Russell from any further decision-making. The result would be disaster.

**The Failed Launch of *Great Eastern* and the Rise of Modern Naval Architecture**

Brunel's rupture with Russell may have contributed to the most significant technical failure of his career, the ill-fated launch of *Great Eastern*. Though the construction of the ship sideways to the river was somewhat novel, Russell had originally intended to launch it traditionally, allowing it to slide freely down the ways into the Thames. However, in 1855 Brunel announced that he was dissatisfied with this approach, arguing that the uncontrolled speed would result in the Great Ship sticking into the mud of the shallow river. He preferred a controlled launch, using a system of great chains, pulleys, and winches to gently lower the ship along the ways into the water. Russell, a highly experienced shipbuilder, objected on the grounds of expense, arguing that the friction between the launch ways and the building cradle that supported the ship would be more than sufficient to slow it down. As usual, Brunel overrode Russell's objections. In late 1856 he caused further problems by insisting that the launching cradle be fitted with iron sliding surfaces and that it slide down the ways on iron rails. Russell faintly complained that such iron-to-iron contact would "bite" and cause the cradle to bind to the ways, but by now Brunel was incapable of hearing anything except his own inner voice.

Although Brunel was convinced that his iron-to-iron launch scheme was correct, he nevertheless turned to his former employee William Froude (figure P.6) to test it, much as Brunel had turned to him a decade earlier on the friction of the atmospheric railway. Soon after the failure of that system, Froude had stopped working—he was then just 38—to care for his ailing father and the family estate in Devonshire. After a decade playing idle country gentleman, Froude was apparently keen to stretch his intellect and accepted Brunel's commission to examine the friction of sliding ways, as well as to investigate the rolling behavior of the Great Ship. Froude first examined the nature of iron-to-iron friction by fabricating a scale model of the launch ways and using a powerful spring-loaded apparatus to record the movement and retarding force of the system. Froude's experiments showed that friction was not independent of velocity, as Brunel and other engineers had thought, but rather *decreased* as velocity increased. Brunel, despite this warning that a controlled, low-speed launch would generate far more resistance, continued with his plans.[25]

Froude's other task would set him on the road to revolutionizing naval architecture by combining theory and scale-model experiments. Brunel was concerned that, when his ship entered the water at launch, its rolling motion would cause it to stick into the river mud. In early 1857 he asked Froude to carry out an investigation of rolling on *Great Eastern*, using mathematical theory and scale-model experiments. In August, Froude wrote Brunel that he would "be able to grind out in something like a tangible and intelligible shape the quest of the movement of a ship on the side of a moving sea."[26] Froude continued to work on rolling through the spring of 1858, well after *Great Eastern* was launched. But by then, Brunel was no longer concerned with those results.

Brunel's lack of interest was due to two grave problems that revealed themselves almost simultaneously. By late 1857 it was obvious that Brunel was suffering from a painful illness, soon diagnosed as Bright's disease (nephritis), a debilitating inflammation in the kidneys. Even as Brunel's strength slowly ebbed, the demands on him mounted,

**Figure P.6**
William Froude.
Credit: Author's collection.

including the preparations for the launch of his Great Ship. Finally, on 3 November, all was set. The great launching chains were wound up around their massive braking drums, ready to check the ship's motion as soon as it started down the ways. A huge crowd had gathered, much to Brunel's chagrin, as he wanted complete silence in order to pass his commands. Just past noon, the directors of the ESN christened the ship *Leviathan*, a name that never stuck—it was later registered as *Great Eastern*—and the order to launch the ship was given. It moved slowly down the ways at first, picking up speed. Then the launch chains grew taut, as the braking drums kicked and spun round, wounding five men, one fatally. Just as Froude had predicted, the slow-speed resistance caused the ship to come to a grinding halt, stuck almost 100 meters from the water. It would not budge.

The ESN had been bleeding money on the ship for over a year, yet now it had to pay the equivalent of another $17 million to bring a flotilla of powerful hydraulic rams to hammer the ship down the slipways and into the water. For three months the rams pounded at the cradle, moving it inches at a time, until in January 1858 it was finally in the Thames. The ESN was broke, and Brunel, personally heavily invested in the ship, made preparations to declare bankruptcy. His health flagging, Brunel followed his doctor's orders and spent much of 1858 and 1859 abroad, spending Christmas in Cairo, Egypt, with his lifelong friend Robert Stephenson, who was by then also gravely ill. Meanwhile *Great Eastern* had lain idle for almost a year while the ESN tried to raise the money needed to complete its fitting out.

By May 1858, Brunel was back in Britain, finding that Russell had made considerable progress while he was out from under the engineer's thumb. Despite his failing health, Brunel visited "his" ship almost daily, hoping to embark on its maiden voyage, scheduled for 7 September. That was never to be. On 5 September, Brunel suffered a stroke onboard the vessel and would spend his final two weeks at home in bed. He received regular reports of the maiden voyage, including the heart-wrenching news that, just a few hours after its departure, it suffered a major steam explosion that killed five men and would lay the ship up for many months. He languished for a few more days, dying at home on 15 September, and was interred at the Kensal Green cemetery in West London.

John Scott Russell never built another ship after *Great Eastern* went into service, instead turning to consulting and writing and occasionally turning his hand to engineering design. He was one of the founders of the Institution of Naval Architects in 1860, the first professional association devoted to the science of ship design. Several years later he led the successful drive to establish a permanent School of Naval Architecture on the site of his Great Exhibition in South Kensington. In 1865 he produced a massive three-volume book, *The Modern System of Naval Architecture*, the size of a coffee table and weighing over 100 pounds, which was priced at the equivalent of $4,000 in today's dollars. It never sold well, and its publisher Day and Son went bankrupt two years later. Russell was left with many unsold copies, which he unsuccessfully attempted to use as collateral for his debts. Russell died in a rented villa on the Isle of Wight on 8 June 1882, poor, without a will, barely remembered and was interred in the Ventnor cemetery.

*Great Eastern* had a short, eventful, but ultimately disappointing career. It never made the voyages to Australia for which it was designed; instead it made just a few round trips to New York, having come into service just when the American Civil War had depressed transatlantic traffic. Despite its widespread fame (Jules Verne immortalized his

# A Bridge Too Far—Brunel and *Great Eastern*

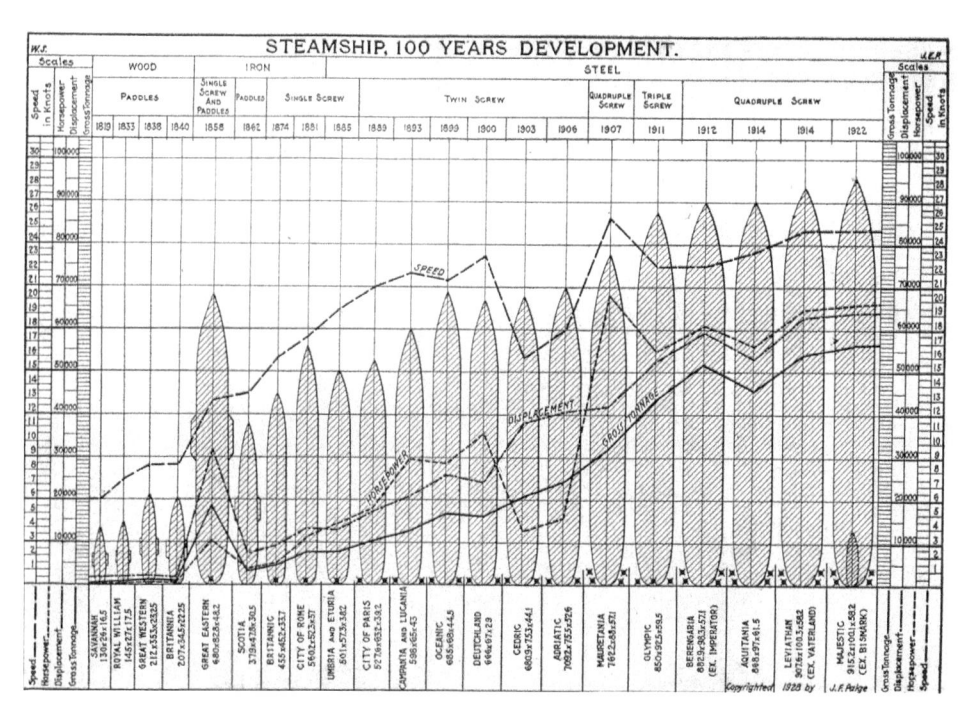

**Figure P.7**
*Great Eastern*'s size would not be surpassed until the twentieth century.
Credit: Author's collection.

own voyage aboard the ship in his novels *A Floating City* and *Propeller Island*), the massive, fuel-hungry vessel proved to be unprofitable as a passenger liner. In 1865 *Great Eastern* was converted into a cable-laying ship for the new network of transatlantic cables, a position it occupied for a decade before being sold as a floating show pier. The immensely strong vessel (in 1862 it survived massive damage from grounding upon a rock) required the shipbreakers to spend a year and a half at their task when it was sold for scrap in 1889. *Great Eastern*'s size would not be surpassed until the stately transatlantic lines of the twentieth century (figure P.7).

Isambard Kingdom Brunel, in the span of a generation, had forever changed the face of Britain, creating a network of railways, tunnels, and bridges that linked the far corners of the nation. He successfully extended that network across the Atlantic, bridging the seas with his ships *Great Western* and *Great Britain*; but his third ship, *Great Eastern*, was simply a bridge too far. The legacy of *Great Eastern* lay not in its voyages but in the advancements in naval architecture that it spawned. To create that one ship, Brunel had assembled the pantheon of engineers who would soon reshape the entire field: John Scott Russell, whose joining of theory and practice found voice when he established the Institution of Naval Architects; William Fairbairn, whose investigations into ship structures, based on his earlier work with bridge design, would provide naval architects with new concepts of structural loading that they would use for over a century; and most importantly, William Froude, who would complete his investigations on rolling and follow up with a novel experimental method using scale models, which would revolutionize the prediction of hull resistance and powering. Together, Brunel and his *Great Eastern* had brought naval architecture firmly into the industrial age.

# 1  Improving Naval Architecture

A ship of the line is one of the most magnificent combinations of the genius of man with the powers of nature. A ship of the line is composed, at the same time, of the heaviest and the lightest of possible matter, for it deals at one and the same time with three forms of matter—solid, liquid and fluid—and it must do battle with all three. It has eleven claws of iron with which to seize the granite on the bottom of the sea, and more wings and more antennae than an insect, to catch the wind in the clouds. Its breath pours out from its hundred and twenty cannons as through enormous trumpets, and proudly answers the thunderbolt. The ocean seeks to lead it astray in the alarming sameness of its billows, but the vessel has its soul—its compass—which counsels it and always points towards the north. In the blackest of nights, its lanterns take the place of the stars. Thus, to oppose the wind, it has its cordage and its canvas; against the water, its timber; against the rocks, its iron, copper, and lead; against the darkness, its light; against immensity, a needle.

—Victor Hugo, *Les Misérables*, 1862

It will always be said of us, with unabated reverence, "THEY BUILT SHIPS OF THE LINE." Take it all in all, a Ship of the Line is the most honorable thing that man, as a gregarious animal, has ever produced. By himself, unhelped, he can do better things than ships of the line; he can make poems and pictures, and other such concentrations of what is best in him. But as a being living in flocks, and hammering out, with alternate strokes and mutual agreement, what is necessary for him in those flocks, to get or produce, the ship of the line is his first work. Into that he has put as much of his human patience, common sense, forethought, experimental philosophy, self-control, habits of order and obedience, thoroughly wrought handwork, defiance of brute elements, careless courage, careful patriotism, and calm expectation of the judgment of God, as can well be put into a space of 300 feet long by 80 broad. And I am thankful to have lived in an age when I could see this thing so done.

—John Ruskin, *The Harbours of England*, 1856

These lines of longing for the bygone age of sail were independently composed almost simultaneously, in the middle of a century that saw the transition from wood and sail to iron and steam.[1] Hugo and Ruskin, both of them poets and authors living at the same time on either side of the English Channel, were enamored of the former and depressed by the latter. Each regarded the enormous social changes wrought by modern technology with a sense of loss, and in their prose sought to preserve in amber the great accomplishments that mankind had achieved without it. They regarded the wooden ship of the line—warships carrying from 50 to 140 guns that had been the mainstay of navies for almost two centuries—as the pinnacle of man's achievements. This view may have been debatable when considering the wider pantheon of cities, roadways, and waterworks, but in the much narrower domain of naval architecture, it was most certainly the case.

Hugo and Ruskin published these tributes to the ship of the line in the mid-1800s, at the same time that Brunel's crowning glory *Great Eastern* came on the scene. Yet even though iron and steam were already clearly in the ascendancy, wooden sailing vessels were just then reaching the summit of their development—larger and faster than any ever before and with a total productivity (economic efficiency of ocean transport) that increased tenfold during the nineteenth century.[2] Modern economists have come to use the term the "sailing ship effect" to describe the phenomenon of old technologies fighting back against competing new technologies by becoming more effective and more efficient. The sailing ship effect, ironically, did not actually apply to sailing ships, whose improvements began long before steamships came of age.

In the early eighteenth century, the decimation of European forests forced shipbuilders to come up with more effective ways of using ever-scarcer timber, with the result that the hull structures were built far stronger and could become larger and more efficient. At the same time, experiments with hull forms—some scientific, some seat of the pants—led to novel designs like the wave-line and the Aberdeen bow, which allowed sailing ships to travel at unparalleled speeds. Many of these advances occurred in Britain, which by then was not only the world's unchallenged industrial power but also the bearer of the mantle of scientific leadership.

## The Transition of Naval Architecture from France to Britain at the Beginning of the Industrial Age

The apogee of the age of sail coincided with the Scientific Revolution (roughly 1600–1800), during which time the discipline of naval architecture—the integration of scientific ship theory into ship design and construction—was born. Naval architecture was largely developed within state institutions, specifically scientific academies and naval administrations, which had the resources to sponsor research; disseminate the findings in books, journals, and memoirs; and methodically apply those results to shipbuilding.

Many of these developments had occurred in France during the ancien régime, which saw ship theory as a strategic force multiplier, a means of augmenting the effectiveness of individual ships to counterbalance the superior numbers in the British navy. The French Academy of Sciences sponsored many of the scientific studies of naval architecture, such as the model experiments on ship resistance conducted in the 1770s that reshaped the existing ideas about fluid dynamics. Naval administrations in turn used ship theory to control the process of ship design and construction, creating systems of standardization and professionalization that ensured warships met uniform criteria across the fleet, epitomized by the designs of Jacques-Noël Sané, that remained in force for almost 40 years. British naval officers, convinced that French warships were better and faster than their own, attributed this superiority to the research and application of ship theory by French naval constructors.[3]

By the end of the eighteenth century, the French revolutionary and Napoleonic Wars (1792–1815) had vaulted Britain to mastery of the world's oceans and also had greatly disrupted French research into ship theory. These same wars, however, also spread French influence in naval architecture across Europe; as Napoleon extended his conquests across the Continent and the Mediterranean, French naval constructors were brought into the occupied dockyards in the Netherlands, Greece, and the Italian states to build and repair warships for Napoleon's fleet, bringing with them both their naval architectural theory and their understanding of how to apply that theory in design and construction. These practices also arrived in Britain, not because of occupation, but in part via a French engineer who took refuge from the worst offenses of the French Revolution.

The civil war that started Napoleon Bonaparte's rise to power during the Revolution devastated much of the French countryside and decimated what remained of many state institutions. The Academy of Sciences was shuttered in 1793, although it was reconstituted in 1795 under the National Institute of Sciences and Arts. Under Napoleon—an artillery officer who never fully trusted the navy—research into the naval sciences was curtailed, while greater emphasis was given to more terrestrial subjects such as chemistry and thermodynamics. This was particularly evident in the rapid change in the annual prizes awarded by the Academy. During the reign of the old Academy, naval and maritime subjects had made up over a quarter of the prizes; as recently as 1791, prizes were offered to examine the nature of ship motions under the action of waves and another on fluid resistance of ships. When the new Academy was reconstituted, there were no prizes in those subjects; in fact, 30 years would elapse before another prize was offered on naval architecture.[4] There was no longer a large-scale, systematic agenda of research into naval architecture carried out under an official Academy.

Instead, there were now just a smattering of studies, done primarily by French naval constructors in their off-hours, when not engaged in the design, construction, and repair of warships for Napoleon's campaign of European conquest.

Most of these studies focused on the issue of hydrostatics and ship stability, which had first been expounded a half century earlier by the French scientist Pierre Bouguer, the Swiss mathematician Leonhard Euler, and the Spanish naval constructor Jorge Juan y Santacilia. The first of these more advanced studies was coauthored in 1798 by Honoré-Sébastien Vial du Clairbois, at the time the chief constructor in the naval dockyard in Lorient. Even though France was then at war with Britain, Vial du Clairbois worked with the British mathematician George Atwood to develop the theory for the righting moments of ships at large angles of heel, identifying for the first time the righting arm, which they titled $\overline{GZ}$ (the term and concept are still used today), that defines in geometric terms the leverage of the buoyancy force against a heeling moment.

Lorient was also the site where, in 1808, a midlevel naval constructor named Antoine Nicholas François Bonjean developed an easy-to-use method for quickly calculating the displacement and center of buoyancy of a ship at any draft or trim. He did so by plotting the sectional-area curves of the hull at various drafts for each section at regularly spaced intervals called stations; Bonjean used 8 stations, while modern naval architects typically divide ships into 20 stations. Bonjean curves are widely used to this day; by drawing the waterline of the ship at the desired draft and trim, then reading off the areas and summing them, designers can rapidly make the required calculations for criteria such as floodable length without spending hours at painstaking integration.[5]

Just as Bonjean was developing his curves in Lorient in 1808, a brilliant 24-year-old constructor named Pierre Charles François Dupin was assigned to one of France's shipyards on the Greek island of Corfu. Although he was the sole supervisor of 100 workers who were building and repairing warships for Napoleon's Mediterranean campaigns, Dupin found plenty of time over the next three years to conduct tests on the strength of wooden beams and to pursue a subject that had fascinated him since a student under the great mathematician Gaspard Monge: the analysis of complex mathematical curves and surfaces, known as analytic geometry. But where Bonjean's curves were eminently practical, Dupin's were purely theoretical: in his memoir "On the Stability of Floating Bodies" (not published until 1822), he fully investigated the general case of three-dimensional stability—that is, the stability of any arbitrary body, rotated around any axis. He defined a set of three-dimensional curved surfaces, formed by the positions of the center of buoyancy and of the waterlines at different inclinations, and demonstrated that stability is a function of the center of gravity in relation to these surfaces. Dupin's elegant theorem, later somewhat improved on by other mathematicians, soon became a fundamental part of the naval architecture curriculum taught in many constructors' courses around Europe. The immense difficulty in solving the complete three-dimensional theorem in numerical terms, however, meant that it was not generally used in the day-to-day stability calculations for heel and trim of ships.[6]

Vial du Clairbois, Bonjean, and Dupin greatly expanded the understanding of hydrostatics and ship stability, but their works were also the only major ones of naval architecture that France had produced in the 30 years since the Revolution. It would be another 30 years before France achieved the level of productivity it had during the ancien régime. Instead, the French were training a new generation of naval constructors across Europe to use ship theory as part of design.

Dupin's assignment in Corfu was hardly unusual for a French naval constructor in the Napoleonic era; he had also worked in the dockyards of Antwerp and Genoa, which along with the dockyards in Naples, Venice, and Trieste came under French control after their occupation by Napoleon's forces. French constructors were put in charge of each of these dockyards to build French warships. These constructors brought with them not only plans but also the know-how of design, calculation, and construction practice, and they taught the local constructors how to draw plans, calculate displacement and stability, and devise estimates for weight and materials. The local constructors, in addition, often visited French dockyards and received their education in the French school for naval constructors.

Thus, in Antwerp, Dupin worked alongside the Dutch constructors Pieter Glavimans and Cornelis Soetermeer, who themselves were trained in the French school for naval constructors that had been temporarily located there. In Genoa and Naples, constructors like Dupin, Joseph Molinard, and Antoine Imbert oversaw the design and construction of French 74-gun warships, the workhorse of the main battle fleets. Under the guidance of the French constructor Jean Tupinier, the Venetian constructor Andre Salvini built French frigates in the celebrated Venice Arsenal, after which he made a yearlong voyage around France to learn more of maritime engineering. Salvini trained his nephew Gaspare Tonello, who took over in the Trieste dockyards after Jean-Baptiste Lefebvre (a classmate of Bonjean) departed there.[7] Allies of Napoleon's regime also called on French expertise, as when Jacques Balthazar Le Brun de Sainte Catherine and Jean-Jacques Sébastian Le Roy overhauled both Russian and Ottoman naval shipbuilding.[8] In each of these cases, the knowledge and practice of theoretical naval architecture remained in those dockyards long after the French constructors had returned to France after Napoleon's downfall in 1814.

But not all constructors and engineers had sworn loyalty to Napoleon. Though Marc Brunel was the most famous royalist engineer to flee the Revolution and emigrate to

Britain, Jean-Louis Barrallier was among the thousands of French who were evacuated from Toulon by a British fleet when the revolutionary army captured the city in 1793. Barrallier was at the time in charge of the Toulon dockyard facilities, but when he arrived in Britain he was brought into the Admiralty (which controlled the design, construction, and operation of ships and fleets) as second assistant under Surveyors of the Navy (i.e., chief constructors) John Henslow and William Rule. Barrallier was effectively the junior naval constructor in the organization. It is not clear where Barrallier, as a civil engineer, learned the elements of naval architecture, ship design, and construction. Nevertheless, he immediately began producing designs for a range of ships, from 44-gun frigates to 80-gun ships of the line.[9]

Barrallier's designs were quite similar to those of French vessels, and his methods of drafting and calculation followed French practice. Both French ships and French methods were already known to British constructors. The British navy had captured dozens of French warships over the years, carefully analyzing them to determine what made them appear to be faster than British ships. The standard British 74-gun ship, in fact, had been copied from the *Invincible*, captured in 1747. French textbooks on naval architecture were widely available in Britain, and several standard British textbooks, like David Steel's *Elements and Practice of Naval Architecture*, carefully explained how to calculate weights, displacement, and stability. What was lacking was the day-to-day understanding of how to translate designs into data for calculations. For example, in 1812 Barrallier provided the Admiralty with calculations of stability and ratio of bow resistance (a measure of speed) for his own 74-gun proposal and for two other vessels, *Black Prince* and *Ajax*.[10]

The Admiralty possibly hoped that Barrallier would provide the inspiration and impetus for British constructors to take a more scientific approach to ship design, emulating their French counterparts to create better ships. This did not happen very quickly. Although the Frenchman had demonstrated the practical methods of naval architecture to the Admiralty, the results were unimpressive; his fellow surveyors found his vessels to be structurally weak, insufficiently stable, and mediocre sailers. Barrallier, originally trained as a civil engineer, quite simply lacked knowledge of many practical aspects of ship design. This cast a pall over the presumed advantages of the French approach to naval architecture, so that when Barrallier returned to France in 1815, the British navy still lacked the technical infrastructure to translate theory into practice.

**The Society for the Improvement of Naval Architecture**

The want of a scientific approach to ship design was not for lack of trying by entities both inside and outside the British navy. John Sewell, publisher of the popular *European*

*Magazine and London Review,* had listened carefully to British sailors complaining of the superiority of French warships to their own. A self-taught expert on shipbuilding—his warehouse on Cornhill Street, near the offices of the British East India Company and other international trading firms, displayed books and models on the subject—Sewell was certain that better naval architecture was the reason for this presumed French dominance. He began printing letters and articles about naval architecture on the blue wrapper that encased the magazine while envisioning a grander scheme to attract public attention.

On 14 April 1791, Sewell called a meeting of 130 men to create an organization that would bring his vision into reality, carefully laying the groundwork for this event. The name of the organization, the Society for the Improvement of Naval Architecture, echoed that of Britain's premier scientific research and advisory body, the Royal Society of London for the Improvement of Natural Knowledge. The meeting place was not in one of coffee shops in Cornhill that serviced financiers and merchants; rather Sewell had strategically chosen the immense, and immensely popular, Crown and Anchor tavern off the Strand—near Whitehall and the Admiralty, the political heart of London.

All this underlined Sewell's vision of his society as not merely a technical gathering but also a means of influencing naval shipbuilding policies. The meeting was a veritable who's who of British powerbrokers, including William Henry, Duke of Clarence (later King William IV, "the Sailor King"); the statesman Charles Stanhope and the famous abolitionist William Wilberforce; the admirals Charles Middleton and Charles Knowles; the scientists Nevil Maskelyne (astronomer royal) and Joseph Banks (president of the Royal Society); and independent ship constructors like Marmaduke Stalkaart. Admiralty constructors like John Henslow and William Rule, whose practices Sewell railed against, pointedly did not attend.

Sewell's opening address minced no words in identifying the threat: "It is but too well known to all who have any skill in Naval Architecture, that the theory is not so well understood as it deserves; and that the French, actually surpassing us in this most important art, have derived many advantages from this superiority in time of war." Sewell's new society, which would eventually have over 300 members from across the globe, resolved to meet every month to discuss technical and policy issues related to naval architecture and shipbuilding. It would publish an annual report containing writings on these subjects. It would also offer a series of cash prizes for experiments, studies, and proposals related to ship resistance, masting, tonnage measurements, and safety. All this would be funded by subscription dues, each member paying two guineas annually (about $150 today, comparable to the dues for modern-day professional societies). The Duke of Clarence, who was elected president, had no trouble obtaining the approval of his father, King George III, for the society (which, however, never became "Royal").[11]

The society and its members lost no time in carrying out experiments and studies on naval architecture. William Hutchison, a dockmaster in Liverpool, floated one-meter-long ship models to determine that their center of pitching motion was identical to their center of gravity, which he had previously located by suspending them in the air (his results were in fact wrong: the center of pitch moves dynamically, generally around the center of floatation). Another society member, Charles Gore, carried out model tests on a series of hull shapes and cross sections to determine which ones had the least resistance and the greatest stability. To no one's surprise, he determined that hulls with the fullest cross section had the greatest stability, a result already known from when the French scientist Pierre Bouguer had first explained the metacentric theory of stability back in 1746.[12]

The most famous series of experiments carried out under the aegis of the society were those of Mark Beaufoy, a wealthy army officer, amateur explorer, and more-than-just-amateur scientist. Beaufoy had already conducted small-scale studies of fluid resistance when he responded to Sewell's offer of a £100 prize to "ascertain the laws of resistance

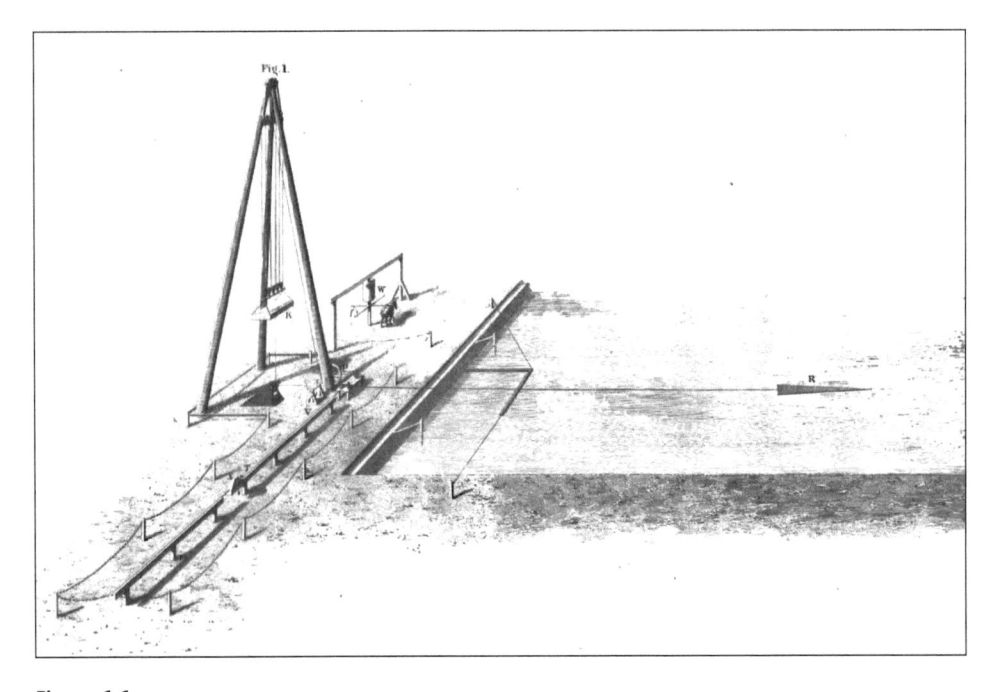

**Figure 1.1**
Beaufoy's experimental setup at the southwestern corner of Greenland Dock (this page) and the site today (opposite page).
Credits: New York Public Library (this page); author's photo (opposite page).

of water to solids of different forms." He took up the challenge in 1793 when William Wells, a member of the society and owner of the Greenland Dock (a huge whaling ship dock in Rotherhithe, just northwest of Greenwich) allowed him free use of his dock to conduct his experiments. Beaufoy set up his outdoor laboratory at the upper end of the dock, out of the way of commercial traffic.

Beaufoy was well aware of previous experiments on the resistance of solids in water, carried out by Jean-Charles de Borda, Charles Bossut, and others in France during the 1770s. Bossut's tests were particularly well known; he had used a tall tower with a falling weight that pulled wooden models of various shapes through the water, timing the passage to determine which shapes had the least resistance, and had decided that bow resistance was the most important factor. Beaufoy noted that Bossut had completely neglected the influence of water friction against the sides, a phenomenon that had been the subject of greater study in the intervening 20 years. On the advice of Charles Stanhope, Beaufoy established his experimental design to examine three sources of resistance: bow pressure, stern pressure, and friction. Although his towing apparatus would be similar to Bossut's—a tall tower with falling weights—he introduced innovations, including an automatic velocity recorder and advanced data-handling techniques to manage the huge amount of test data (figure 1.1).

Beaufoy conducted a methodical series of 1,671 recorded trials of geometric and ship shapes over the space of five years, from 1793 to 1798, eventually spending £30,000 ($2 million today) of his own money in the process. Beaufoy determined that ship resistance was divided into three components, each of which could be measured separately:

- bow pressure (positive)
- stern pressure (negative)
- friction

He towed bodies at and below the surface to establish the effects of end shapes on pressure and towed thin planks (effectively without pressure-causing end shapes) to determine laws of frictional resistance. He did not, however, consider wavemaking in his studies.

Beaufoy published early results with the society in 1794 and 1800, although it would not be until 1834 that his son Henry Beaufoy fully tabulated them. Beaufoy never extrapolated these model results to full-scale ships, although other engineers would soon do so to calculate the power required for the first steam vessels, which is discussed in chapter 2.[13]

The Society for the Improvement of Naval Architecture proved to be short-lived; it closed its books in 1801, shortly before Sewell's death.[14] Its influence was modest at best, primarily through the legacy of Beaufoy's experiments and the later careers of some of its members.

**Strengthening the Wooden Ship**

Wooden ships are complex structures of many hundreds or thousands of pieces held together with myriad fasteners. A wooden hull does not behave as a modern steel hull does. Steel plates and girders in modern ships are fairly uniform in strength and ductility. In a wooden hull, the frames and planks are made of different timbers—usually dense, hard oak for the frames and softer pine, fir, or cedar for the planking—which behave differently under different loads. In steel hulls, the individual pieces (plates and beams) are rigidly connected and respond to loads as a continuous girder. In a wooden hull, the strips of shell planking are caulked with hemp rope and tar for watertightness but not otherwise rigidly connected to each other, so they slide relative to one another and rotate around their fastenings to the frames. As ships age, these connections tend to work loose. This results in a wooden structure that is subject to large flexures and distortions under loading.

Over the centuries, European shipbuilders evolved different methods of ship design and construction to achieve watertightness and strength. Many early vessels were clinker

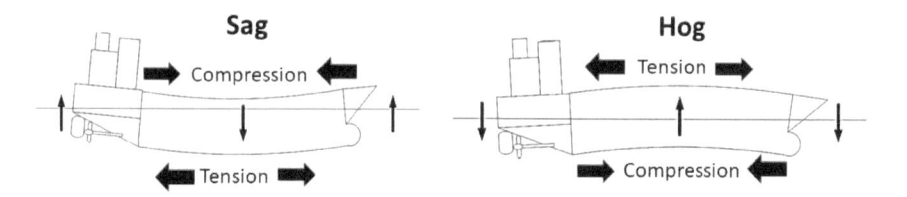

**Figure 1.2**
Hogging and sagging of hull girders, and corresponding directions of tension and compression in deck and keel.
Credit: Author.

built, meaning that the hull was constructed from overlapping strakes (similar to the way roof shingles are laid one atop the other), with caulking applied at the overlap. By the late Middle Ages, carvel-built ships—with planks joined and caulked flush at the edges—became more common for longer oceanic voyages. At the same time, the method of construction changed from shell first, in which the hull strakes were built up into the final hull shape before frames were cut and fitted inside, to the stronger frame-first construction, in which internal frames were erected along the length of the hull to develop the shape, after which the strakes were fastened to the frames and caulked. By the end of the eighteenth century, almost all European oceangoing ships were built frame first using the carvel technique, which provided the greatest resilience against the perils of the sea.[15]

A sailing wooden ship is subjected to many loads, even before it is afloat. The process of launching imposes enormous stresses as the hull enters the water, transferring the support from the keel and bracing to the buoyancy of the hull itself. The standing rigging, already pretensioned to keep the masts upright, imposes sudden localized loads under wind gusts. The ship's rolling and wind and waves push and pull the hull structure side to side, a process known as racking. The greatest loads come from longitudinal bending of the hull, both in the static condition of calm water and under dynamic wave conditions, causing the hull to flex up in the middle (hogging) and down (sagging) as shown in figure 1.2.[16]

The problems of racking became more pronounced as ships spent increasingly greater lengths of time at sea. This was especially true for English (later British) warships, which could spend many months or years in permanent overseas squadrons. As early as the 1600s, constructors such as Mathew Baker and Anthony Deane were using diagonal pillars set athwartships to stiffen the hull against racking. The use of diagonal pillars became more widespread when the British constructors who went to work in Spain in the 1750s took the practice with them (figure 1.3).[17]

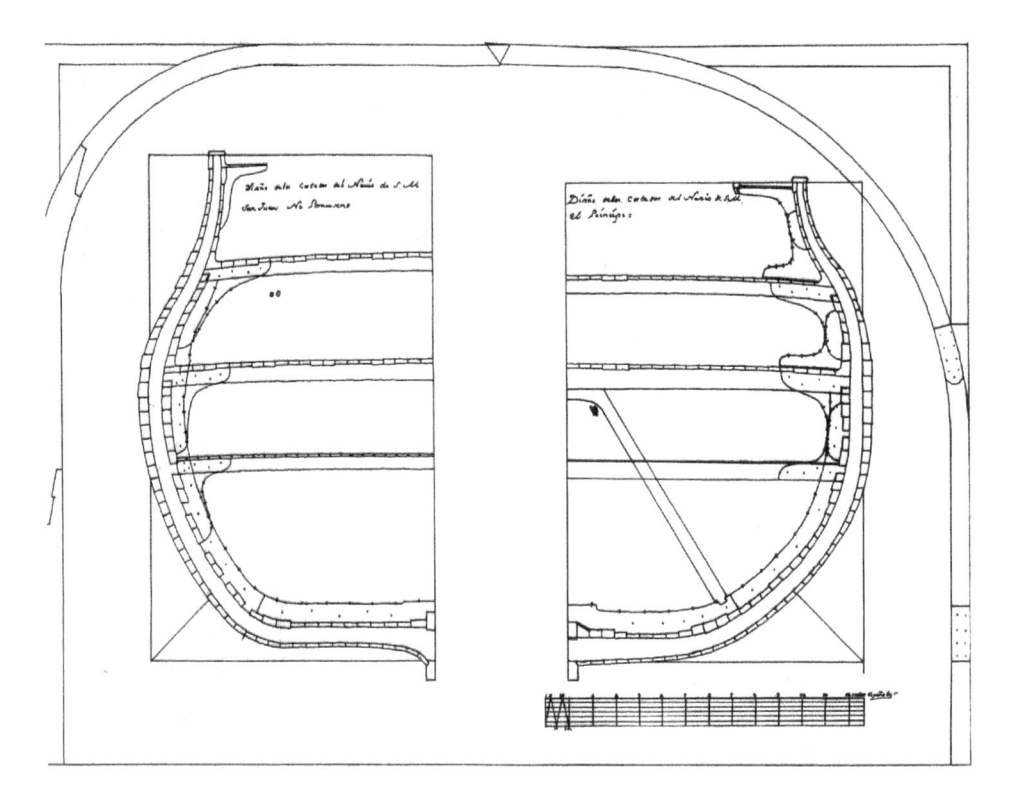

**Figure 1.3**
British-style diagonal pillars in the Spanish warship *Principe* (right, 1759), compared with later French-style construction on *San Juan Nepomuceno* (left, 1766).
Credit: Armada Española, Colección de Dibujos Técnicos y Planos de Buques.

This method of construction reached its apogee in the merchant ships of the East India Company. In the 1780s, the company's chief constructor, Gabriel Snodgrass, developed a comprehensive system of diagonal pillars and iron braces to build these immensely strong vessels, which were often of similar size and armament to line-of-battle warships. Many of these East Indiamen were built not in Britain but in the great Indian port city of Bombay (today Mumbai). There were two reasons for this: first, the East India Company's Bombay dockyards were staffed by a highly skilled Indian workforce under the masterful direction of the chief constructor Jamsetjee Bomanjee of the Wadia dynasty of constructors; and second, the ships were built of teak, an immensely strong hardwood that was in many ways superior to oak, cedar, or fir. The combination of Snodgrass's system, the workmanship of the Bombay constructors, and the use of

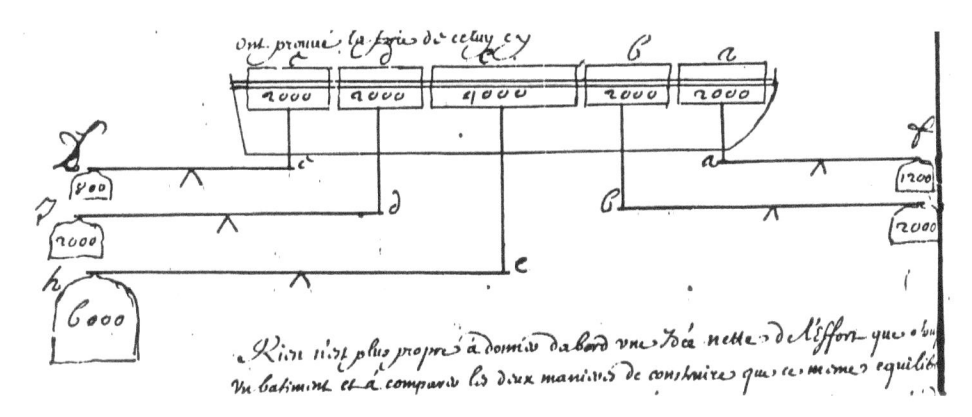

**Figure 1.4**
Alexandre Gobert's analysis of hogging due to differences in weight and buoyancy along the hull. Credit: Bibliothèque Nationale de France/Gallica.

teak resulted in unusually long-lived ships. The British navy took full advantage of this shipbuilding prowess by ordering a number of warships from Bombay, including HMS *Trincomalee*, which survives to this day in the British port of Hartlepool.[18]

The use of diagonal pillars had addressed side-to-side loading; but as ships increased in length, the problem of hogging due to longitudinal loading became the controlling factor in structural design. Wooden sailing ships would often acquire a permanent hog soon after launch, with the keel drooping as much as a meter at each end compared with amidships. The underlying reasons were well understood by naval constructors of the seventeenth century. Part of the problem was due to improper construction techniques, as the French mathematician Paul Hoste explained in 1697. His colleague Biaggio (Blaise) Pangalo presented a more fundamental reason when he stated that "a ship is wider in the middle so that this part displaces more than it weighs … whereas the two ends, which are much narrower, weigh more than they displace, so they push down into the water." His fellow constructor Alexandre Gobert graphically demonstrated this, as shown in figure 1.4, by modeling each section of a vessel as a balance, with weight and buoyancy on each side of a fulcrum. In his diagram, the middle of the vessel weighs 4,000 *livres* (1,960 kilograms), but displaces 6,000 *livres* (2,930 kilograms), with a net upward force of 2,000 *livres*. The ends, however, have net downward forces. This unequal loading for each section results in large bending loads—and thus hogging—along the hull.[19]

Gobert's solution to hogging was to arrange the planking diagonally (figure 1.5). This distributed the loads more evenly along the length and also stiffened the overall

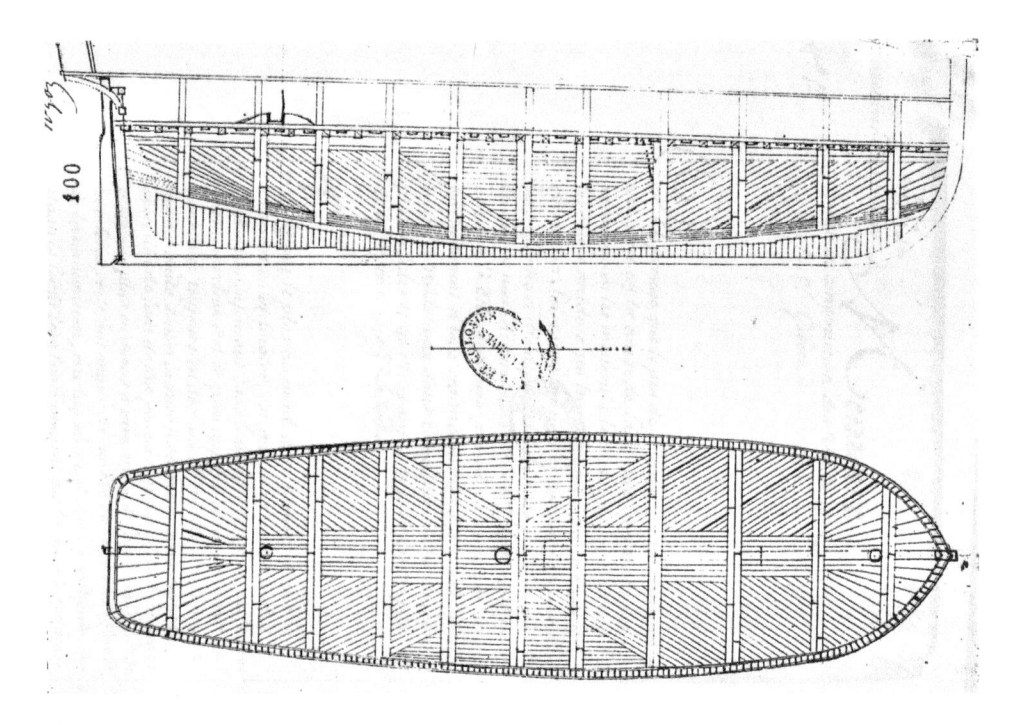

**Figure 1.5**
Alexandre Gobert's diagonal planking.
Credit: Bibliothèque Nationale de France/Gallica.

structure, in the same way that a square wooden frame is stiffened by adding diagonal slats from corner to corner. Gobert successfully used diagonal planking in several refits and new constructions, including the 70-gun *Saint-Michel* in 1706. The Gobert system, as it became known, was widely used until the mid-1740s, notably in the ships of Joseph and Blaise Ollivier such as the 64-gun *Fleuron*, although after 1745, Gobert's system was largely abandoned because of its higher costs and greater difficulty of fabrication.[20]

In 1775 the Royal Danish Academy of Sciences took up the question of ship structures for its annual prize. A young naval constructor named Ernst Vilhelm Stibolt submitted the winning entry, which was published a decade later. In it, he expanded on Gobert's analysis, by showing how to divide a ship into 11 sections, calculating the displacement of each slice and distributing the ship's weight to match the displacement of each section, thus avoiding the possibility of breaking the keel. Stibolt did not discuss diagonal framing, although in his ships he used additional knees to strengthen the joints, a practice followed by other Danish constructors.[21]

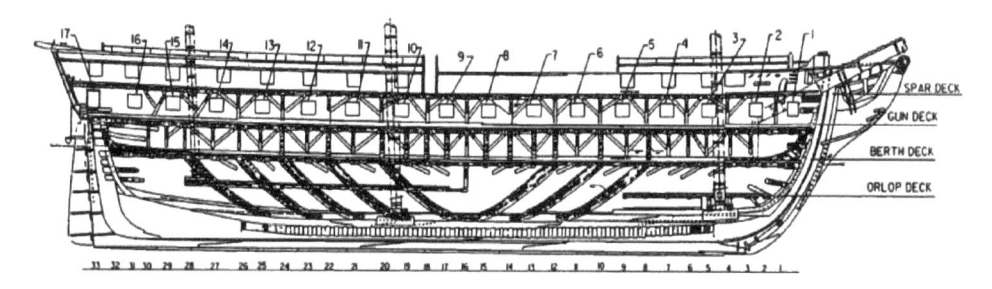

**Figure 1.6**
USS *Constitution*, showing original diagonal riders.
Credit: Robert Wasalaski.

The notion of diagonal structures came independently to constructors in several other nations who had little need of ship theory but instead were solving quite practical problems. In the newly born United States, Joshua Humphreys had been given the contract in 1794 to design a fleet of six frigates to counter pirates in the Mediterranean and to protect against any potential European threat. Humphreys, although apparently unfamiliar with prior theoretical work on hogging, knew that his design for long, heavily armed and heavily built frigates would be subject to enormous longitudinal stresses. He devised a system of diagonal riders (figure 1.6) to distribute the bending loads and stiffen the hull against hogging. These riders were installed on three 44-gun frigates, *Constitution*, *United States*, and *President*, but not on the remaining ships, owing in part to objections by Humphreys's assistant Josiah Fox.[22]

Meanwhile, in the Netherlands, a Dutch constructor of British descent, William May, was trying to solve the problem of corrosion in iron bolts and nails. His solution—wooden treenails—weakened parts of the structure, so he developed the idea of using short diagonal trusses in the hold to bolster its strength (figure 1.7). In 1785, May constructed the 68-gun *Leiden* with these same diagonal trusses. He also sought to reduce the vulnerability of the warship's stern—normally a large, flat expanse adorned with windows and easily penetrated by gunshot—by rounding it off, enclosing it in wood and adding gunports. May's friend the Danish constructor (and contemporary of Stibolt) Henrik Hohlenberg picked up on this idea and designed his vessels, such as the 90-gun *Christian VII*, with narrow sterns variously called ducks' asses or pinched sterns. Within a few years, both May's and Hohlenberg's ships would be captured by the British navy and help launch the most important advances in wooden warship construction of the nineteenth century.[23]

**Figure 1.7**
Sections of William May's 1784 ship model with diagonal trusses.
Credit: Rijksmusem, Amsterdam.

The reason the May and Hohlenberg ships were captured was that the Netherlands and Denmark, along with much of the rest of Europe, were at war with Britain. As the French revolutionary and Napoleonic Wars extended France's domination and influence across the Continent, Britain was increasingly engaged in battle with nations either occupied by France (like the Netherlands) or allied with it (Denmark). Britain's navy, the vaunted "wooden walls" that protected the nation and its empire, was soon overstretched from engagements that ranged across the globe. It grew from 276 ships at the outbreak of hostilities to almost 400 at its height.[24] Dockyard constructors were constantly at work building new ships, refurbishing captured vessels, and repairing battle-damaged ones.

Britain had begun the war with something of a timber crisis. Its own forests had long ago been cut down, and the loss of its American colonies meant the loss of their steady supply of critically important oak and fir. This had largely been replaced by Baltic timber until France's blockades threatened to strangle that supply. Now, the overwhelming requirements of wartime construction and repair forced administrators and constructors to develop innovative methods of economizing shipbuilding timber, in particular the long, straight pieces and heavy curved knees (called compass timber) that were increasingly hard to find.[25]

Robert Seppings was the master shipwright at the Chatham dockyard in May 1805 when HMS *Kent*, a 74-gun ship that had been weakened by two years of heavy wartime cruising in the Mediterranean, was brought into dock for a complete refit. That very same month, HMS *Leyden*—the British name for the Dutch warship *Leiden* that had been captured in 1799—was brought into Chatham for refit. Seppings had already

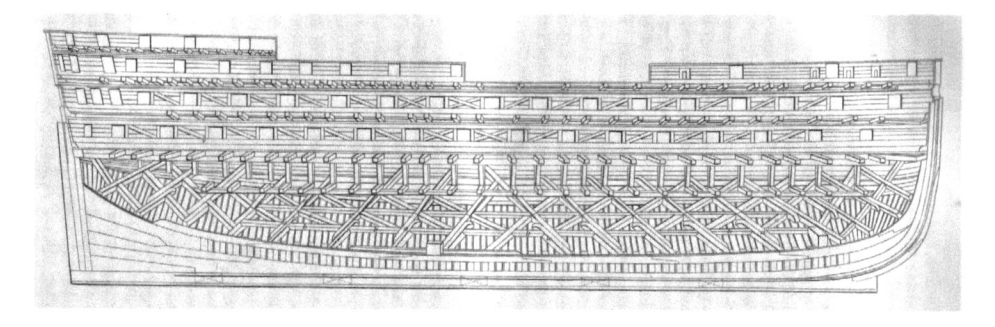

**Figure 1.8**
Robert Seppings's complete system.
Credit: Author's collection.

been thinking for some time about ship hogging and timber shortages, and it is almost certain that when he saw William May's diagonal trusses in the former *Leiden* it crystallized for him the potential solution to both problems. Seppings rejected the Admiralty's recommendation to refit HMS *Kent* with Snodgrass's system, instead taking (without attribution) William May's concept of diagonal trusses to straighten out and strengthen the hull while saving on timber.[26]

The HMS *Kent* refit was the first step that Seppings took to evolve his system. After his initial success, he added more elements to his refits of *Warspite* and *Tremendous* in 1806 and 1810, eventually replacing individual diagonal trusses with sturdier X-shaped trusses. By 1811 he had worked out all the elements during the refit of HMS *Albion*. Seppings's complete system (figure 1.8) consisted of the following:

1. Inserting X-shaped trusses to stiffen the hull against hogging (Seppings's analogy was that of a rectangular gate with diagonal braces); the trusses were short and thus economized on long timbers.
2. Filling in the voids between the lower transverse frames with old timber; this reinforced the keel against the compressive stresses due to hogging.
3. Connecting deck beams to the transverse frames using continuous shelf pieces and waterways instead of heavy knees; this economized on increasingly hard-to-find oak compass timber required for knees.
4. Using diagonal deck planks instead of planks running fore and aft; this also economized on timber and strengthened the deck against transverse loads.

This integrated approach to structural design, developed by a systematic series of experiments and observations, went far beyond anything that had been attempted previously. It marked Seppings, although not classically educated, as a clear scientific thinker on par with members of the Royal Society. That did not stop the Admiralty from turning to the Royal Society itself to help them decide on whether to wholly adopt Seppings's system. In 1814, Seppings submitted a long explanation of his system to the society. Thomas Young, a Royal Society fellow who had already made a name for himself by defining what is now called Young's modulus of elasticity (the relationship between stress and strain for a given material), then examined Seppings's system using the latest developments in beam theory.

Young's carefully constructed analysis considered several different loading scenarios, including still water, waves, and end loads due to water pressure. His ultimate support of Seppings's system, later endorsed by the French naval constructor Charles Dupin during one of his many study trips to Britain, validated what the Admiralty had already decided—that Seppings's system would be the basis for all future construction and refit of British navy ships. Seppings, who was subsequently made a fellow of the Royal Society, went on to add several features to his designs, including using diagonal iron straps as well as wooden trusses and incorporating rounded, enclosed sterns that were in part based on Hohlenberg's ships captured after the 1807 attack on Copenhagen. Seppings's complete system is visible today aboard the museum ship HMS *Unicorn* in the port of Dundee.[27]

In 1815 the Battle of Waterloo ended the Napoleonic Wars, and previously hostile nations began looking to the British navy as a model to redevelop their own fleets. Among the first to adopt Seppings's system were France (via Dupin's memoirs) and the Netherlands, whose ships had helped inspire Seppings in the first place. In 1818 several Dutch constructors, including Cornelis Soetermeer and Cornelis Jan Glavimans, visited Britain to study Seppings's system, and by 1822 the Dutch navy had adopted it. By 1827 the Russian navy had also begun using the system, first on the 110-gun *Imperator Aleksander* and later on a series of 74-gun ships. In some navies, only parts of the Seppings system were adopted. For example, when the U.S. constructor Samuel Humphreys (son of Joshua Humphreys) built the 120-gun USS *Pennsylvania* in 1837, he used diagonal trusses only in the upper decks.[28]

Merchant shipbuilders were also quick to adopt Seppings's system, in whole or in part. Seppings himself had proposed a modified system for merchant ships, relying primarily on iron and wooden diagonal trusses. A number of constructors developed and patented diagonal framing systems of their own, generally based on Seppings's ideas.[29] Sailing merchant ships quickly grew in length and capacity as a direct result of the increased strength and rigidity afforded by the Seppings system and its derivatives, which were often coupled with iron framing (discussed in chapter 2). Early transatlantic packets of the 1820s and 1830s, built with traditional framing, were typically about

50 meters long and 700 tuns. By the 1850s, clippers built with diagonal iron trusses were measuring over 70 meters and 2,500 tuns. The largest wooden sailing ship ever built, the four-masted bark *Great Republic* launched in 1853, measured 122 meters and 4,500 tuns. Its extensive diagonal-truss framing system kept the hull straight and true over its nearly 20-year life.

Ironically, of all these structural improvements, the most influential derived from a 1,500-year-old invention. As long ago as the Jin dynasty (265–420), Chinese junks had been built with strong internal bulkheads that provided structural rigidity and separated a ship into compartments. Although these bulkheads had limber holes at the bottom to allow the free flow of water, they apparently could be made watertight to prevent progressive flooding in the case of damage.

Marco Polo had described bulkheads in his fourteenth-century book *Travels*, but it was not until 500 years later that Westerners seriously considered their use. The American statesman and polymath Benjamin Franklin, for example, wrote his celebrated "Maritime Observations" in 1785 after returning from serving as America's first ambassador to France. In it he suggested dividing ships up into "separate apartments after the Chinese manner … caulked tight to keep the water out." John Schank, a naval officer and member of the Society for the Improvement of Naval Architecture, came up with the same idea in 1792. This idea was first acted upon by the British naval constructor Samuel Bentham, who undoubtedly saw firsthand Chinese junks fitted with bulkheads during his sojourn in far eastern Siberia. In 1795, after his return to Britain, he built two sloops, *Arrow* and *Dart*, with watertight bulkheads, the first time Western ships had been so fitted. Watertight bulkheads soon became standard fixtures on ships, a practice that continues to this day.[30]

### The Search for Speed under Sail

The growth in size of wooden sailing ships during the nineteenth century was matched by a corresponding increase in speed. This was due to a combination of practical improvements such as coppering and advances in theoretical hydrodynamics that overturned previous concepts and paved the way for a more complete understanding of ship resistance.

The Scientific Revolution had spawned great leaps in the understanding of ship resistance and hydrodynamics. Unfortunately, neither theoretical developments nor experimental testing had resulted in any practical improvements to the actual performance of ships. Hydrodynamic theory had been the focus of extensive research by Pierre Bouguer, Leonhard Euler, and Jorge Juan, but the only practical result in ship

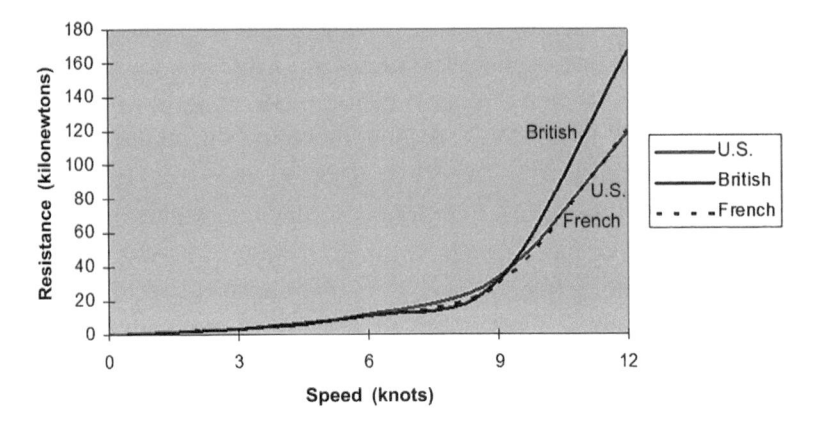

**Figure 1.9**
Ship resistance versus speed for U.S., British, and French frigates, circa 1805.
Credit: Author, after Dirk Böndel.

design was a coefficient called the ratio of bow resistance used by French constructors, which gave only comparative results with other vessels and could not predict ship speed. John Sewell's Society for the Improvement of Naval Architecture had been founded in part on the belief of British naval officers that French ships were faster than British ships and that this superiority was due to the French use of hydrodynamic theory. As figure 1.9 shows for a series of normalized U.S., British, and French hulls, they appear to have been correct that French ships *were* faster, but this had nothing to do with the use of hydrodynamic theory; the nascent American navy also had faster ships than the British, but without any resort whatsoever to such theory.[31]

Experiments in towing small wooden prisms and ship-shaped models, conducted by French scientists such as Jean-Charles de Borda and Charles Bossut and the Swedish naval constructor Frederik Henrik af Chapman, gave conflicting results that were almost impossible to apply to full-scale ships. Chapman, in fact, ceased using theory to design his hulls and instead advocated his method of distributing a hull's underwater volume in the shape of a parabola; this technique made the hull easier to define in mathematical terms but owed nothing to hydrodynamics.[32]

These mathematical theories came under increasing criticism during the early nineteenth century. In the Netherlands, a mathematics professor named Jan Frederick van Beeck Calkoen condemned the works of Leonhard Euler and Jorge Juan for inaccurately defining metacentric stability (although in fact Calkoen himself was wrong) and developed a series of geometric curves to define hull shapes that were never used in practice. A fellow countryman, Folkert Nicolaas van Loon, favored model tests over mathematics for improving the hull forms of sailing yachts and steamboats (primarily by avoiding any straight lines), and his designs did come into common use on Dutch waterways. In Sweden, the mining engineer Pehr Lagerhjelm, as part of a larger series of experiments on hydraulics and water power, meticulously towed bodies of various shapes through water and arrived at experimental results that were so similar to Beaufoy's that Beaufoy had intended to publish them side by side with his own but died before being able to do so.[33]

Sail theory fared little better during this time. Bouguer, Euler, and Juan had exhaustively studied the placement of masts on ships, but only Bouguer's concept of the *point vélique* (windage point), which defined where the center of sail effort should be located for maximum effect, was routinely incorporated into ship construction plans. However, the concept was never built upon with further research, and it disappeared from practical use by the mid-1830s. These same three scientists also calculated the wind force on sails as a measure of propulsive power. That work was extended somewhat during the 1820s by a French naval officer named Louis-Adrien Thibault, who measured the force of air on inclined cardboard surfaces, using a whirling-arm apparatus similar to that of another by Borda, then scaled the results up to full-sized sails. His work, regrettably, received almost no notice from scientists or constructors. Instead, most enhancements in sail design, such as split topsails and topgallants, were intended to make sail handling less manpower intensive.[34] By the early nineteenth century, therefore, new ideas for improving the speed of ships under sail were sorely needed.

Many inventors worked on improving the sailing qualities of ships. The aforementioned John Schank developed the now-popular sliding keel, essentially a centerboard that can be raised and lowered through a slot in the hull and that allows deepwater vessels to enter shallow waters. Richard Gower, a ship's master for the East India Company, tinkered with rigging and masting to improve speed and handling.[35]

Despite all this tinkering, the only innovation that universally improved the speed of sailing ships was copper sheathing. European navies and shipowners had tried for centuries to protect against the teredo worm, which ate through wooden hulls, by using various methods including lead sheathing. During the Seven Years' War, the British navy, after several tentative trials using copper plates in 1758–1759, applied full copper sheathing in 1761 to the hull of the 32-gun frigate HMS *Alarm*, bound for the West Indies. In addition to the immediate cessation of worm damage, the navy noted that, after two years on station, the ship's bottom stayed relatively free of barnacles and other fouling (copper is poisonous to marine life). Since hull fouling greatly increases skin friction, copper sheathing immediately translated to sustained higher speeds throughout a ship's lifetime.

The practice of coppering was halted when the iron spikes holding the copper plates began eroding at a rapid pace, endangering the watertight integrity of the hull. The problem was traced to the galvanic action between the copper plates and the iron spikes that held them to the hull. The British navy tried several means of isolating the two metals, using felt and other materials, but the problems of corrosion remained. Other navies learned of the advantages of coppering and immediately began sending naval officers to Britain to learn more about it and to transfer the new technology to their own shores. In France, the navy noticed the problem of galvanic corrosion almost simultaneously with the British and, like them, quickly halted the practice.

The War of American Independence (1775–1783), which brought Britain, France, and Spain into conflict, revived the practice of coppering. By the middle of the war, Britain had developed a temporary fix using tarred paper insulation, which allowed Britain, at enormous expense, to copper its entire fleet, while French and Spanish ships remained largely uncoppered. This disparity helped the British prevail in several encounters, including the Battle of the Saintes in 1782 and the Battle of Cape Spartel in 1783. By the end of the war, Britain had developed copper and bronze spikes forged from a harder alloy, which could be substituted for the ironwork to largely eliminate the galvanic action. France quickly established its own domestic supply base for copper plating and spikes. It sent agents into British factories to copy their manufacturing methods and to entice British coppersmiths to come work in France. Spain brought British workers into its factories and dockyards for their copper expertise. All three fleets were fully coppered when they met in 1805 at the Battle of Trafalgar, which ended with an overwhelming British victory over the French and Spanish. Coppering quickly spread to merchant fleets; by 1820, almost all East India Company ships were coppered, dramatically reducing both sailing times and overall costs.[36]

Coppering was an expensive process, and hardened copper spikes were both more costly and weaker than iron fasteners, so navies and shipowners continued to use iron in certain applications to save money. However, this continued to cause problems with galvanic corrosion, finally driving the British Admiralty in 1823 to request that the Royal Society investigate the problem. Humphrey Davy, the noted chemist, set about testing configurations of copper sheets and fasteners, along with iron and zinc plates, submerged in seawater. He concluded that the corrosion, caused by the oxidation of the copper, could be significantly reduced by attaching blocks of zinc or iron to the hull, which would corrode instead of the copper. These sacrificial anodes, as they are now called, soon came into widespread use around the world and are used to this day to protect hulls from galvanic corrosion.[37]

Another practical development, the introduction of yacht-like hulls for naval vessels, became enmeshed in British politics. Large sailing warships were generally built with wide U-shaped hulls (as seen in cross section amidships), which provided sufficient displacement and hull volume to carry large guns with adequate stability and to contain plenty of stores, ballast, and armament. It was well known at the time that such full-hull forms were slower than V-shaped hull forms with sharp deadrise, used by yacht builders.

William Symonds was a naval officer who, after the Napoleonic Wars, took to designing racing yachts having particularly sharp deadrise and wide beam that required little or no ballast to remain stable. One of these yachts, which sailed exceptionally well, caught the attention of some yachtsmen who happened to be high-ranking politicians of the Whig party. Symonds then designed yachts for his influential friends (most notably Robert Vernon), and they in turn persuaded the Admiralty to build several warships along the same lines, convinced that Symonds had discovered a new principle of naval architecture. In fact, the combination of wide beam and sharp deadrise was already used by American naval constructors to build small, fast warships like revenue cutters and sloops of war.

Symonds, by now a Whig partisan, rode the wave of reform when in 1830 the previously dominant Tory Party lost the British Parliament to the Whigs. The new party swept away the vestiges of the old one, including fixtures of the Admiralty such as the Navy Board, which had previously decided on ship designs. Symonds was appointed surveyor of the navy in 1832, and without the Navy Board looking over his shoulder, he was given free rein to design warships to his liking. Assisted by his more experienced deputy surveyor John Edye, who performed the calculations, Symonds developed over a dozen yacht-like vessels that made excellent speed in fine weather, most notably his 78-gun HMS *Vanguard*. However, during the 1844–1845 series of so-called Experimental Squadrons that matched Symonds's vessels against more traditional ships, Symonds's warships performed badly in rough weather, slowing considerably, rolling and pitching uncomfortably, and proving to be unsteady gun platforms. Much of the criticism of his designs, however, was actually prompted by the political nature of his appointment. By 1847 the weight of these political critiques forced Symonds to resign from office.[38] By that time, a new concept in hull forms, based in geometric practice and not on a yachtsman's sea sense, had come to dominate the field of naval architecture.

The original notion of basing ships' hull forms in mathematical theory came from Isaac Newton, for the simple reason that he invented the type of mathematics (the calculus) necessary do so. In his groundbreaking 1687 work *Principia Mathematica* (Mathematical principles) he developed the concept of a flat-nosed solid of least resistance

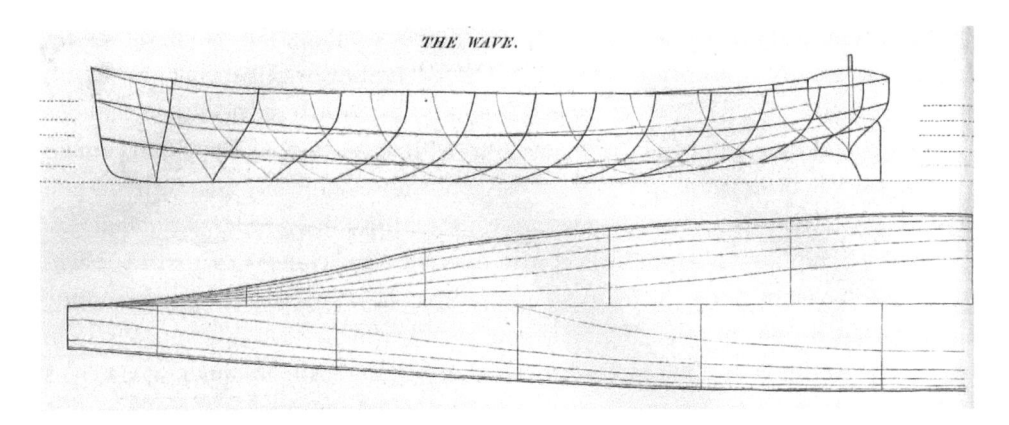

**Figure 1.10**
John Scott Russell's *Wave* with concave waterlines.
Credit: Author's collection.

based on the notion that fluid resistance was caused by the shock (impact) of fluid particles against a surface. When John Scott Russell began his experiments on canal boats in 1834–1835 (described in the prologue), he immediately thought of Newton's solid as the basis for his own wave-line form of least resistance, although he substantially altered it from a flat nose to a parabola. As he reported in 1835 to the British Association for the Advancement of Science (BAAS), his 23-meter boat *Wave* was built with parabolic, concave waterlines forward (figure 1.10). Russell claimed the hull "appeared to give no motion to any particles of water" but rather passed through the water "without ruffling the surface," which was "the criterion of minimum resistance."[39]

It was no oversight on Russell's part to present his results to the BAAS and not to the Royal Society, despite the society's notable work on ship structures and hull coppering. The BAAS had been created just a few years before, in 1831, as a direct response to the perceived decline in British science in general and the Royal Society in particular, compared with nations such as France and Germany that provided official patronage for their scientists. Noted British scientists such as Charles Babbage (who would later help Brunel with his railroad work) railed against the Royal Society for having descended from its previous heights into "malpractice [and] incompetence." Babbage and others felt that the Royal Society had become a club run by amateurs, whose leaders had no real interest in science but instead promoted its officers on the basis of nepotism and not on their scientific contributions.[40]

The BAAS was intended to change all that. Like the Royal Society, the BAAS did not receive any direct government patronage but instead raised funds through membership dues and other fees. Unlike the Royal Society, whose members could weigh in on any subject whether they had expertise in the field or not, the BAAS was organized around seven sections—mathematics, mechanical science, and so on—each with its own professional standing committees whose members identified subjects in need of investigation, evaluated research proposals, and allocated grants to carry out experiments and analysis.

Right from the start, naval architecture was at the top of the BAAS's list for requiring systematic research, and Russell's interest in waves and ship forms fit right in. From 1837 to 1844 the BAAS granted him £1,132 (over $1 million today), the second-highest sum paid out by the BAAS in that period, to study the nature of waves at sea and the proper form of ships to reduce wavemaking. Even as he was working as an engineer at the Caird marine engine works on the Clyde, he was also playing scientist by examining how different shapes of solids generated waves and developing some of the earliest observations on how waves traveled in groups.

The twin interests of science and engineering came together for Russell when he built and tested over 100 different hulls, ranging in size from 1-meter models to actual oceangoing ships of 60 meters' length, to confirm that his wave-line hull form generated the least wavemaking resistance. As with his 1834–1835 canal boat experiments, Russell tested the smaller models using falling weights to tow them in canals and basins. At the same time, he designed several larger vessels (such as the 80-tun steamer *Flambeau*, built in the shipyard of Robert Duncan), which were tested at sea using a tug and a dynamometer. Russell claimed to have conducted over 20,000 trial runs on these vessels, but unfortunately he never published their particulars. Rather, he simply wrote a short notice in the 1843 BAAS report saying that his discovery of the wave line "demonstrated a remarkable law by which it appears that each velocity [of the hull] had a corresponding form and dimension peculiar to that velocity." In other words, Russell claimed to have uncovered the fundamental principle by which *any* ship's hull should be designed, regardless of size.

Because Russell's wave-line theory dominated much of mid-nineteenth-century ship hydrodynamics, it is worth examining in some detail.[41] On the basis of his observations of wavemaking during the trials of canal boats described in the prologue, Russell suggested that waves come in four varieties, or "orders": (1) waves of translation, like the soliton, which carry mass and are responsible for ship resistance and ocean tides, (2) oscillatory waves, like wind-driven ocean waves, (3) capillary waves caused by surface tension, and (4) corpuscular waves, like sound waves. He focused most of his research on the first type of wave, which became the basis for his wave-line theory. He suggested, following the research of previous mathematicians, that the wave of translation had a vertical profile in the form of a sine curve. The length of this wave profile, he said, followed the formula $L = 2\pi V^2 / g$. For example, a wave moving at 10 knots would be 16.1 meters long. The form of this wave would determine the optimal hull shape to generate the minimum resistance.

Russell was well aware of the earlier experiments of Mark Beaufoy, which identified bow pressure, stern pressure, and friction as the primary components of ship resistance. Russell, however, showed that wavemaking resistance was a critical additional component of overall resistance that had been overlooked by previous researchers, and he determined to find the proper hull form that would minimize this component. In the first trials of his boat *Wave* in 1835, he used parabolic curves as his preferred horizontal hull lines (i.e., waterlines), noting that this shape cut through the water cleanly. After 1844, when he had fully defined his wave-line theory, he argued that the waterlines should *not* follow parabolic lines but rather should follow the shape of the wave of translation—that is, a sine curve.

Russell's reasoning went as follows: (1) A particle of water resting at the surface will, when disturbed, follow the path of least resistance to return to its original state of rest. (2) This path of least resistance is the sine curve, which forms the free surface of the wave of translation in the vertical plane. (3) To follow the law of least resistance, this particle would necessarily follow the same sine curve, but in the horizontal plane, to return to its original state of rest. (4) Thus, the waterline of a vessel should also follow the same sine curve as the vertical wave of translation to give it the minimum wave-making resistance. To confirm this theory, he sailed a wave-line vessel in a field of small floating balls and observed that they did not strike the hull but were instead simply nudged aside.

Russell argued that the sine curve was applicable only for the shape of the forward waterline, since the bow pushing through the water created a wave of translation. For the stern of the ship, he imagined that there was also a "wave of replacement" that filled in the space vacated by the ship as it passed through the water. This wave, he argued, was similar to ocean waves, which according to the prevailing theory of the time (discussed in chapter 3) had a cycloidal shape. This cycloid—which is a type of trochoidal curve, or generated by a point on the circumference of a rolling circle—would be only two-thirds the length of a wave of translation for the same speed, so a 10-knot wave would be 10.8 meters long. Consequently, he argued, the aft waterline should be cycloidal and not a sine curve.

Russell's formula for creating a wave-line hull form was as follows: first, set the length and beam of the ship and its speed; next, establish the length of the forward (entrance) waterline on the basis of the speed of the wave of translation, and trace the sine curve from the centerline at the bow out to the maximum beam; then establish the length of the aft (run) waterline on the basis of the wave of replacement, and trace the cycloid from the centerline at the stern out to the maximum beam; and finally, connect the ends of the two curves with a parallel middle body. For example, a 30-meter-long ship with a 10-knot speed would have a forward sine-shaped entrance of 16.1 meters, a 3.1-meter parallel middle body, and an aft cycloidal run of 10.8 meters (figure 1.11).

It is important to note what the wave-line theory did *not* do. Most importantly, it did *not* provide any means of estimating resistance; Russell simply assumed (quite incorrectly) that the wave-line hull form had zero wave resistance, so that the overall resistance of a wave-line hull would be based on the same factors (friction, pressure) that Beaufoy had identified. The wave-line theory did *not* have a basis in physics; despite Russell's claim of 20,000 trial runs, he actually had little in the way of experimental data that described the mechanism of resistance. With its insistence on perfect sine curves and cycloids, the wave-line theory was less of a theory and more of a geometrically descriptive concept. Finally, the wave-line theory did *not* provide an ironclad rule for designing hulls, and its use was more honored in the breach than in the observance. The required sinusoidal and cycloidal shapes were often insufficient to provide the hull with the displacement and water-plane area required for flotation and stability. One yacht designer, finding that the wave lines provided too little displacement for his hull,

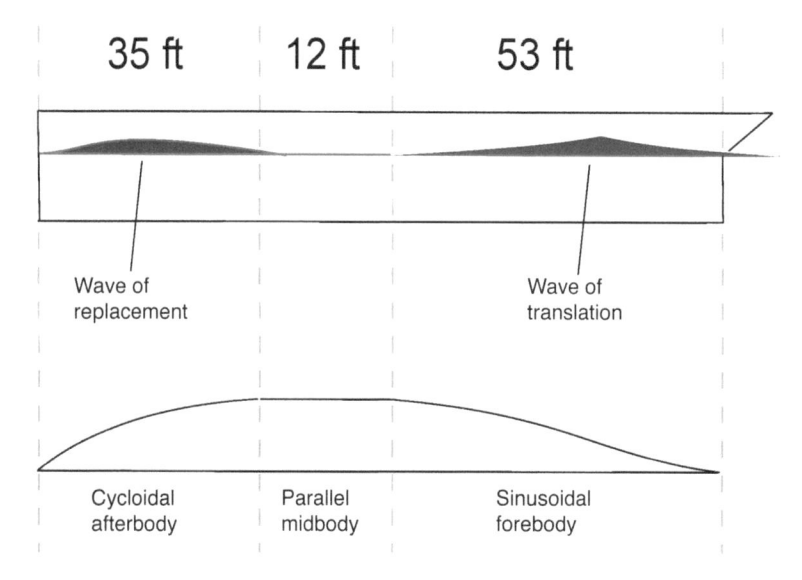

**Figure 1.11**
Wave-line theory.
Credit: Alex Pollara and author.

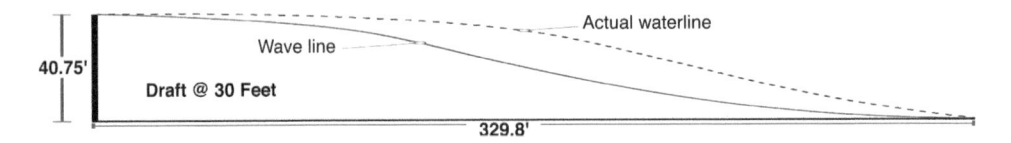

**Figure 1.12**
Russell's *Great Eastern* hull lines compared with wave line.
Credit: Alex Pollara and author.

was counseled by Russell himself "to fill out the lines, and by the time this was done … there was very little of the wave-line left." Russell's own *Great Eastern*, which he claimed to have been built according to the wave-line theory, had waterlines that were much broader and flatter than the perfect sine curves he espoused (figure 1.12)

Russell may have ignored some of the principles of the wave line for his own ships, but he ensured that the wider world knew of them through a constant stream of BAAS reports and his own lectures. He memorably described the work of the ship's hull as "excavating" a channel through the water. This visual representation of the ship as a plow captured the public's imagination. The wave-line theory was widely discussed in trade journals such as *Mechanics' Magazine* as well as more popular publications like *Literary Gazette*. The British naval officer Edmund Gardiner Fishbourne publicly explained the technical details of the wave line in his *Lectures on Naval Architecture* (1846). The wave line even drew the attention of poets and novelists. Henry Wadsworth Longfellow, in his 1849 paean to clippers, *The Building of the Ship*, alluded to the stern wave of replacement in his stanzas, "Broad in the beam, but sloping aft / with graceful curve and slow degrees / that the currents of parted seas / closing behind, with mighty force / might aid and not impede her course." Jules Verne, in *20,000 Leagues Under the Sea* (1869), was clearly referring to the wave-line theory when he described his fictional submarine *Nautilus* as having "lines sufficiently long and its run extensive enough for the displaced water to escape easily and provide no obstacle to headway."[42]

Shipbuilders who built wave-line ships noted that they had less resistance and thus were faster than older, more traditional hulls. The wave line became the basis for many steamships and sailing ships, although frequently modified by naval architects to suit specific needs. For example, packet ships built to cross the English Channel underwent a rapid evolution from 1830 to 1845 as a result of the wave-line theory, as shipbuilders quickly adopted the sharper, hollow-bowed hull form as being ideal for speed. A wave-line hull did not, however, guarantee a successful ship. The Scottish shipbuilder James R. Napier built several vessels to Russell's theories, but they were uniformly poor performers and Napier lost money on each.[43]

The wave-line theory found its ultimate expression in the clippers and yachts of the mid-to-late 1800s. Clippers were built for fast transport of passengers and perishable goods; yachts were built to win races. For both those types of ships, speed was paramount and almost every aspect of design and construction was bent to that goal. The clipper was originally developed in the United States in the 1840s to take advantage of the rapidly expanding tea trade with China. Its developer, a 30-something New York ship designer named John Willis Griffiths, was quite conversant in the latest theories of naval architecture, having apprenticed under the renowned New York shipbuilder Isaac Webb and trained as a naval constructor in the Gosport Navy Yard in Norfolk, Virginia. He had studied the various BAAS reports on Russell's wave-line theory, and in 1843 he apparently took those lessons to heart when he exhibited, at a New York City trade fair, his model for a fast ship having hollow waterlines at the bow. In 1844, while employed at the Smith & Dimon shipyard in lower Manhattan, he designed a new type of ship for the trading firm Howland & Aspinwall to ply the China tea routes. Named *Rainbow*, it was the first clipper to have hollow waterlines forward, although the waterlines were only slightly concave and not true wave lines. *Rainbow*'s first voyage to China in 1845 was so fast that the firm quickly ordered a second clipper, *Sea Witch*, which Griffiths designed with hollow sine-shaped waterlines that closely followed the wave line (figure 1.13). *Sea Witch*, launched in 1846, proved to be one of the fastest sailing ships ever built and greatly influenced the subsequent development of the clipper.[44]

Griffiths's very public demonstration of the benefits of the wave line, which had been until then confined to a few British yachts and channel steamers, influenced a new generation of shipbuilders on both sides of the Atlantic. After his third and fourth clippers, *Memnon* (1847) and *Universe* (1850), Griffiths never built another; instead, he went on to design fast steam packets, such as *Georgia* and *Ocean Bird*, and the gunboat *Pawnee* for the U.S. Navy, all sporting Russell's wave-line hull form. Other shipbuilders produced a

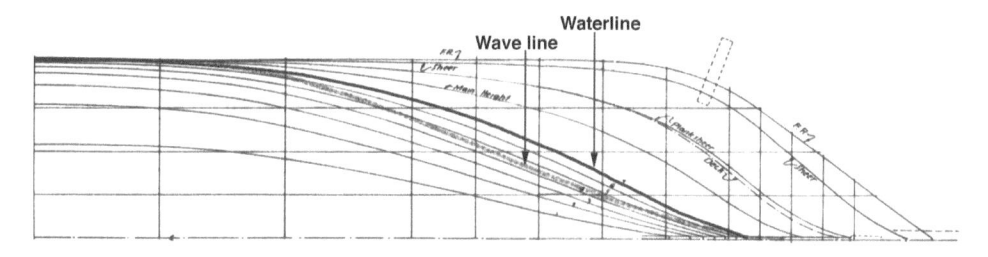

**Figure 1.13**
Griffiths's clipper *Sea Witch* hull lines compared with wave line.
Credit: Alex Pollara and author.

slew of wave-line clippers to meet the enormous demand for fast sailing ships, not just for the China tea trade but also for the sudden eruption of passenger traffic that followed the discovery of vast quantities of gold in California in 1848 and in Australia in 1851.

These shipbuilders, in the United States and Britain, were steadily shifting their work from slow and sturdy transatlantic packets to lighter and faster clippers. Packets were generally built with bluff bows, flat bottoms, and full waterlines to maximize the cargo capacity, but they could travel at only about 6 or 7 knots. Clippers had narrow V-shaped hulls and hollow waterlines, which gave them great speed (10 knots was commonplace, 20 knots was not unheard of) but reduced their cargo capacity. This was of secondary importance because their cargoes—generally tea and passengers—were both comparatively light and highly perishable, requiring the fastest possible delivery time.[45]

In the United States, New York and Boston became the hubs of clipper construction, with the hollow wave line forming the basis for their hull design. The most famous clippers were built by Donald McKay, one of a family of shipbuilders that moved from Nova Scotia to New York. McKay, who knew Griffiths from his own apprenticeship under Isaac Webb, adopted the wave line when he established his yard in Boston, building such legends as *Stag Hound*, *Flying Cloud*, and *Sovereign of the Seas*. Robert McKay (Donald McKay's uncle) put it succinctly when, during a visit to London, he told John Scott Russell, "I have adopted the wave principle in the construction of all my American clippers, and that is my secret. I first found the account of the wave line in the publications of the British Association." Other constructors also followed suit, including William Webb (Isaac Webb's son), whose extreme clipper *Challenge* broke the record for the passage to California.[46]

British clipper designers, surprisingly, were slower to adopt the wave-line principle. The wave-line bow was in fact preceded by a few years by the Aberdeen bow, an early development by Alexander Hall and Sons. In 1839, the Scottish shipyard came up with the idea of extending the upper deck to create a raked, or elongated, bow, giving ships more cargo capacity for the same registered (taxable) tonnage. Hall also demonstrated, by means of testing small towed models, that a raked bow was faster. The marriage of the Aberdeen bow with Russell's wave line apparently began with Hall's 1848 clipper *Reindeer* and was continued by other constructors around Great Britain. Alexander Stephen of Glasgow, for example, was convinced by his own experiments, carried out in 1853, to "try the Wave Line" for his clipper *Storm Cloud*. After the 1860s, the wave line lost its luster; none of the ships that later ran the great tea clipper races—*Ariel* and *Taeping*, followed by *Thermopylae* and *Cutty Sark*—had hollow bows. In the Netherlands and France, however, the shipyards of, respectively, Frederik François Groen and Jean-Lucien Arman continued to build wave-line clippers through the 1860s.[47]

The wave line became a far more prominent and longer-lived fixture in yacht design. Boat constructors began building wave-line yachts in 1844, a year after Russell's BAAS reports outlined his model test results.[48] Russell himself built the wave-line yacht *Titania* for the railroad engineer Robert Stephenson, which helped smooth Stephenson's entry into the prestigious Royal Yacht Squadron. In 1851, the squadron invited its counterparts at the New York Yacht Club to Cowes on the Isle of Wight to celebrate the Great Exhibition that Russell was helping oversee.

Unbeknownst to the Royal Yacht Squadron, the members of the New York club were readying their own wave-line yacht to compete against Britain's finest. George Steers was a young constructor who had apprenticed at Smith & Dimon with Griffiths and had undoubtedly learned the wave-line principle from him. Steers used it on one of his early yachts, *Mary Taylor*. Its performance so impressed the New York Yacht Club that a syndicate (led by Commodore John Cox Stevens) commissioned Steers to build a larger yacht on the same wave-line principles (figure 1.14), which they would sail across the Atlantic to race against the best of Britain.

On 22 August 1851, *America* soundly defeated a fleet of 14 British boats in the Hundred Sovereigns Cup regatta around the Isle of Wight. *America*'s victory was so resounding that a cartoon of the time showed Queen Victoria asking which yacht came in second, only to be told, "Ah, your Majesty, there is no second." A separate race a week later pitted *America* against Russell's own wave-line yacht *Titania*, which *America* again handily won. Steers's lifelong friend John Griffiths crowed, "The wave line principle, carried out in the construction of the *America*, had been [an] abundant success." John Scott Russell blamed his defeat on arcane government regulations that forced him to compromise his wave-line principles, whereas "Mr. Stevens' yacht, *America*, was a pure wave-line vessel built without the trammels of measurement tonnage." Russell claimed that *America*'s victory forced British regulators to drop their onerous requirements: "America," he said, "reaped a crop of glory; England reaped a crop of wisdom.... It was worth the loss of a race to gain so much."[49]

*America*'s victory had even wider repercussions for naval architecture. The wave line was no longer seen as a mere scientific curiosity but as a fundamental principle for designing yachts and other fast boats. For example, when the conventionally hulled U.S. revenue cutter *Joe Lane* was badly damaged in 1851, it was completely rebuilt with

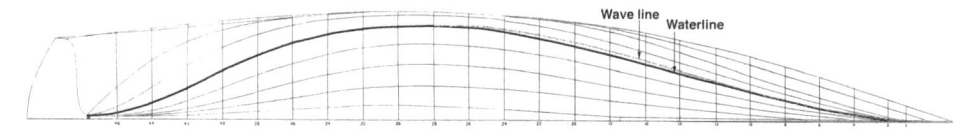

**Figure 1.14**
The hull lines of Steers' yacht *America* compared with wave line.
Credit: Alex Pollara and author.

a wave-line hull. Entire chapters of boatbuilding texts were devoted to the wave line. Other designers quickly took note of the wave line's success and tried to take credit for the concept. Thomas Assheton Smith, a wealthy British sportsman, claimed that he had built wave-line yachts several years before Russell. Another wealthy British dilettante, Robert Montagu, developed the vague notion that water particles follow a geometric dividing line that can be used to form the hull shape. Colin Archer, a far more experienced Norwegian naval architect who went on to design Fridtjof Nansen's polar ship *Fram*, argued in 1878 that the wave line applied not to the shape of the hull but to the shape of the sectional-area curve.[50]

Not all naval architects believed in the wave-line theory. The great yacht designer and builder Nathaniel Herreshoff, who was trained as an engineer at the Massachusetts Institute of Technology, explicitly rejected the wave line (and all other scientific theories) in favor of his seat-of-the-pants approach to hull design. His engineering intuition proved unerring; from 1893 until 1920 he designed and built five consecutive America's Cup defenders—including his 1903 masterpiece *Reliance*—none of which had a whit of hollow in the bows.[51] Herreshoff's skeptical intuition about the wave line also proved correct, because continued advances in hydrodynamic theory showed that Russell's concept was based on flawed principles. Nevertheless, the fascination with the geometric simplicity of the wave line continued well into the twentieth century.

**The Sailing Ship Effect**

The wooden sailing ship reached its apogee during the mid-to-late nineteenth century—built larger and faster and capable of carrying more cargo or armament than anything in the previous three centuries. This happened at exactly the same time that the technologies of iron and steam were being developed at a breakneck pace, yet the wooden sailing ship persisted for almost half a century before being completely sidelined. The reason that shipowners continued to order and operate wooden sailing vessels was that their increasing size and speed made them profitable even in the face of competition from steamships.

Sometime in the 1960s economists started using the term "sailing ship effect" to describe industries that were faced with new, disruptive technologies threatening to sideline existing products and that responded by improving and innovating their old technologies to save them from extinction. The term came from the perception that wooden sailing ships fought back against iron and steam, that shipbuilders and shipowners developed innovations directly in response to the threat of the iron steamship. This was a complete misreading of the chronology; in fact, the most important innovations that gave wooden sailing ships their second life—diagonal truss framing, coppering, waveline hulls—all arose well before the oceangoing iron steamship had become a feasible venture. Instead, these sailing ship innovations—which also later found a place in iron steamships—were the result of the dogged, scientifically informed study and experimentation that would become the hallmark of modern naval architecture.[52]

For some time, these improvements in naval architecture, coupled with handier sail designs, helped stave off the rapid advance of iron steamships, but the pace of industrialization, the decline of older wood-building trades, and the expansion and integration of manufacturing industries inexorably sounded the death knell for the stately, craft-built windjammer. The nostalgia felt by Hugo and Ruskin for the bygone era of sail persisted well into the industrial age and may explain why economists chose its imagery to explain one of their more cherished theories. This longing was perhaps best expressed in John Masefield's 1916 poem "Ships," whose final lines may be found today on the dry dock walls surrounding the legendary clipper *Cutty Sark*:

> They mark our passage as a race of men—
> Earth will not see such ships as those again.[53]

## 2  Steam, Iron, and Steel

The nostalgia that Victor Hugo expressed in *Les Misérables* for old-fashioned sailing ships was equaled only by his disdain for the newfangled steam-powered ones: "A thing which smoked and clacked on the Seine, making the noise of a swimming dog, came and went beneath the windows of the Tuileries. It was a worthless mechanism, a sort of toy, an inventor's pipe dream, a utopia: a steamboat. The Parisians looked upon the useless thing with indifference." John Ruskin, meanwhile, complained that steam was "the very curse and unmaking of us" and had little use for iron as a structural material.[1] Poets might have railed against the intrusion of such technologies into their carefully constructed aesthetic, but societies as a whole adopted steam, iron, and later, steel, for railways, river and canal boats, and ocean shipping, all owing to one overarching reason—*predictability*. Predictable transportation greased the skids of commerce, leading to lower market costs, increased throughput, higher profits, and thus greater confidence by business owners for making future investments.

The rise of New York City in the early 1800s to become America's center of trade was a microcosm of this drive for predictable transportation. In the early years of the American republic, Philadelphia was both the nation's most populous city and its largest trading port, in part because, as a freshwater port, fouled ships would see their barnacles and other marine growth die and slough off while loading and unloading cargo, without the expense of hauling down and scraping the hull. Coastal trade between Philadelphia and other cities was also brisk. In 1790, the clockmaker and inventor John Fitch sought to take advantage of this by introducing the world's first successful steamboat service, among Philadelphia, Trenton, and other cities along the Delaware River. The money-losing venture lasted less than a year before Fitch pulled out for lack of funds, and no one else stepped in to take his place. In New York City, by contrast, the artist and inventor Robert Fulton initiated a longer-lived steamboat service along the Hudson River (then called the North River) to Albany in 1807. Although Fulton and his business partner Robert Livingston theoretically held a monopoly on New York steam navigation, in practice other inventors like John Stevens also introduced ferry services in the region.[2]

After 1815, which marked the end of both the Napoleonic Wars in Europe and the War of 1812 between Britain and the United States, a flood of pent-up trade was released. By then, much of America's commerce was internal, carrying overseas goods to inland ports, bringing goods from the nation's interior for export to Atlantic ports, and most of all, trading among the expanding numbers of states. Philadelphia's trade network had not developed much beyond the days of sail, with ferry and cargo sailings not departing on fixed dates and times but instead subject to the vagaries of wind and tide. By contrast, the merchants of the New York port had already adapted to the advantages of steam. For the first time, steamboat timetables were being printed for ferry and cargo service to New Jersey, Long Island, Connecticut, up and down the Hudson, and eventually to the nation's heartland via the Erie Canal. Steam propulsion enabled the predictability of routes and times, so that commerce could now be carried on to fixed shipping schedules. This was soon extended to the packet lines that set sail across the Atlantic on specified dates, independent of the weather. At the same time, the decreasing availability of dependable timber supplies catalyzed many industries (especially shipbuilding) to substitute iron as their primary building material, now being mined and produced in predictably increasing quantities. Thanks to the rapid adoption of steam and then iron, New York went from being a regional hub to the nation's center for trade, manufacturing, and finance, accounting for two-thirds of all U.S. imports, one-third of its exports, and 70 percent of its immigrants.[3]

On both sides of the Atlantic, the changeover from wood and sail to iron and steam played havoc with the ship design and shipbuilding industries. The initial assumption was that wood craftsmen would become iron forgers and riveters and that boilermakers and engine makers would replace sailmakers. Workers would simply trade in one set of skills for another, and yards that had formerly constructed wooden sailing ships would begin making iron vessels. Although this transition did occur in some shipyards, for the most part this was not the case; there was almost always a clean break between the wood shipwright and the iron shipbuilder. Most wood-building shipyards went out of business, and new iron shipyards sprang up on greenfield sites or alongside locomotive works. The overwhelming evidence shows that the iron-and-steam ship was not the product of the wood shipyard; it was the product of the machine shop.[4]

This chapter describes the development of the technologies—steam engines, paddle wheels, propellers, iron, and steel—that reshaped shipbuilding in the nineteenth century. Chapter 3 explores the theoretical advances in naval architecture that accompanied iron, steam, and steel shipbuilding as engineers and scientists attempted to understand the fundamental laws that governed their behavior and allowed them to predict their performance.

**Human-Powered Mechanical Propulsion**

The earliest systems of mechanical propulsion were paddles and oars. They were undoubtedly used in Paleolithic times for waterborne voyages, not just in lakes and rivers but across oceans: out of Africa, to Australia and the Pacific Islands, and along the icebound North American coastlines. However, the first direct evidence of oared boats so far found is in a series of Bronze Age rock carvings in Norway dating from approximately 2000 B.C.[5] For millennia, the paddle and the oar evolved and were perfected by cultures across the globe, with no recourse to mathematics or mechanics.

The Mediterranean Sea was home to one of the most famous types of oared ship, the rowed galley, and this became the subject of the first theoretical treatment of naval architecture to ship propulsion. In 300 B.C., a scholar from the school of Aristotle wrote *Mechanical Problems*, a treatise of how various practical problems could be solved with the correct application of the lever. Several of those problems concerned the action of oars and tillers in a large rowed galley, most notably the need to have more of the oar on the inside of the ship to provide the greatest leverage for propulsion. Although the mechanical principles of the lever were not well understood at the time (the assumption was that the product of force times velocity was equal on either side of the fulcrum, as opposed to the product of force times distance), it provided the impetus for later scholars to suggest improvements to the oar. The most famous of these were Vettor Fausto and Galileo Galilei, both of whom worked with the Venice Arsenal in the sixteenth century, and the eighteenth-century Swiss mathematician Leonhard Euler. Fausto had translated *Mechanical Problems* into Latin and came away with the insight that the then current Venetian galley, the trireme (three rowers per bench), could be improved by adding two additional rowers inside the first three. In 1529 he created the larger and faster quinquereme (five rowers per bench), some of which served as Venetian flagships. In 1592, Galileo used the same Aristotelian text to propose his own theory of rowing to the Arsenal commissioners, but with conclusions opposite Fausto's. Instead of recommending more rowers inside the vessel, Galileo suggested more of the oars be outside for improved leverage. The commissioners pointed out that this would cause oars to break and that the required cross-sectional area for longer oars would make them heavy and unwieldy, which gave Galileo the impetus to research the strength of materials. This later resulted in his famous *Two New Sciences*, which overturned many classical theories in favor of mathematically based rational mechanics. A century later, in 1747, Leonhard Euler was also inspired by *Mechanical Problems* to apply more advanced mathematical analyses to rowing to calculate the speed and efficiency of galleys (arriving, for example, at the very modern conclusion that the best way to improve rowing efficiency is to increase the blade area of the oars). With oared galleys already largely replaced by sailing ships, Euler's analysis of rowing would not be improved upon until the twentieth century.[6]

By Euler's time, the attention of both inventors and scientists was not focused on rowing but on other means of propulsion, notably the paddle wheel. The undershot waterwheel, which appeared across the globe by A.D. first century, may have inspired the idea of a sort of rotating bank of oars to propel a ship. The first recorded appearances of the paddle wheel date to about the fifth century, in Roman and Chinese texts. Whereas the European paddle wheel appears to have fallen out of favor soon after the fall of Rome and for almost a millennium thereafter, Chinese shipbuilders continued to build larger and more elaborate paddle wheelers, with either side or stern wheels (or both) driven by men on treadles, primarily for river and inland navigation. In the Song dynasty (960–1279), ships with 11 paddle wheels on each side and a stern wheel carried hundreds of amphibious troops into battle. In fact, human-powered paddle wheelers were in operation as late as 1842 by the Qing dynasty navy against British steamships during the First Opium War.[7]

By contrast with China, knowledge of paddle wheel boats appears to have been lost in Europe through most of the Middle Ages. They began to reappear during the Renaissance, as evidenced in drawings of Leonardo da Vinci and paintings by Raphael. Spain at this time was at the center of paddle wheel development, because it needed a reliable means of getting the hundreds of warships and armed merchantmen that sailed annually to its new colonies in Asia and the Americas into and out of port, without having to wait for favorable winds and tides. Many sailing ships were equipped with a line of oars to accomplish this, which had to be rowed by experienced crew. Paddle wheels held the promise of greater efficiency without requiring highly trained oarsmen. The most famous of the many Spanish paddle wheel developments were those conducted from 1539 to 1543 by Blasco de Garay, a naval officer and inventor. With financing from King Carlos I, Garay conducted a series of six carefully measured experiments in Málaga and Barcelona, using several mechanical arrangements and with geared crankshafts or capstans driving up to three pairs of paddle wheels. However, the paddle wheel ships proved slower than oar-powered ones, so were abandoned. Another widely reported experiment in 1702, involving a boat equipped with a paddle wheel system created by François Du Quet, a prolific French inventor, met the same fate. Other paddle wheel trials in Spain and France during the seventeenth and eighteenth centuries gave similar results, with the upshot that human-powered paddle wheels never caught on in Europe the way they did in China.[8]

The failure of paddle wheels did not stop navies, scientists, or inventors from pursuing the dream of navigating without oars or sails. The French Academy of Sciences was the nexus of those three parties during the eighteenth century. In addition to funding the ship model experiments mentioned in chapter 1, the Academy also sponsored annual prizes on subjects of national interest. The 1753 prize was titled "On the Most Advantageous Way to Supplement the Action of Wind on Large Vessels." Nine entries were received, of which that of Swiss mathematician Daniel Bernoulli was declared the winner and Leonhard Euler's received an honorable mention. The two entries had much in common, not surprising given the close working relation (largely by correspondence) between the authors. Both men proposed and analyzed several mechanisms for moving boats through the water, including oars, paddle wheels, and feathering boards. Of greater interest was that Bernoulli and Euler proposed waterjets as a propulsion device, with the reaction coming from the expulsion of water from the after end of the ship by either human-powered pumps or by raising water to an elevated tank and letting it flow out the back. Euler admitted, however, that the waterjet would be inefficient compared with oars or paddle wheels and instead suggested a novel device inspired by windmills. As shown in figure 2.1, Euler proposed a rotating vane projecting from either the bow or stern, with paddles set at oblique angles to propel the ship—in other words, the first written description of a screw propeller.[9]

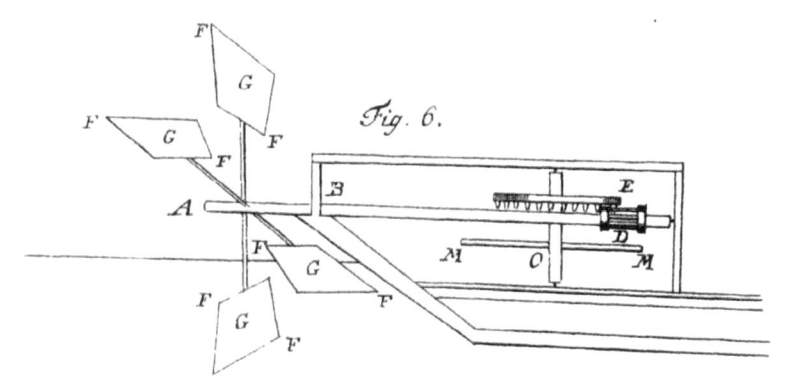

**Figure 2.1**
Leonhard Euler's propeller, 1764.
Credit: Berlin-Brandenburg Academy of Sciences.

Just over 20 years later, near the start of the War of American Independence in 1776, an American inventor named David Bushnell built the human-powered submersible *Turtle* to attack British warships. He, too, was inspired to fit his boat with a pair of windmill-shaped screw propellers, the first known to have ever been successfully used on a vessel.[10] By the end of the eighteenth century, two promising methods of propelling ships—the paddle wheel and the screw propeller—had come into use. What was needed was a more powerful source of motive power than humans. That motive power would be steam.

**The Evolution of Steam Power**

Daniel Bernoulli's 1753 French Academy prize memoir on waterjet propulsion included a short discussion of a novel means of raising water to an elevated tank—using steam instead of humans to power the pumps.[11] Bernoulli was generally dismissive of the concept, arguing that the method was inefficient. He based this observation on a pair of Newcomen pumping engines, built in 1726 to raise water from the Thames to supply London with fresh water. Thomas Newcomen, an ironsmith in Devonshire, had developed the first practical steam engine for pumping water in 1712. Newcomen was attempting to solve the problem of dewatering tin and coal mines, which were prone to frequent flooding. His device could raise water by admitting steam from a boiler into a cylinder that pushed up a sliding piston, which then condensed the steam, creating a vacuum that drew down the piston. The alternating action of the piston moved a beam up and down, which in turn operated a water pump.

Newcomen's "atmospheric engine," as he termed it, quickly found its way across Britain and into Europe and the Americas, not only to dewater mines but also to pump water into cities and towns. But Newcomen's original concept, as Daniel Bernoulli correctly observed, was inefficient. The problem was that Newcomen's concept required the cylinder itself to be alternately heated and cooled to generate work, which because of thermal inertia of the cylinder walls (usually copper) was a slow process that wasted a lot of heat.[12]

As inefficient as the Newcomen engines were, for many years they were the only ones available. In 1736 a British inventor named Jonathan Hulls took out a patent for a tugboat powered by a Newcomen engine that was attached by ropes and pulleys to a paddle wheel at the stern (it is not clear if it was ever built). In 1776 a member of the French nobility, Claude François Dorothée, marquis de Jouffroy d'Abbans, also used a Newcomen engine to propel a small boat named *Palmipède* (Webfoot) on the river Doubs, using a rod attached to pivoting oars on each side of the vessel, in a motion similar to a duck paddling. The experiment was an unhappy one, so in 1783 he built a larger boat in Lyons, *Pyroscaphe* (Steamboat), with a more powerful two-cylinder

Newcomen engine that drove a pair of paddle wheels. However, a patent dispute kept Jouffroy d'Abbans from further developing his steamboat until 1816.

Across the Atlantic, Newcomen-style engines figured in the fierce competition between two American inventors, James Rumsey and John Fitch. The men had independently developed and demonstrated steam power to propel riverboats but by using very different methods of propulsion, which became the subject of a patent dispute. Both men had learned about Bernoulli's waterjet and paddle wheels from Benjamin Franklin's "Maritime Observations," mentioned in chapter 1. Franklin, who met Rumsey and Fitch, was enthusiastic about waterjets but deeply critical of paddle wheels, which appears to have influenced both men to discard paddle wheels in favor of other devices. Rumsey's boat, built for George Washington's new Potomac Company, which built canals to the nation's interior, used a jet propulsion system in which the steam piston connected directly to a pump that alternately sucked water through valves in the hull and then expelled it from a copper tube at the stern. After a single successful trial on the Potomac River in 1787, however, Rumsey dismantled the boat.

John Fitch, by contrast, favored pivoting oars for his steamboat *Perseverance*, which he demonstrated that same year before delegates of the Constitutional Convention in Philadelphia. After another three years' experimentation, in 1790 he built an 18-meter boat equipped with articulated duck-leg paddles at the stern. To verify the boat's speed, Fitch set up along the Philadelphia waterfront what was apparently the very first measured mile—sets of flagpoles set exactly one mile apart, which were sighted on to precisely determine speed. At a respectable eight miles per hour, Fitch's unnamed vessel began plying the Delaware River as the world's first successful steamboat service. The company survived less than a year, because both he and Rumsey had been granted patents on their respective engines, but their competing claims—and lack of monopoly status—meant that investors shied away from both projects. The fallout from the Rumsey-Fitch dispute led to the creation of the Patent Act of 1793, which swept away much of the previous confusion.

Although Newcomen engines were in wide use elsewhere, in Britain they were already on the wane. In 1765, James Watt, an instrument maker at the University of Glasgow, lit on the idea of connecting a separate condenser to the Newcomen piston-cylinder arrangement, which would allow the steam to evacuate and condense without having to alternately heat and cool the cylinder. The dramatic improvement in efficiency meant that more power could be generated with smaller, lighter machinery. By 1775 Watt had worked through the mechanical details of his new engine, after which he partnered with Matthew Boulton, a manufacturer of metal goods, to build them in quantity in the Soho Manufactory outside Birmingham. Boulton and Watt engines (and their derivatives after Watt's patent expired in 1800) were soon widely used for dewatering mines and pumping water to cities, as well as for running mills and factories. The idea of propelling ships and boats was never far behind.

A British engineer working in Glasgow was the first to successfully use a Boulton-and-Watt-style engine on a boat. William Symington had experimented on several variations of steamboats throughout the 1790s, with little to show for it. In 1803, Thomas, Baron Dundas, commissioned Symington to build a steam towboat for his Forth and Clyde Canal Company. Symington's 17-meter paddle wheel boat, named *Charlotte Dundas* (figure 2.2) after the baron's daughter, was equipped with a single-cylinder engine that generated a respectable 10 horsepower (7 kW). It was able to pull two heavily laden barges along the canal in strong headwinds. Word of the machine spread quickly and attracted visitors from far and wide to witness it in action. The owners of the canal company were less impressed, ostensibly concerned that the wash from the paddle wheels would erode the canal banks, and cut off further funding.

Among the visitors to *Charlotte Dundas* were Robert Fulton, an American painter; the British hotelier Henry Bell; and a 12-year-old boy named David Napier. Each would go on to make major strides in maritime steam propulsion. Robert Fulton had moved to London in 1786 to study under the Anglo-American painter Benjamin West. Fulton was also an inveterate tinkerer with a solid understanding of engineering. He was familiar with steamboat developments in both Europe and America and petitioned several

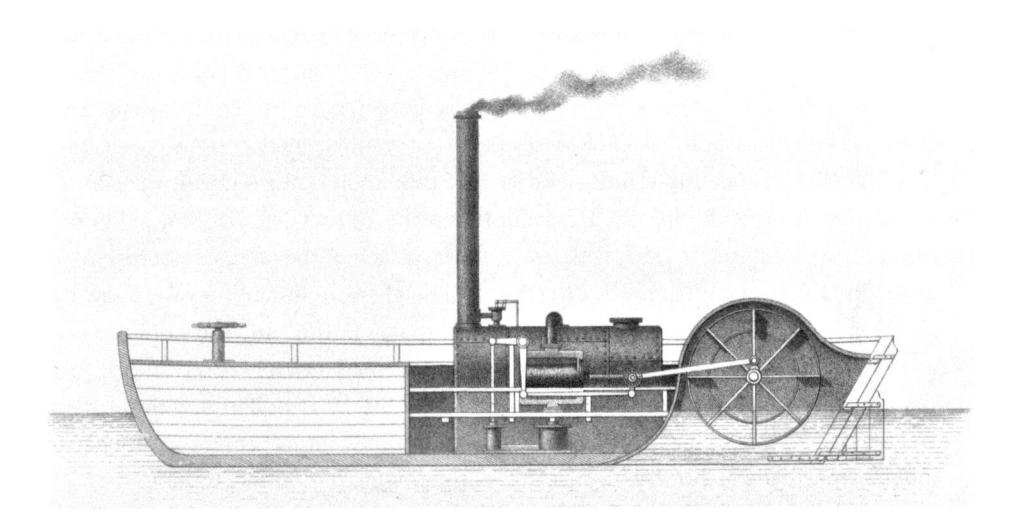

**Figure 2.2**
*Charlotte Dundas* of William Symington, 1803.
Credit: National Maritime Museum, Greenwich, London.

times for a scheme to improve canal navigation, without success. In 1797 he moved to Paris to try his hand, where he developed canal schemes as well as naval weaponry, notably a crude torpedo and a small manned submersible called *Nautilus*, again with little success. His opportunity for success came in 1802 when he met Robert R. Livingston, the new American ambassador to France.[13]

Livingston had been sent by President Thomas Jefferson to negotiate the purchase of the port city of New Orleans, which France had just taken over from Spain. Napoleon, however, offered to sell the entire Louisiana territory for a bargain price. Livingston immediately saw that this would give the United States complete control over the Mississippi River, the commercial lifeline of America's heartland, and that the steamboat would be the way to exploit it. Back in the United States, Livingston had been granted a 20-year monopoly on Hudson River steamboat traffic, contingent upon meeting a service speed of four knots. However, he and his brother-in-law John Stevens failed in their numerous attempts to build a working steamboat. When Livingston met Fulton, they realized that together they had the keys to building not only a steamboat empire on the Hudson but also an empire on the biggest prize of all, the Mississippi River.

Backed by Livingston, Fulton built his first steamboat, driven by a chain of paddles, but its trials on the Seine were disappointing. After his visit to *Charlotte Dundas*, however, Fulton had a clear idea of what success looked like—a paddle wheel boat driven by a Boulton and Watt engine. By 1807 he had bought a 20-horsepower (14 kW) model from the Soho Manufactory and had it shipped to New York City, where it was installed in a 46-meter boat. In August the *North River Steamboat* (later referred to as *Clermont*) began regular service between New York City and Albany, with newer boats added to the line. In 1811, Fulton's steamboat *New Orleans* inaugurated service along the Mississippi River. In the meanwhile, John Stevens had not been idle. Now operating independently of Livingston, he and his sons developed a series of screw propeller and paddle wheel steamboats to ferry passengers around New York and Philadelphia.

In Britain, Henry Bell also applied his observations from *Charlotte Dundas* to build a small steamer in 1812 (named *Comet* not for its speed but after the Great Comet of 1811) that would ferry passengers along the Clyde from Glasgow to his small hotel near Greenock. *Comet*'s boiler was built by David Napier, by then 22 years old and on the cusp of turning his father's foundry into one of the great marine engineering companies in Britain. In 1818, Napier commissioned his cousin William Denny to build the 90-tun *Rob Roy*, which plied the Irish Sea between Greenock and Belfast and thus became the first steam vessel to provide regular oceangoing service. The pace and extent of construction accelerated; by 1834 there were 70 shipyards around Britain building steam-powered vessels, although Clyde shipyards and marine engine builders dominated the market.[14]

At first, steam power was used primarily to supplement the inability of sailing ships to move against wind and tide or in confined spaces. J. M. W. Turner's painting *The Fighting Temeraire* on this book's cover shows a classic use—one of the last of Admiral Lord Nelson's sailing warships under tow by a steam paddle wheel tug. The first steamship to cross the Atlantic Ocean, *Savannah*, used its steam-driven paddle wheels only to move in and out of harbor or in adverse winds. Gradually, steam would come into its own as the primary source of propulsion power for all types of vessels.[15]

At this time, the development of steam-powered vessels was concentrated in two nations—the United States and Great Britain. However, geographic differences caused those developments to take widely divergent paths. With the Louisiana Purchase, the United States had become a vast nation of rivers and inland waterways whose trade would be primarily internal. The great steamboat empires begun by Fulton and Livingston were soon crowded out by dozens of competitors. On the Mississippi, Henry Shreve was both inventor and pilot of numerous steamboats that connected the rapidly growing urban areas along the great river and its tributaries. In New York, the steamboat ferries begun by Fulton and the Stevens family were soon challenged by an up-and-coming entrepreneur, Cornelius Vanderbilt. Steamboats spurred the rapid development of canals into the west, most famously the Erie Canal, which opened for business in 1821. For a time, steam even spurred a novel lower-cost competitor, the horse-powered ferry, on the more popular routes.[16]

In comparison with the United States, Britain was a small country with few major rivers and rapidly becoming a nation of exporters. Almost from the start, its steam activities were turned toward ocean transport and coastal trade. Soon after *Rob Roy*'s first voyage, steamships were regularly crossing the Irish Sea, North Sea, and the English Channel. In 1825, the East India Company established its first steamship voyages to India and began populating India's rivers with steam tugs and ferries. Early steam engines were fairly inefficient, with boilers operating at pressures of around one atmosphere (100 kilopascals, or 15 pounds per square inch), and the use of simple expansion engines meant that coal consumption was very high. The lack of coaling stations around the world also limited the utility of marine steam engines. For this reason, steam was initially used as auxiliary propulsion for oceangoing sailing ships to come in and out of harbors and in unfavorable winds. Such "mixed" ships were often equipped with screw propellers that could be hoisted into the ship to reduce drag under sail and funnels that were retractable to get them out of the way of the sails. When making ready to switch from sail to steam, the command "up funnel, down screw" was passed. In fact, pure steamships accounted for less than 20 percent of the total registered tunnage as late as 1869. After that date, the opening of the Suez Canal drastically cut the sailing distance (and thus reduced coaling costs) from Europe to East Africa and Asia, which led to a dramatic increase in steamship tonnage at the expense of sailing ships.[17]

A plethora of designs for boilers, engines, and condensers emerged from manufacturers, whose merits were hotly debated in the pages of trade journals like *Mechanics' Magazine*, even if fundamental terms like "horsepower" were still only loosely defined. The British and American navies closely watched these developments, allowing private companies to develop steam technologies and adopting the most promising ones only after careful consideration.[18] Navies had every reason to be cautious, for these early days of steam power were marked by rapid and widespread experimentation as shipbuilders and engine builders tried new ways to increase steam pressure, gain horsepower, and improve engine efficiency, often with disastrous results. Boiler explosions were a common occurrence in British and American waters, yet for many years both nations maintained a hands-off policy for fear of interfering with the free market. In 1830, the U.S. government funded the Franklin Institute, an industry-sponsored mechanic's institute, to conduct wide-ranging research into the causes of boiler explosions. This resulted in the 1838 Steamboat Act, which required regular boiler inspections. Britain followed suit in 1846 with the Steam Navigation Act.[19] By the 1850s, navies across the globe had universally embraced steam as the primary motive power and were actively involved in improving marine engineering.

As steamships began crossing the Atlantic, the competition increased between rival companies, pitting improved hull design against advances in marine engineering. As the prologue describes, in 1838 Isambard Kingdom Brunel built *Great Western* as the extension of his London-to-Bristol railway all the way to New York, two years after which Samuel Cunard extended Canada's steamship network from Montreal to Halifax all the way to Liverpool. Whereas Brunel continually advanced the state of hull design by subsequently constructing *Great Britain* and *Great Eastern*, Cunard poured his investments into Robert Napier's marine engineering company (Robert was a cousin of David Napier) to produce more efficient and more reliable steam engines built in very conventional hulls. In 1850 Edward Collins began service with his New York and Liverpool United States' Mail Steamship Company (known as the Collins Line) with high-speed ships having almost twice the installed power of the Cunard liners. The emphasis on marine engineering won the day; by 1858, Brunel's companies had gone bankrupt, and Cunard and Collins were leading the race to create a transatlantic empire.[20]

Although Britain and the United States dominated steamship development, other nations quickly followed suit, often starting with imported technology and know-how to jump-start their own indigenous capability. France, although the site of the first steamboats of Jouffroy d'Abbans and Robert Fulton, lagged well behind Britain. After the end of the Napoleonic Wars in 1815, France began buying its steam engines from British companies, and Aaron Manby, a British engineer, set up his engine works just outside Paris. To gain in-depth knowledge of how steamboats operated, the French navy in 1819 sent the engineering officer Jean-Baptiste Marestier on a mission around the United States. His memoir on steam propulsion provided a blueprint for France's rapid development of this technology for both civilian and military use. This experience led the French navy in 1828 to create its own marine engine works on the site of the gun forges at Indret, near Nantes.[21]

Other European nations also based their technological development on British steam engineering. Gerard Maurits Röntgen, a Dutch naval officer, visited British dockyards in 1818, and came back to found the Netherlands Steamboat Company in 1823. In 1843 a Danish naval officer on an industrial tour in London persuaded a young British engineer named William Wain to work at the Danish Naval Dockyard, after which he joined with Danish engineer Carl Christian Burmeister to form one of Europe's great marine engineering firms, Burmeister & Wain. Spain and the Italian states bought their first steamboats from Britain (in 1817 and 1834, respectively), and for many years afterward their navies and merchant shipbuilders bought engines from British manufacturers before eventually developing indigenous manufacture. The same held true for Russia (first steamboat purchased in 1818) and the Ottoman Empire (1822). The expansion of steam marine power in Asia happened along these same lines many years later, first in Japan in 1855, then in 1868 in China.[22]

Apart from the opening of the Suez Canal, the largest boost to the worldwide adoption of marine steam propulsion was development of the compound steam engine, which made more efficient use of the steam by allowing it to expand in multiple stages. It was first adopted in textile mills and locomotives before being applied in ships during the 1850s by marine engineering firms like John Elder of Glasgow and Henry Maudslay of London. The compound steam engine reached its apogee in the triple-expansion engine, developed independently in the 1870s by Charles-Benjamin Normand in France and Alexander Kirk in Britain. This engine type, coupled with improvements in the design and construction of fire tube boilers that boosted steam pressure (up to 10 atmospheres by the 1880s), further improved overall efficiency and reduced coal consumption. Although the sailing ship had reached its peak by the 1890s, the combination of lower coal costs due to the widespread adoption of the triple-expansion engine, lower shipbuilding costs, and reduced manning, coupled with steamships proving to have far fewer losses at sea than sailing ships, had tipped ocean freight rates in favor of steam over sail and marked the end of the stately, craft-built windjammer.[23]

At the time of the widespread adoption of the triple-expansion engine, three other revolutions in marine engineering were also taking place that would crystalize the future of shipping. The first was the development by the British engineer Charles Parsons of the compound steam turbine, in which steam drove a series of turbine wheels at very high speed, extracting an even greater proportion of energy than piston engines. Parsons famously demonstrated his invention in 1897 by running his launch *Turbinia* at the unheard-of speed of 34 knots through Queen Victoria's Diamond Jubilee Fleet Review off Portsmouth. The second, pioneered by the Italian naval constructors Nabor Soliani and Vittorio Cuniberti in the 1880s–1890s, was the use of "liquid hydrocarbon" (petroleum fuel oil) as a replacement for solid coal, on the grounds that it had higher heating power than coal (meaning greater speed), removed the need for coal stokers, and was also becoming more widely available than previously and thus less expensive. The third was a novel internal-combustion engine invented by the German engineer Rudolph Diesel at the end of the nineteenth century, which was fitted to ships by the beginning of the twentieth century. By the 1940s, the typical oceangoing ship was fitted with oil-fired steam turbine or diesel engines, a paradigm that lasted through the end of the industrial age.[24] This complex evolution in ship engines was only half the story; at the other end of the drivetrain, the evolution of the propulsor was just as convoluted.

**Paddle Wheels to Propellers**

In the 1753 French Academy of Sciences prize "On the Most Advantageous Way to Supplement the Action of Wind on Large Vessels," both Euler and Bernoulli roundly dismissed the paddle wheel, calling it an ineffective propulsion device. This did not stop the first hands-on inventors like Jouffroy d'Abbans, Symington, Fulton, and Bell, or their successors from using it on their steamboats. Indeed, for almost 30 years, paddle wheels (like those shown on *Charlotte Dundas* in figure 2.2) were the uncontested means for transforming steam power into speed, first on rivers and lakes and later on the oceans. Even after the introduction of the screw propeller in the 1830s, paddle wheels remained widely used for another two decades.

Steamboat operators had noticed early on what Euler and Bernoulli had argued decades before—paddle wheels were inefficient compared with oars, in large part because the floats (the blades of the wheel) entered and exited the water surface at a steep angle, slapping the water on entry and pulling up water on exit. These actions drew considerable energy from the steam plant without providing any forward thrust. In the 1830s, two British engineers, Elijah Galloway and Joshua Field, independently proposed a fix to this problem—a paddle wheel fitted with stepped floats arranged in a cycloid, so that each part of the float entered the water more or less vertically. Field, who by then was the senior partner in the firm Maudslay, Sons, and Field, which provided the engines for Brunel's *Great Western*, also fitted the ship with cycloidal paddle wheels. These stepped floats sustained considerable damage in service and were replaced with conventional floats within a few years. Other ships experienced similar problems, so cycloidal paddle wheels soon fell out of favor.[25]

Another, more successful mechanism to improve efficiency was the feathering paddle wheel, also invented independently by two engineers—Elijah Galloway (again) and François Cavé of Paris. In these systems, an eccentric center wheel was linked by articulating levers to hinged floats, so that they remained almost vertical throughout the bottom of the rotation and thus improved efficiency. Cavé's invention was not particularly successful, and the French navy rejected it in 1839 following a series of disappointing trials. By contrast, Galloway's invention saw greater acceptance after it was sold to William Morgan, a British engineer working in the Austrian Empire, who first mounted an improved version in 1829 on ferries between Trieste and Venice. These proved so successful that the Morgan wheel became widely used on river and intracoastal steamers. A theoretical and experimental analysis by British civil engineer Peter Barlow showed the Morgan wheel to be about 7 percent more efficient than conventional paddle wheels. During the U.S. Civil War, Confederate blockade runners slipping through Union lines at night preferred feathering paddle wheels because the floats slapped the water with less force and were more silent. However, the fragility of the articulating levers meant that feathering paddle wheels were rarely used on oceanic steamers.[26]

Paddle wheels in general were problematic for many oceangoing vessels. As mentioned earlier, steam was initially used as auxiliary propulsion for sailing ships, and paddle wheels were notoriously difficult to dismount to reduce drag under sail. Under steam, the roll of the ship would alternately raise and lower the rotating paddle wheels in and out of the water, which was inefficient and stressful on the machinery. As coal was burned and the ship drew less water, the floats would be less immersed, which affected efficiency. Variations in speed also caused the wave pattern along the hull to change, which could place the paddle wheels at a wave crest or trough, further reducing efficiency.[27] Naval officers disliked the paddle wheel boxes taking up so much valuable real estate at the middle of the ship at the expense of broadside cannon. In battle, the wheels were vulnerable to gunfire and to being fouled by masts being shot away. Finally, the paddle wheel displaced their command position from the time-honored quarterdeck at the aft of the ship, to a bridge that spanned the paddle wheel boxes (the term "bridge" stuck long after paddle wheels were gone). For these reasons, navies relegated paddle wheels to second-line warships and auxiliary vessels but never to major fighting ships.

Some inventors tried to improve on the vertical paddle wheel by laying it on its side. The most famous of these was a system developed by a U.S. naval officer, William Hunter, who in 1841 patented a pair of rotating horizontal wheels that protruded underwater from the sides of the ship. Although avoiding the pitfalls of the vertical wheel, the Hunter wheel lost much of its power pushing water through the enclosed casing. A series of experiments conducted from 1841 to 1847 on U.S. Navy sloops and Revenue Cutter Service cutters led to its abandonment. In Britain, a different horizontal wheel system, closer to the waterjet first proposed by Bernoulli and Euler, was developed by the civil engineer John Ruthven and installed in 1866 on the iron gunboat HMS *Waterwitch*. Trials showed that Ruthven's hydraulic propeller, which drew water from the ship's bottom into the encased horizontal pump wheel and ejected it via tunnels through two nozzles in the sides of the ship, was more complex than the screw propeller without providing any greater effectiveness. The system was abandoned, but successful waterjets in a different form would reappear in the twentieth century and be used for high-speed boats, ferries, and warships.[28]

The screw propeller eventually became the propulsor of choice to replace the paddle wheel, but only after more than a half century of experimentation did a standard configuration emerge. From 1785 until 1850, engineers and inventors (primarily in Europe) produced over 100 widely different patents and models for screw propellers, only a handful of which made it past the prototype stage and into large-scale production (figure 2.3). These screw propellers fell into two broad categories: those inspired by windmills, such as the first human-powered screw propellers of Euler and Bushnell mentioned earlier, and those inspired by the Archimedes screw, although these categories included many variations such as the British civil engineer George Rennie's spiral propeller modeled on a fish's tail.[29]

The windmill was perhaps an obvious choice to serve as a propeller model, because it spun freely in open air. Robert Fulton, who was familiar with both the French Academy of Sciences and Bushnell's *Turtle*, put a hand-cranked two-bladed propeller on his submersible *Nautilus* in 1800. Two years later, a British engineer named Edward Shorter fitted a similar propeller to a long pole at the stern of a merchant transport ship to demonstrate that a team of men working a capstan could move the ship around the Bay of Gibraltar at a very slow speed.

A windmill-shaped propeller was used by John Stevens, the first person to pair steam power with the screw propeller. At the time, he was attempting to start a steam ferry service across the Hudson River. From 1802 to 1804, Stevens experimented with a 25-foot

*17*  *18*

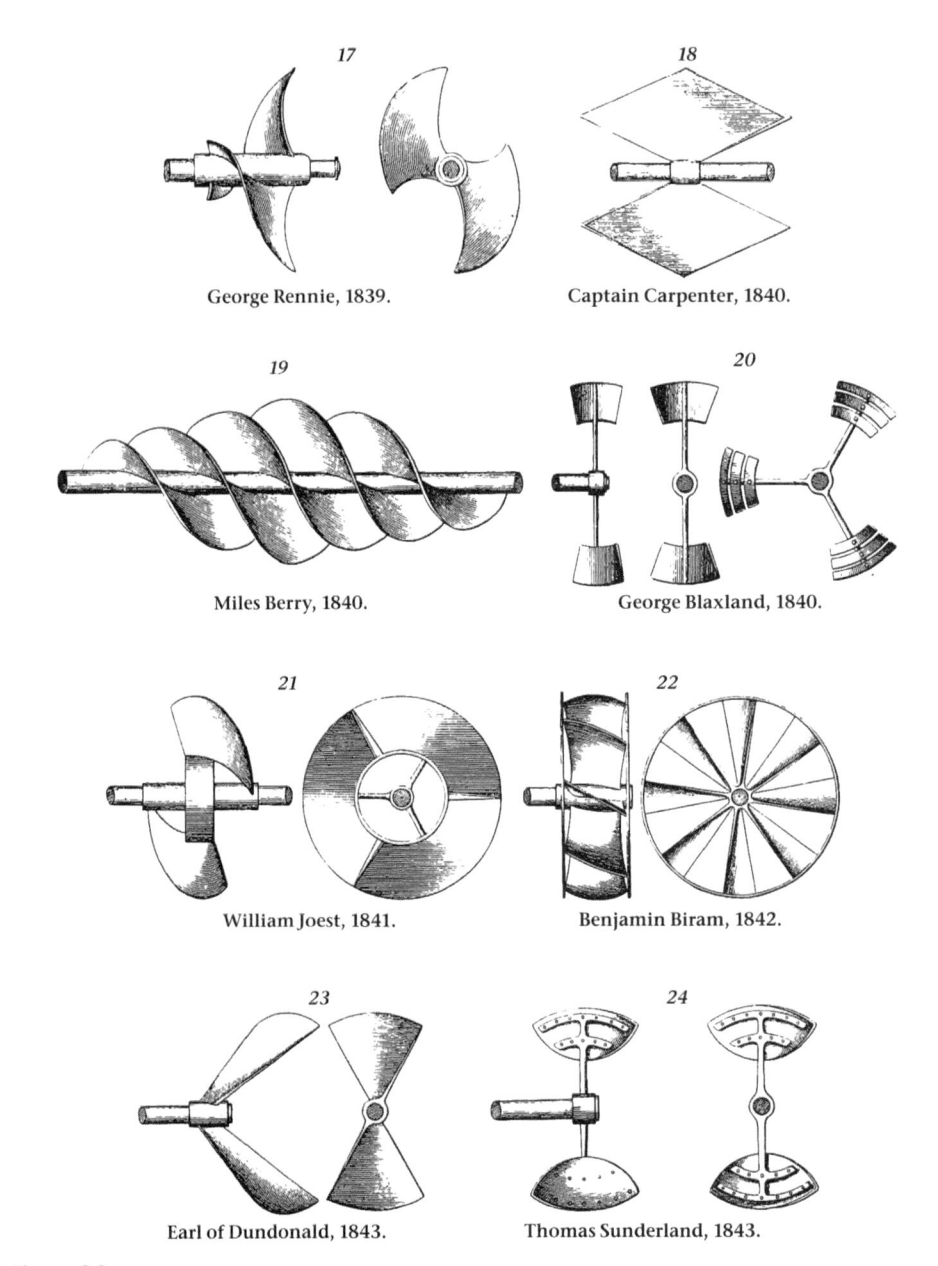

George Rennie, 1839.   Captain Carpenter, 1840.

*19*  *20*

Miles Berry, 1840.   George Blaxland, 1840.

*21*  *22*

William Joest, 1841.   Benjamin Biram, 1842.

*23*  *24*

Earl of Dundonald, 1843.   Thomas Sunderland, 1843.

**Figure 2.3**
Some early screw propeller configurations.
Credit: Cornell University.

steamboat, named *Little Juliana* for his daughter, fitted with a single propeller whose blades were "like those on the arms of a windmill." He noted that the torque of the propeller caused the boat to pivot uncontrollably, so in 1805 he developed a gear mechanism that enabled him to fit a pair of screws, fitted port and starboard and rotating in opposite directions, that solved the problem. Although his little screw steamboat was successful, Stevens reverted to paddle wheels as he built larger and larger ferries.[30]

Rather less obvious as a model for the propeller was the Archimedes screw, which for almost 2,000 years had been used for irrigation and drainage. The screw acts like rotating bucket, turning inside a hollow tube to scoop water from the bottom and raising it to spill out the top (figure 2.4). François Du Quet, the same inventor who developed an early paddle wheel, was the first to describe use of the Archimedes screw in a free stream. His 1729 patent to haul boats upriver without being towed by animals or humans consisted of an Archimedes screw fixed between two boats, which were attached by cables along a stretch of river. The screw was attached to a system of pulleys, so that as the current turned the screw it would be pulled upstream. It is not clear that this device was ever built.

With the advent of steam power, several inventors began modifying the Archimedes screw to serve as the propulsion device, instead of a paddle wheel. One of the first was Josef Ressel, a forestry inspector in the Austrian Empire and serial inventor. In 1827 he patented the idea of placing an Archimedes screw, which he called a spiral, in the deadwood just forward of the rudder (figure 2.5, top; Ressel predated the drawing to 1812 because of a priority dispute). The following year he contracted with the Panfilli shipyard in Trieste to build the 15-meter steam yacht *Civetta* equipped with a screw propeller, which in 1829 he tested in Trieste harbor. A faulty steam pipe stopped the

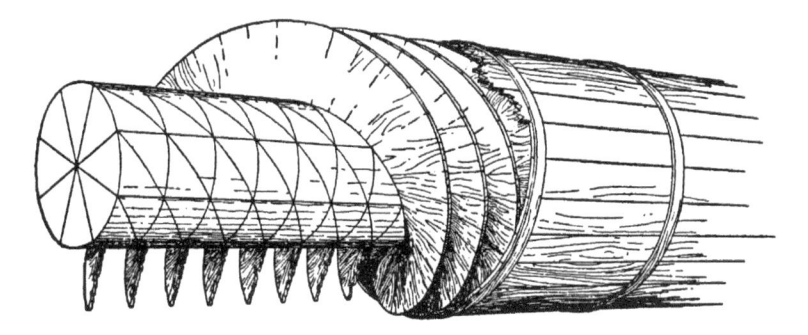

**Figure 2.4**
Archimedes screw.
Credit: Cornell University.

tests after just a few minutes, and further trials were precluded when William Morgan (who had just developed his feathering paddle wheel) successfully asserted his monopoly privilege over all steamships in Trieste. Ressel had already demonstrated his propeller in Paris and shopped his patent around Europe, but he received no contracts or royalties since other inventors simply copied his idea.[31]

One of the inventors who copied Ressel's idea was an émigré London merchant named Charles Cummerow, who in 1828 patented "certain improvements in propelling vessels, communicated by a foreigner"—that is, Ressel. Cummerow's patent, identical to Ressel's, was almost certainly known to a young farmer in Herndon outside

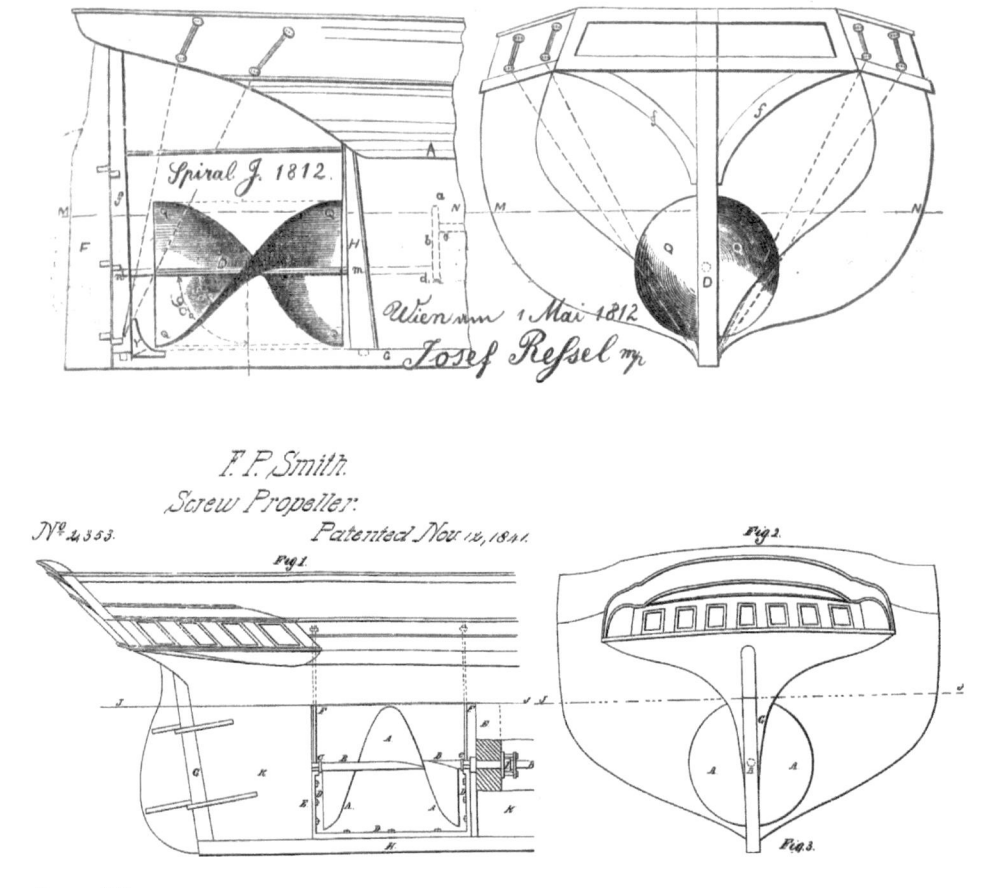

**Figure 2.5**
Patents for screw propellers by Josef Ressel, 1827 (top), and Francis Pettit Smith, 1841 (bottom). Credits: New York Public Library (top); University of California (bottom).

London. Francis Pettit Smith had been experimenting on his farm's pond with small boat models driven by a long spiral screw of two and a half turns. In May 1836 he applied for a patent for the device (like Ressel, Smith patented the placement of the screw in the deadwood, not the screw itself). With financial backing from a consortium of bankers, engineers, and aristocrats, Smith built a 10-meter steamboat fitted with the screw, which he tested on the Paddington Canal and the Thames River. In February 1837, the boat struck debris in the water, which broke the screw in half. Smith immediately noted that the shorter screw gave the vessel higher speed, so he refitted his trial boat with a short screw of a single turn, whose placement he patented a few years later (figure 2.5, bottom). Smith experimented with his boat for another year before attempting a demonstration to the British navy, aware that another inventor had also attempted a demonstration of a screw propeller and failed utterly to convince the Admiralty of its utility.[32]

Another inventor was John Ericsson, an émigré Swedish engineer who had worked on railways and steam boilers in Britain before he turned his attention to the screw propeller. In July 1836, just six weeks after Smith had filed for his patent, Ericsson patented a contrarotating propeller that, although inspired by Archimedes, bore no resemblance to Ressel's screw. It was composed of two hoops fitted with paddles at the circumference, each of which rotated in opposite directions. Ericsson had anticipated the benefits of contrarotation (higher efficiency, canceling out torque) some 50 years before its widespread use on underwater torpedoes. Like Smith, Ericsson installed his device on a small model to test the concept, then on a 14-meter steamboat he named for his financial backer Francis B. Ogden, the American consul in Britain. In August 1837, he put his propeller-powered boat through its paces for the British navy in the hopes of selling his invention, towing a barge packed with admirals and captains down the Thames. William Symonds, the surveyor of the navy who favored yacht-like warships, was unimpressed; as he correctly argued, placing the propeller aft of the rudder (see figure 2.6) would render the vessel difficult to steer. Ericsson did not receive a contract from the Admiralty Board. Unable to recoup the investment he had plowed into his prototype, Ericsson went bankrupt and, as a foreigner, was sent to debtor's prison.[33]

That same group of naval officers would have an entirely different reaction when Francis Pettit Smith demonstrated his propeller-driven steamboat just six months later. Compared with Ericsson, Smith had a better propeller arrangement (forward of the rudder, so that the propeller wash impinged on the rudder and improved steering), a more powerful consortium of financial backers, and the added advantage of being British. This time, the Admiralty Board was cautiously optimistic about the advantages of the screw propeller, but it was not about to adopt such a novel technology without further

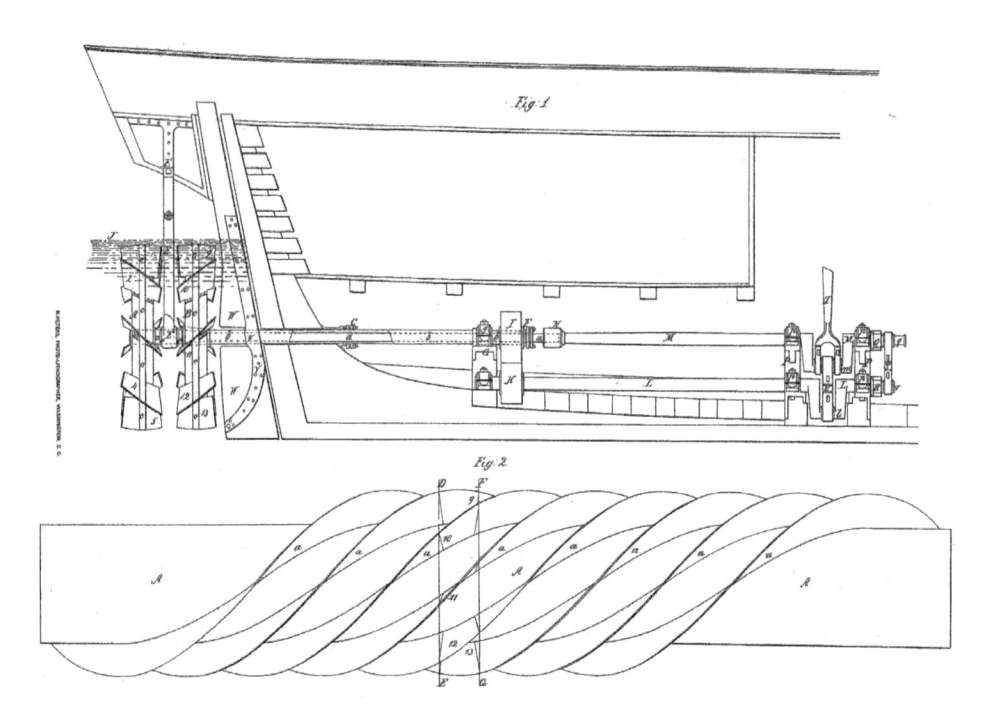

**Figure 2.6**
Original Ericsson propeller patent, 1838.
Credit: U.S. Patent Office.

proof on a larger vessel. Smith and his backers formed the Screw Propeller Company, which built the 200-tun screw steamer *Archimedes* in 1838. For the next several years, Smith sailed *Archimedes* around Britain to show off its potential, all the while testing a variety of propeller configurations, and comparing its performance against the fastest paddle wheel packets. As described in the prologue, one of these trips in 1840 brought *Archimedes* into Bristol harbor, which convinced Isambard Kingdom Brunel to test, and ultimately adopt, the screw propeller for his ship *Great Britain*. The Admiralty was suitably impressed with Brunel's diligence and in 1841 requested his help in assessing the propeller for naval use. As is related in chapter 3, Brunel carried out several additional trials in conjunction with Smith and the Screw Propeller Company, leading to the construction of the screw sloop HMS *Rattler* and its famous tug-of-war trial with the paddle wheel sloop HMS *Alecto* in 1845. The British navy by then had decided that the screw was superior to the paddle wheel but did not buy *Archimedes*, as Smith had hoped, which bankrupted his Screw Propeller Company and left Smith almost penniless.[34]

As for John Ericsson, he did not stay in the notorious Fleet Prison for long, because Francis Ogden helped bail him out. Ogden's American business colleague Robert F. Stockton, a former naval officer then in Britain, had witnessed Ericsson's experiments. Convinced of the utility of the screw propeller for American commercial interests, Stockton contracted the William Laird and Son shipyard in Birkenhead to install an Ericsson contrarotating propeller on the iron steamer it was building for him, and in 1839 he sailed it to the United States. Ericsson himself soon followed and began building propeller steamers. Stockton had by then resumed his naval commission and helped Ericsson get the design contract in 1842 for the steam corvette USS *Princeton*, which had an Ericsson screw propeller, but this time simpler and fitted forward of the rudder for improved steering.

The events surrounding the development of the screw propeller in France happened simultaneously with, and generally mirrored, the developments in Britain, even to the point of a comparable cast of characters. The part of Francis Smith was played by a moody inventor named Frédéric Sauvage; that of Brunel by Augustin Normand, a Le Havre shipbuilder of impeccable lineage; that of the Admiralty Board by the French Ministry of Finances; and John Ericsson played himself.

Pierre Louis Frédéric Sauvage had worked in his father's shipyard in Boulogne during the Napoleonic Wars, but with the subsequent French defeat and economic depression, he turned his attention to marble cutting and inventing. He may well have learned about the Ressel propeller demonstration in Paris, for just a few years later, in 1832, Sauvage demonstrated his patented model of a hand-cranked screw propeller that was quite similar to Ressel's. He did not, however, have the money to develop a full-sized steam-powered version. Augustin Normand was a fifth-generation shipbuilder (and father of the previously mentioned Charles-Benjamin Normand, inventor of the triple-expansion steam engine) who at the time had teamed with British marine engineer John Barnes to compete for steam packets for the French mail service to Corsica, under contract with the Ministry of Finances. Normand saw potential in the screw propeller and made a deal with Sauvage to develop it in exchange for his patent rights. Normand and Barnes received the shipbuilding contract in 1841. They built the propeller-driven steam packet *Napoléon* (later renamed *Corse*) the following year and tested several types of screws before putting the ship into postal service in the Mediterranean. By this time, Frédéric Sauvage had been pushed completely out of the picture, and like Francis Smith, was left destitute.[35]

The French navy also became interested in the screw propeller, and shortly after *Napoléon* was put into service it began contracting for screw steamers. One, *Pomone*, was built with engines and propeller designed by John Ericsson (who was now in the United States; his agent in France handled the order). Another, *Chaptal*, was built by

François Cavé soon after his unsuccessful experience with feathering paddle wheels. Ericsson's design featured the propeller fitted aft of the rudder, the same as on his British trials, which the French navy, like the British navy earlier, rejected for making the ship unsteerable. Cavé's design, with the propeller forward of the rudder, proved a success and became the model for future French screw steamers.

By the mid-1840s, the major navies and shipowners in Europe and the United States had firmly committed to the screw propeller. Shipbuilders settled on a relatively standard configuration, referred to as the common screw, which combined two, three, or four separate blades of the windmill-type screw with the helical structure of the Archimedes screw. Inventors made improvements or additions to this configuration, such as variable-pitch designs by Bennet Woodcroft and Robert Griffiths.[36] As engine power increased, single screw propellers could no longer handle the greater thrust. In the 1860s, the British navy began building twin-screw warships. Tugboats soon followed suit, but it would be another 20 years before the first twin-screw passenger liners crossed the Atlantic. By that time, the last vestiges of sail had been removed from passenger liners as well as warships, as improvements in steam engine efficiency, coupled with the widespread availability of coaling stations, finally made trustworthy the oceangoing steamship.[37]

**From Wood to Iron**

If the adoption of steam propulsion allowed shipping to occur on predictable schedules, then the adoption of iron provided predictability for shipbuilding materials. As noted in chapter 1, timber shortages routinely plagued most European nations in the eighteenth century. Britain and France, for example, often ran short of the critical oak compass timber needed to fashion the right-angle knees that braced the hull and deck frames. Iron fittings were seen as a suitable replacement for these hard-to-find timber components as early as 1670, when Anthony Deane fitted them in the 102-gun *Royal James* as he had "not one [timber] knee in the yard." Inspired by Deane's innovation, Alexandre Gobert began to do the same for French warships, starting in 1706 with *Saint Michel*, whose diagonal planking was described in chapter 1. Iron knees became part of the Gobert system that was in wide use in France through the 1740s, and they continued even after diagonal planking went out of favor. Similarly, British constructors throughout the eighteenth and early nineteenth centuries, such as Gabriel Snodgrass and Robert Seppings, made systematic use of iron for knees, braces, riders, straps, trusses, and other structural fittings on wooden ships, as their access to oak compass timber worsened after the loss of the American colonies. Spanish constructors, by contrast to those in France and Britain, did not appear to use iron in any systematic way for wooden hull reinforcement, in part due to the comparative lack of foundries and forges in Spain.[38]

While timber shortages may have explained the increasing use of iron fittings in ships, they cannot explain the sudden appearance of iron canal barges in the British Midlands at the end of the eighteenth century. John Wilkinson was a Shropshire industrialist who had pioneered the use of cast iron in the early days of the Industrial Revolution. In part to promote the region's nascent iron industry, he was the driving force behind the Iron Bridge in Shropshire in 1778. In 1787, Wilkinson built the world's first iron-hulled cargo vessel, apparently named *Trial*, to serve his iron foundries on the Birmingham Canal. Fabricated of cast-iron plates bolted to elm frames, it was about 20 meters long and weighed 8 tonnes empty. He appears to have built *Trial* for the same reason as the Iron Bridge, as a showcase to promote the region's iron industry. No other rationale fits the circumstances: there was no timber shortage for the elm and other hardwoods needed to build inland barges and canal boats; timber was far cheaper than iron; although iron barges proved to have shallower drafts than wooden ones, a great advantage in canals that often ran nearly dry, and were far more durable than wooden boats, these benefits were not understood until much later; and the structural benefits for supporting steam engines would not become apparent until well into the next century. Wilkinson built several more iron-hulled barges and boats in the following year before abandoning the idea for good. Nor did his promotional efforts bear immediate fruit; for over 20 years, no other iron vessels appeared in the Midlands or elsewhere in Britain.[39]

Even though iron boat construction had begun at roughly the same time that steam propulsion came on the scene, its adoption was far slower and more sporadic. It was not until 1818, near Glasgow, that Thomas Wilson built the barge *Vulcan*, the first vessel constructed entirely from iron plates and framing, for Scottish canal service. In 1821, Aaron Manby, managing partner of the Horseley Coal and Iron Company near Birmingham, began work on the first vessel that would marry iron construction with steam propulsion. Manby had formed a French company with the British naval officer Charles Napier (no direct relation to the Napier family of marine engineers), who had seen extensive combat in the Napoleonic Wars and was now advocating a series of reforms, including steam, for the British navy. The 32-meter, 30-IHP (20 kW) paddle wheel vessel *Aaron Manby*, as it was named, was built at the Horseley works but dismantled and then reassembled on the Thames River. Manby and Napier sailed it across the English Channel and up the Seine River to Paris. Although their company soon went out of business, the vessel and its sisters operated for several decades carrying cargo on French rivers and canals (these were the boats Victor Hugo railed against). By the 1830s, a growing fleet of iron steamers was operating regularly on inland waterways in Britain, France, the Netherlands, and other parts of Europe, as well as in British India.[40]

These first iron canal boats and barges were built in not wood shipyards but machine shops that also fabricated bridges, building and canal structures, and boilers and locomotives. This pattern would repeat itself in the coming decades for the builders of oceangoing iron ships such as William Laird and Son of Birkenhead, originally established as a boilermaker in 1824. The skills for wood shipbuilding were not easily transferable to iron, which is what British naval architect Westcott Abell had suggested in his 1948 book *The Shipwright's Trade*, wherein he imagined a wood shipwright, an ironsmith, and a boilermaker quickly coming to a meeting of the minds on how to build a ship.[41] The truth was quite different—for example, the processes of rolling, bending, and riveting iron plates had no equivalent in hewing, forming, and nailing timber. Managing the more specialized workforce also required a more hierarchical structure not found in wood shipyards.[42] Even the nature of shipyard injuries changed in scope and scale; whereas in the wood era a relatively few moderately serious injuries occurred, often hernias and lacerations, the advent of iron brought a large increase in hospital admissions and more severe hurts such as eye injuries from sparks and disfiguring burns from molten metal.[43]

Nor did the equipment needed to fabricate and join riveted iron plate and girders have much in common with wood shipbuilding practice or even with the shipyard forges that had made nails and fittings for wooden vessels. Wood shipwrights frequently brought their own tools to the yard—adzes, saws, chisels—and the shipyard's capital cost for equipment was fairly small. By contrast, the metal-working tools used in the iron shipbuilding industry were very different in power, size, and cost. A rolling mill cost $4,000 ($5 million in today's money), punching machines $3,000 each, and furnaces $2,500. While wood shipyards were often unable to amortize the initial equipment cost for converting to iron construction, machine shops could spread this capital outlay across many product lines like ships, bridges, factories, and railways. Admittedly, some wood shipyards did make the transition to all-metal shipbuilding, notably Portsmouth Royal Dockyard in Britain during the Crimean War (1853–1856) and the Charles Cramp shipyard in Philadelphia during the Civil War (1861–1865), both of which had to make the change to accommodate wartime requirements for armored iron warships. These, however, tended to be the exception rather than the rule.[44]

Instead of making the technological leap from all-wood to all-iron ships, many shipbuilders and shipowners opted for a mixture of the two. Composite construction, as it was known, took many forms on either side of the Atlantic and was noteworthy in the construction of fast clippers in the 1850s and 1860s. In the United States, where wood shipbuilding dominated the commercial market through the nineteenth century, it generally meant the use of heavy diagonal iron trusses to support and stiffen wooden hulls. The *Great Republic* by Donald McKay, discussed in chapter 1, had extensive cross bracing of inch-thick (25 millimeter) iron trusses that extended from the floor to the main deck. "Composite construction" meant something quite different in Europe, in general denoting wooden hull planks bolted over iron frames and bracings. This was the system used in the British clippers *Thermopylae* and *Cutty Sark* and by the Bordeaux constructor Lucien Arman in his clippers like *Maréchal de Turenne*. Another Bordeaux native, Jean Cathérineau, devised a complex system of wooden and iron frames and stiffeners that was abandoned after one trial.[45]

Even when the technology of all-iron construction became routine, there remained an enormous disparity between Britain's ironworking capacity and that of other nations. In 1860, Britain was producing 2 million tonnes of pig iron annually, compared with 800,000 tonnes in France and 820,000 tonnes in America.[46] So while the British navy would be able to order iron ships from a large and well-established production network, the lack of reliable networks of iron producers in France and the United States led their navies to order warships of composite construction. In France, the existing foundries were unable to produce large amounts of plate to the high quality required. Thus, of the six first ironclads in the fleet, five of them—the *Gloire* and *Magenta* classes—were built in timber, with 120 millimeter iron armor bolted to the sides. Only one ship, *Couronne*, was constructed entirely of iron at Lorient, the only dockyard in France that had been specially equipped for iron shipbuilding. In America, the first Union ironclads of the Civil War, notably USS *New Ironsides*, *Galena*, and *Monitor*, had wooden-framed hulls with iron armor bolted on. Even with the limited amount of iron used in their construction, the states that built those ships—Pennsylvania, Connecticut, and New York—had their local networks of ironworks and foundries stretched to their limits.[47]

The relatively slower adoption of iron (compared with steam propulsion) for shipbuilding, and the persistence of wood and composite construction, was due to more than just lagging industrial capacity or capability. This was the result of not the sailing ship effect discussed in chapter 1 but rather several serious problems with the use of iron having to be overcome before it was fully accepted by the maritime industry. These "reverse salients"—which the historian of technology Thomas P. Hughes calls the parts of a system that, owing to insufficient development, prevent the system from realizing its full potential—included the problems of hull fouling, iron brittleness, and compass correction, all of which had to be resolved before iron could be widely accepted as a shipbuilding material.[48]

The problem of hull fouling, which had been largely addressed by the coppering of wooden ships, returned when iron hulls came into wider use in the 1840s. Copper plates simply could not be attached to iron hulls, since the same galvanic action that corroded iron spikes in wooden ships would also cause iron hull plates to corrode and waste away. In response, shipbuilders tried solutions such as sheathing the iron with zinc alloys or Muntz metal (a copper-based alloy) or using antifouling compositions of tar, copper, and other materials. These experiments were at first largely unsuccessful because galvanic corrosion continued to occur; the 1844 gunship HMS *Grappler* suffered such severe hull corrosion that it had to be scrapped after just five years' service, and the 1848 sloop HMS *Triton* while in dry dock was found to have some hull plates "reduced to the thickness of writing paper." Workers trying to scrape off barnacles punctured the hull. Starting in 1860, the development of reliable, noncorroding antifouling paints removed one of the obstacles to the widespread adoption of iron.

The second obstacle was both more pernicious and harder to overcome. The first iron bridges, railways, and canal boats had been built from brittle cast iron, but by the end of the 1840s, wrought iron had replaced cast iron as the favored material for railways, factories, bridges, and ships. Wrought iron is made by first melting the iron ore in a puddling furnace, letting it cool, and then rolling and hammering the blooms into bars, plates, and beams. The resulting change in the metallurgy means that wrought iron is more ductile than cast iron and therefore better able to resist bending and tensile loads before breaking. However, the quality of manufacture at that time was quite variable between foundries, and wrought iron could become brittle if poorly made.

The problems of brittleness, as well as those of corrosion, led the British navy to roundly condemn iron-hulled warships and to stop building them from 1847 until the end of the Crimean War. During the 1840s, the British navy had experimented widely with wrought iron, building iron-hulled frigates, sloops, and gunboats, and carrying out a series of ballistic tests on armor plate. But the operational experience of the iron warships was not happy; several of the gunboats developed leaks due to corrosion. At the same time, the ballistic tests carried out from 1845 to 1846 at Woolwich Arsenal revealed that wrought iron plate, when struck by shot, could spall and splinter very badly, to more deadly effect than wood splintering. The rising tide of government opposition to iron warships—encouraged, ironically enough, by Charles Napier, now an admiral who had been bankrupted by his early investments in iron vessels—led the Admiralty in 1847 to cancel all further orders for iron ships and to convert the iron frigates already on the ways to troopships. One of those troopships, HMS *Birkenhead*, struck a submerged rock off South Africa in 1852, resulting in the deaths of 450 troops and civilians (the famous Birkenhead drill, "women and children first," stems from this accident). Even though brittle fracture was not to blame, this accident added to the general prejudice against iron hulls, which was further cemented in 1859 when the iron steam clipper *Royal Charter* wrecked off Wales with a loss of 400 lives. One member of Parliament, Douglas Howard, went so far as to declare that "vessels constructed entirely of iron were utterly unfit for all the purposes of war." However, the Crimean War, in which both the French and the British navies used floating ironclad batteries to supplement their steam-powered wooden warships, had already demonstrated to the naval authorities the need for iron armor to protect against the exploding shells that had just come into common use. Thus, although some popular sentiment continued against iron and in favor of Britain's time-honored "wooden walls," a steady series of manufacturing improvements in iron plate in France and Britain, culminating in new and successful armor trials in both nations from 1856 to 1857, removed the second obstacle to the widespread adoption of iron.[49]

The third obstacle, that of compass correction, was addressed early on, but the proposed fixes became caught up in debates over political authority that required cultural as well as technical solutions. It was well known that the presence of iron causes magnetic compasses to deviate from magnetic north. Even the iron fittings on wooden ships interfered with navigation, a problem noted by the British naval officer Matthew Flinders during his voyages to Australia. In 1805 he proposed placing an unmagnetized iron bar aft of the compass (normally on the poop deck) to compensate for the ironworks forward. While the Flinders bar, and the associated Barlow plate, did work for wooden vessels with only a few metal fittings, the advent of iron construction demanded new solutions. Compass deviation was especially problematic for the iron ships destined to serve Britain's far-flung empire, whether at sea or on inland waterways

In 1838 the British navy asked George Biddell Airy, the astronomer royal at Greenwich, to experiment with compass correction on the iron steamer *Rainbow* at Deptford. Airy understood that the process of building the ships caused ships' hulls to become magnetized, and he compensated with a combination of magnets and iron bars. But this turned out to be only a partial fix, and in the ensuing years several wrecks due in part to faulty navigation cast doubt on Airy's methods. This mistrust came to a head at the 1854 British Association for the Advancement of Science meeting, where the scientist William Scoresby, also wearing the mantle of a former ship's captain, questioned the scientific underpinning of Airy's work. While the arguments between Airy and Scoresby over scientific authority continued in the popular press, Airy reexamined the mathematics of magnetism recently developed by Carl Gauss, Wilhelm Weber, and Siméon Poisson and recommended several additions to his system, including a Flinders bar. Airy's improved system became widely adopted; in 1863, during the U.S. Civil War, the new National Academy of Sciences conducted tests of the system on the Union monitor USS *Passaic* and pronounced it "most effective" for use on ironclads. However, errors due to rolling, pitching, and induced magnetism still crept into the navigation. It was not until 1876 that William Thomson (later to become Lord Kelvin) developed the now-famous binnacle, with a gimbaled compass and spherical compensating magnets, which provided a complete solution to the problem of compass deviation in iron ships.[50]

By 1860, the three reverse salients of fouling, corrosion, and compass correction had been largely overcome, and the advantages of iron construction over wood—shallower draft for inland vessels, superior strength and durability, wider and more predictable availability, and greater structural stiffness to mount steam engines and propellers—were widely recognized. Even so, public debates over the merits of wood and iron continued apace.[51] These arguments became increasingly moot as the pace of iron shipbuilding noticeably accelerated; by 1864 the annual shipbuilding tunnage in Britain of iron ships had surpassed that of wooden ships.[52] The British navy reenergized its iron shipbuilding program with the first all-metal ironclad, HMS *Warrior*, coming off the building ways in 1860. The French all-iron *Couronne* followed in 1862. Other European navies followed suit, although they too had to overcome their own industrial salients. Spain's first ironclad *Tetuán* was launched in 1863, but another six years would pass before it could deliver another. Italy built its first iron warship in 1866, the sloop *Vedetta*, but another decade elapsed before it could build another.[53]

As noted in chapter 1, wood shipbuilding persisted in many areas long after the obstacles to iron had been surmounted. In the case of the United States, a series of incredibly high tariffs on iron (much of which was still produced in Britain) kept shipbuilders wedded to wooden ships throughout the nineteenth century—in fact, American iron tunnage did not surpass wood tunnage until 1899. In other cases, notably in the smaller ports around the world, wood shipbuilding continued well into the twentieth century. This was often driven by the persistence of family-owned shipyards having close ties to local shipping and business communities, which still found profit in the wood-and-sail ship trade even as the larger shipping world became increasingly globalized.[54]

**The Spread of Iron as a Shipbuilding Material**

When Brunel conceived his *Great Eastern* in 1852 to have the same longitudinal structure as William Fairbairn's Britannia Bridge, he was following an already well-established practice of adapting bridge, factory, and railroad construction methods to iron shipbuilding. The earliest structural shape was the L-shaped angle iron, either hammered into shape or rolled, commonly used in smaller buildings and boilers; the steamboat *Aaron Manby* was transversely framed with angle iron. The rise of the factory economy in Britain led to the development of larger and stronger iron beam shapes. As large multistory mills sprang up to produce cotton, flax, and other textiles, it became increasingly difficult to find large timber to support heavy steam-driven machinery across wider floors. Cast-iron beams, followed by wrought iron, filled that need, with the added benefit of being fireproof. The T-shaped beam, stronger and stiffer than the angle iron, was first used in 1801 to construct a steam-driven cotton mill built and furnished by Boulton and Watt and then in 1803 in a flax mill built by Charles Bage in Leeds. William Fairbairn, who in 1831 expanded his civil engineering practice to include iron steamboats, framed his first boat, *Lord Dundas*, and subsequent vessels, with T beams. The I-shaped beam, which began as a novel profile for railroad rails, was widely touted as the strongest and stiffest shape in the popular 1842 edition of the handbook *A Practical Essay on the Strength of Cast Iron* by the civil engineer Thomas Tredgold. Fairbairn extensively used I beams in his Britannia Bridge, as did Brunel and Russell for *Great Eastern*. Each beam shape spawned variations—sections built up from simpler shapes, flanged sections, bulbs, and so on, each with different degrees of strength, stiffness, and above all, cost (figure 2.7). The framing and riveting practices, which also varied, were shared between shipbuilding and civil construction. Individual engineers and shipbuilders used a personal mix of shapes and connections and their own manufacturing tools, machinery, and methods to achieve the required strength and stiffness at the most effective cost.[55]

Each shipbuilder and ironworks had a unique combination of material, machinery, and methods and kept them closely guarded against industrial espionage and theft. Technical know-how was still developed primarily through apprenticeships and on-the-job experience, and it was commonplace for factory managers to poach one another's more skilled workers and bribe others to steal machinery to gain a technological advantage. This threat extended beyond British firms to European and American governments and industries bent on catching up with Britain's acknowledged superiority in industries ranging from textile manufacturing to iron production. From 1780 to 1820, Parliament passed laws prohibiting any skilled artisan from leaving Britain to carry on his trade, while the customs office seized vital machinery, such as iron-working rollers, bound for export. Starting in the 1820s, after the end of the Napoleonic Wars allowed the government to lift wartime restrictions on exports and migration, British shipbuilders, railway builders, and ironworkers opened their doors to visitors from Austria, Germany, Sweden, and France, intending to open new markets for their goods. Thus, from 1823 to 1829, the French inventor Marc Seguin and members of his extended family of engineers made numerous trips to the factories of Marc Brunel, Robert Stephenson, and Philip Taylor to bring back to France the understanding of how to build railways, suspension bridges, and steamships. In 1824, a French navy captain, Jacques Philippe Mérigon de Montgéry, published *Memoir on Iron Ships*, detailing his examination of iron

**Figure 2.7**
Photo of built-up bulb T beam and deck-hull girder connection on *Star of India* (1863) in San Diego.
Credit: Author's photo.

steamships such as *Aaron Manby* and recommending the French navy begin shifting its ship construction from wood to iron.[56]

The most significant visit to British shipbuilders was made in 1842 by a young French naval constructor, Stanislas-Charles-Henri-Laurent Dupuy de Lôme. The visit gave rise to a global arms race that would span half a century. Dupuy de Lôme (figure 2.8) was born to a naval family in Ploemeur, Brittany, on 15 October 1816. He entered the École Polytechnique (Polytechnic School) in 1835 and then the École d'Application du Génie Maritime (School of Naval Engineering) in 1837, and he became a naval constructor in Toulon in 1839. In 1841 he began carrying out experiments on ship resistance in addition to his regular duties of operating and maintaining the dry docks and was marked by his superiors as someone who could combine theoretical insights, practical engineering, and hands-on management of complex problems.[57]

In June 1842, the French navy sent the 26-year-old Dupuy de Lôme, at his own request, to Britain to "examine the state of technologies then in use for the construction of ships, study the recent improvements made in this industry ... and to publicize this knowledge in [French] ports." Although unable to converse fluently in English, he spent the next 10 months touring shipyards in London, Liverpool, Bristol, and Glasgow, making copious notes and drawings. He was especially impressed with the William Laird and Son shipyard and considered the steam paddler *Guadalupe*, under construction there for the Mexican navy, to be the epitome of iron warships. He carried on a lively correspondence with the owner John Laird on details of the shipbuilding process, Dupuy de Lôme writing in French, Laird replying in English, and both sets of letters translated by Laird's staff.[58]

On his return to France, Dupuy de Lôme compiled his notes and sketches to produce the monumental *Mémoire sur la Construction des Bâtiments en Fer* (Memoir on the construction of iron ships) and accompanying *Atlas* of drawings, which were published 1843–1844. This was a comprehensive, objective examination of the state of iron shipbuilding and a vision of its future. Dupuy de Lôme admitted that there were "many prejudices, reasonable doubts and real difficulties" with iron, but he argued that the benefits outweighed those drawbacks. One of the greatest advantages of all-iron construction, he said, was that it avoided the complicated truss systems then in use for wooden hulls, which made iron ships lighter, roomier, and easier to build. Another benefit was the inherent watertightness of riveted connections compared with caulked wooden joints, which, coupled with the use of watertight bulkheads, made iron vessels stronger, safer, and longer-lived than wooden ships (figure 2.9). The report concluded with a step-by-step guide on how to establish the machinery and production facilities necessary to sustain an iron shipbuilding industry in France.[59]

**Figure 2.8**
Stanislas-Charles-Henri-Laurent Dupuy de Lôme.
Credit: Bibliothèque Nationale de France/Gallica.

Steam, Iron, and Steel

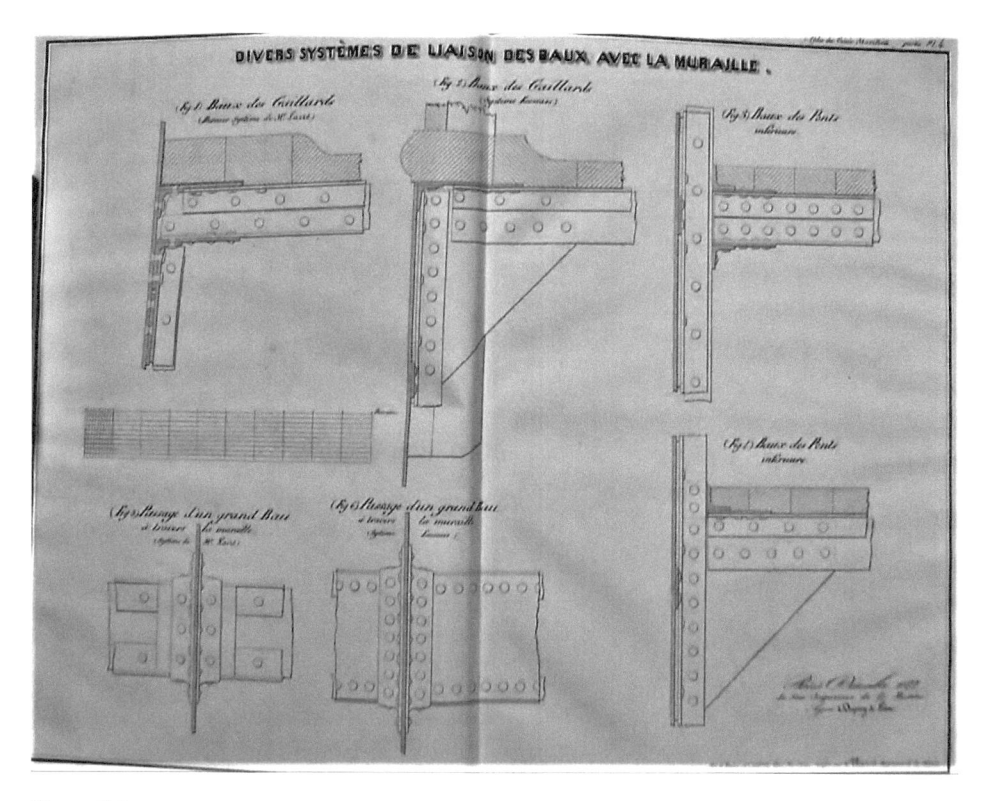

**Figure 2.9**
*Atlas* of the *Mémoire sur la Construction des Bâtiments en Fer* (Memoir on the construction of iron ships) showing different systems of connecting the deck and hull girders.
Credit: David K. Brown.

Dupuy de Lôme wasted little time before trying to implement his newfound knowledge. In 1845 he proposed the construction of a large iron warship that would combine armor protection with screw propulsion, but the French navy judged the project premature given the state of the industry. Two years later he successfully proposed a series of even larger 90-gun screw-propelled warships, built of wood, the first of which was christened *Napoléon* after it was launched in 1850. *Napoléon* served in the Crimean War, where its speed and maneuverability helped turn the French navy away from paddle wheels in favor of screw propulsion. But as mentioned earlier, the war also convinced both the French and the British navies of the utility of iron hulls. Although the development of the exploding shell by Henri-Joseph Paixhans in the 1820s hinted that the end of the wooden warship was in the offing, it was not until the 1840s that shell-firing guns became widespread in Europe. At the Battle of Sinope in 1853 at the start of the Crimean War, a Russian fleet used shell guns to destroy an Ottoman squadron. This battle brought France and Britain into the war on the side of the Ottomans and also alerted them to the need for armor against shellfire. Dupuy de Lôme, by then a senior naval constructor, exchanged visits with his British counterpart Thomas Lloyd to each other's dockyards and factories. Together they came up with a plan to jointly build, test, and deploy a combined fleet of steam-powered ironclad batteries, which bombarded and destroyed Russian fortifications at the Battle of Kinburn in Crimea with little damage to themselves. Soon after the peace was declared in 1856, the short-lived French-British alliance would turn back to competition for control of the seas and trigger the arms race that led to the dreadnoughts of the twentieth century.[60]

This competition for sea control began in 1857 when the French emperor Napoleon III named Dupuy de Lôme as both director of matériel and director of naval construction, placing him in charge of the design, construction, and maintenance of the entire navy. Dupuy de Lôme now had the authority to realize his vision of an all-iron, propeller-driven fleet that could stand toe-to-toe with the much larger British navy. The *loi de programmation navale* (navy budget) of 1857, followed by subsequent budgets, represented an enormous increase in ships, port facilities, and arsenals and included provisions for six large ironclads. As noted earlier, Dupuy de Lôme understood that the French shipbuilding industry was still not capable of producing large quantities of iron plate, so when the first ships of the *Gloire* class, displacing 5,600 tonnes, were launched in 1859, they were timber-framed with iron armor; only *Couronne* was all-iron. Even so, the appearance of these ironclads on the high seas—*Gloire*'s sister *Normandie* would be the first ironclad to cross the Atlantic Ocean—caused the British government to reconsider its own naval budget.[61]

Even though Parliament was notoriously tightfisted, it decided that the British navy needed to regain mastery over the French. Funds were allocated to a build a class of two 9,000-tonne ships, HMS *Warrior* and HMS *Black Prince*, which were fabricated entirely of iron. Designed by Chief Constructor Isaac Watts and Chief Engineer Thomas Lloyd, each of these ships displaced almost twice as much as their French counterparts when they were launched in 1860 and 1861.[62] In 1861, the *Provence* class became France's first class of all-iron major warships. In Britain, the high cost of *Warrior* (£377,000, almost $1.4 billion today) drove the government to order a series of smaller, less expensive ironclads. This was followed by a major buildup of ironclads in the British and the French fleets; by 1869, Britain had 19 ironclads compared with the French navy's 15. Dupuy de Lôme began equipping his warships with large iron rams, like Greek and Roman triremes of the past, arguing that the ram would "rip open by the shock, at even a moderate speed, any armored ship it attacked." This triggered his British counterparts, and soon navies around the world, to develop iron ram warships as part of their fleets.[63]

The ironclad race was on. The 1861 Civil War duel of the American ironclads USS *Monitor* and CSS *Virginia*, in which both ships emerged relatively unscathed, gave most other navies the confidence to build in iron. The particular success of *Monitor*, which had only two guns mounted in a revolving turret, also led to the replacement of the broadside gun battery by revolving gun mounts and turrets. The 1860s and 1870s saw a spate of wars in which these new ironclad designs were tested in battle, notably the Danish-German War of 1864, the Italian War of Independence of 1866, the Franco-Prussian War of 1870–1871, and the War of the Pacific from 1879 to 1883. Yet even as iron was receiving its trial by fire at sea, a new material, steel, was being developed for the railroad, civil construction, and arms industries, and within a decade it would become the material of choice for shipbuilding that would last until the present day.[64]

**From Iron to Steel**

Steel differs from wrought iron primarily in its carbon content. Wrought iron typically has about 0.3 percent carbon, while mild steel—the most common alloy in shipbuilding—has between 0.3 and 2 percent carbon. The higher carbon content makes mild steel stronger; the steel of the mid-to-late nineteenth century typically had an ultimate tensile strength before rupture of about 26 tons per square inch (360 megapascals), which was about 30 to 50 percent higher than wrought iron. This meant that a steel structure could be lighter than a wrought iron structure for a given application. Steel had been manufactured in small quantities for many centuries, but in 1856 a British inventor named Henry Bessemer, looking to improve gun manufacture, developed a new furnace that forced air into molten pig iron to raise its temperature and blow off impurities. This process could be accomplished in just minutes, which meant that large quantities of steel could quickly be produced for industrial use, but its quality was highly variable. British boilermakers quickly turned to Bessemer steel to obtain higher operating pressures, but British shipbuilders were at first wary of its unreliability and used it sparingly for masts, yards, and local hull stiffening while retaining tried-and-true wrought iron for the primary structures. At exactly the same time that Bessemer was working on his furnace, a German-born engineer in Britain, Carl Wilhelm Siemens, developed an open-hearth process that, while slower, gave better control over the chemistry and thus the quality of the steel. In 1865, the French engineer Pierre-Émile Martin adapted Siemen's reverberatory furnace, and the more reliable Siemens-Martin process soon replaced the Bessemer process in the major steel firms, such as Schneider-Creusot in France, Krupp in Germany, and John Brown in Britain.[65]

As with the development of ironclads, it was France, the weaker naval power, that led the innovation toward steel warship construction, in the expectation that the more advanced technology would be a force multiplier, or a means of augmenting the effectiveness of individual ships as a counterbalance to Britain's superior numbers. Only later would Britain catch up to and then overtake France in steel construction by dint of its superior industrial capacity.[66] In 1860, while both France and Britain were building their first ironclads, the École d'Application du Génie Maritime produced an in-depth study of steel manufacture and, while acknowledging the problems in the Bessemer process, predicted that it would "more and more be used in construction."[67] However, it would not be until 1873 that Louis de Bussy introduced Siemens-Martin steel into the production of the battleship *Redoubtable* built at Lorient, which he estimated would reduce the hull weight by 10 percent. The success of steel also extended into the manufacture of shipboard guns, armor, and machinery. Other navies quickly took notice. After an 1874 visit to Lorient, the British chief constructor Nathaniel Barnaby convinced the Admiralty to build in steel. Italy followed suit in 1876 and Austria-Hungary in 1879. Germany was a relative latecomer due in part to industrial turmoil after unification in 1871; its first steel warships were not launched until 1882.[68]

Commercial shipowners also took notice of the advantages of steel for shipbuilding. Compared with the decades-long transition from wood to iron, shipyards made the transition from iron to steel quite quickly, in just a few years, since the skills for ironworking were more translatable to the steel industries. In the British shipyards on the Clyde, Tyne, and Wear Rivers, steel construction went from just few percent of the total commercial tunnage built in 1883 to almost 100 percent by 1888. The shipyards on the "American Clyde," the Delaware River near Philadelphia, also changed over from iron to steel production in just a few years. For this reason, the U.S. Navy turned to these commercial shipyards when it began building its steel fleet, starting with the ABCD ships (*Atlanta*, *Boston*, and *Chicago* cruisers and *Dolphin* dispatch ship) in the 1880s, instead of relying on the naval dockyards it had traditionally turned to for warship construction.[69]

**"The Ship That Found Herself"**

The new technologies of steam, iron, and steel may have seemed abhorrent to Victor Hugo and John Ruskin, but they found their herald in the British author and poet Rudyard Kipling, who in poems about marine steam engines and tramp ships, proclaimed them as the icons of the mighty, world-dominating British Empire. In 1895—ironically enough, while he was living in Vermont—he elegantly captured the importance of these otherwise mundane artifacts in his short story "The Ship That Found Herself." A newly built cargo steamer *Dimbula* leaves Liverpool for New York, and along the way its personified parts—the plates and stiffeners, the steam engines and machinery, all grumble and argue with each other. The ship's captain explains to a passenger that "rivets and plates put into the form of a ship" did not mean the vessel was yet whole, because "the parts of her have not learned to work together … she has to find herself yet."

And indeed, the ship does. In Kipling's narrative, forces and strains and structural details escape from the naval architect's notebook and are imbued with life:

> "Can't you keep still up there?" said the deckbeams. "What's the matter with you? One minute you weigh twice as much as you ought to, and the next you don't!"
>
> "It isn't my fault," said the capstan. "There's a green brute outside that comes and hits me on the head."
>
> "Tell that to the shipwrights. You've been in position for months and you've never wriggled like this before. If you aren't careful you'll strain us."
>
> "Talking of strain," said the deck stringers, "are any of you fellows—you deckbeams, we mean—aware that those exceedingly ugly knees of yours happen to be riveted into our structure—ours?"

As the waves twist and slam the ship in the seaway, the parts begin to settle into place and work together. It was Steam, the most experienced part of the vessel, who best understood: "When a ship finds herself, all the talking of the separate pieces ceases, and melts into one voice, which is the soul of the ship."[70]

# 3   The Quest for Accuracy

As far back as ancient Greece, engineers and scientists had tried to understand the behavior of wooden ships sailing on the ocean, seeking to predict how they would operate and how to improve them. The introduction of steam, iron, and steel meant that the knowledge accumulated over centuries was now almost useless and that new rules for designing and building ships had to be developed and verified. For example, on sailing ships the rolling motions are heavily dampened by the sails themselves, so when steamships began shedding their masts in the 1850s, neither ship designers nor ship captains knew what kind of motions to expect at sea; new scientific and engineering tools had to be developed to fill those knowledge gaps.

This new era of research into the fundamental principles of naval architecture began with the most basic of steps—measurement. New tools and methods provided engineers and scientists the means to gather data on speed, power, force, and material strength, which allowed them to build equations and models and in turn allowed them to predict the characteristics and performance of these new steam, iron, and steel technologies. Although prediction had been the rationale for navies and governments to sponsor naval architecture research in the eighteenth-century Scientific Revolution, during the nineteenth-century industrial age much of that effort was carried out by commercial shipbuilders and engineers.

The most common pattern at this time was that industries would develop new technologies (e.g., steam power and the screw propeller) long before any deep theoretical understanding of the subject was available. After that, these developments spurred localized scientific investigations into the underlying theories (e.g., theory of resistance and the momentum theory of propeller disks). These discoveries would lead builders to improve the technologies themselves (more streamlined hulls, more efficient propellers) and to standardize those products across the production line (model testing for speed and power, criteria for propeller performance).

The well-known story of the development of the steam engine and concurrent rise of thermodynamics is illustrative of this pattern. In Britain, Thomas Newcomen invented the steam engine in 1712, and James Watt improved on it in 1775 with a separate steam condenser. Both these inventions came well before French scientist Sadi Carnot devised his theory of the efficiency of heat engines; in fact, Watt's engine provided the underlying model for Carnot's cycle involving a high-temperature source (cylinder and piston) and a low-temperature sink (condenser). By the same token, the hot-air engine of Robert Stirling (1816) spurred later investigations by William Thomson (Lord Kelvin) into the theoretical relationship between heat and mechanical motion. Carnot's and Kelvin's theories led to more efficient steam engines and the development of performance standards for those engines.[1]

Chapter 2 describes the steam, iron, and steel technologies that shaped nineteenth-century shipbuilding. This chapter explores these theoretical developments in naval architecture—ship motions, speed and power, propellers, maneuvering, and structural design—that gave engineers and scientists an accurate understanding of their behavior and performance. Chapter 4 discusses how shipbuilding industries and organizations used these theoretical foundations to control and improve ship design and construction in order to meet the demand for standardized shipbuilding processes and products.

**Rolling and Dynamic Stability**

William Froude is the name most often associated with this period in the development of theoretical naval architecture and is uniquely attached to his development of scale-model experimentation that revolutionized the prediction of hull resistance and powering. Every first-year naval architecture student around the world does calculations using the Froude number, the dimensionless number used to relate speed and power for ship forms whether model scale or full size. They also learn the story of how he developed his experimental methods in the 1870s at his now-famous model basin at Torquay and how he verified his speed-power theory at full scale on the sloop HMS *Greyhound*. Yet as mentioned in the prologue, Froude's first forays into using scale models to develop ship theory occurred over a decade earlier, when he was asked by Isambard Kingdom Brunel to investigate the rolling behavior of *Great Eastern*. The methods that he pioneered for these investigations, combining scale-model experiments, full-scale measurements, and mathematical theory, would serve as a template for his later speed-power research and indeed as the model for future researchers in the field of naval architecture.

William Froude was born on 28 November 1810, the sixth of eight children. The Froudes (in Old English the name means "prudent") were an ancient line of farmers mentioned in the Domesday Book, but William's immediate family was deeply connected to the Anglican church; his father Robert was archdeacon of the Devon parish of Totnes; his older brother, Hurrell, was a church leader; and his younger brother, James Anthony, was a noted historian and religious polemicist. As noted in the prologue, William Froude became a railway engineer under Brunel in 1837. Under Brunel's tutelage, Froude learned the importance of accurate measurements in the field to support his engineering calculations. The year following his appointment, he could not have failed to be impressed by Brunel commissioning the mathematician Charles Babbage to experimentally verify the ride quality of the Great Western Railway using a device he (Babbage) created to automatically record the movement of the railway carriage with a pen tracing a graph on a continuous sheet of paper. Even though Froude took early retirement in 1846 to care for his ailing father, he continued to dabble in engineering, frequently emulating Babbage's methods by employing pen-and-graph-paper devices to record the performance of farm machinery and the efficiency of model screw propellers.[2]

Brunel asked Froude in 1857 to come out of retirement to examine the launch friction and rolling of *Great Eastern*, the latter to ensure it would not stick in the mud after launch. By then, the subject of rolling had been examined by researchers for over a century. Daniel Bernoulli initially tackled the problem in 1738 and was the first to mathematically define the three degrees of freedom of a ship—roll (rotation side to side), pitch (rotation fore and aft), and heave (movement up and down). Pierre Bouguer and Leonhard Euler had addressed the motions of vessels in 1746 and 1748, respectively, as an outgrowth of their theories of ship stability. The problems of rolling, pitching, and other motions were so important that the French Academy of Sciences commissioned five separate prizes on these subjects in the 1750s and 1760s. In 1771 Jorge Juan y Santacilia developed methods for evaluating a ship's roll period in still water and in waves that were based on the theory of the pendulum. But all these investigations were of kinematics, examining the only motions of the ships themselves, and did not tackle the dynamics, or the forces that underlay the kinematics.[3]

The greatest obstacle confronting these early theoretical developments was the lack of observational data to back them up. The metacentric theory of a ship's initial stability could be verified by the inclining experiment, using a pendulum to measure the angle of heel while hanging a static weight over the side. By contrast, it was almost impossible to use a standard pendulum to accurately measure the angle of inclination while a ship was dynamically rolling or pitching, due to inertial effects that would cause the pendulum to regularly overshoot or undershoot. This was the same problem that early clockmakers faced using a pendulum balance to build a marine chronometer until John Harrison perfected the use of the spring balance in 1759 that enabled navigators to calculate longitude at sea.[4] A few individuals invented devices intended to accurately measure the angles of roll and pitch at sea—the oscillometer, developed in 1800 by the French naval officer Nicolas Baudin, and the nauropometer, built in 1829 by British naval constructor Henry Chatfield—but none of these caught on.[5]

It would not be until 1850, in a paper read before the Royal Society, that Henry Moseley made the first attempts to develop a comprehensive theory of the dynamics of rolling and to verify it experimentally. Moseley was a mathematician, a schools inspector, and a prominent Anglican clergyman (which is why he was often referred to as Canon Moseley), who also wrote highly regarded textbooks on engineering and gave examinations on the subject to naval constructors at the short-lived Portsmouth Central School of Mathematics and Naval Construction. It was therefore unsurprising that this polymath chose to examine the arcane subject of dynamic stability in a moving seaway at a time when the state of naval architecture was limited to studies of initial stability in calm waters. Moseley's breakthrough was to think about ship motions not just in kinematic terms but also in terms of dynamics, or work done. Partly on the basis of Dupin's "On the Stability of Floating Bodies" from 1822, Moseley developed the formula

$$U(\theta) = W(\Delta H_1 - \Delta H_2),$$

where $U(\theta)$, the work done by rotating a floating body through an angle $\theta$, is equal to the work done raising $W$, the weight of the body, through the difference in heights of the center of gravity $H_1$ and center of buoyancy $H_2$. To verify this formula, Moseley worked with two of his colleagues at the Portsmouth school, Robert Rawson and John Fincham, to build two models, each one meter long, of ship forms with circular and triangular cross sections. These models were floated in an experimental apparatus (figure 3.1), and when a sudden force was applied at the end of an attached crossarm (thus creating an inclining moment), the resulting angle of roll and period of oscillation were plotted on a graph. The results of the experiments correlated well with those predicted by the formula. Moseley concluded that the theory of ships at large angles of heel, developed in 1798 by Vial du Clairbois and Atwood, underestimated the angle of roll because it did not account for the momentum due to inertia of the rolling body.[6]

In 1857, while *Great Eastern* sat high above the Thames River being prepared for launch, Froude and William Bell, one of Brunel's engineers on the Dartmouth railway, were at work trying to estimate the rolling behavior of *Great Eastern* after its launch. Bell built a small (probably two-meter) model of the ship at the launch displacement of 11,600 tonnes, which he and Froude tested in a water tank fitted with a hand-cranked wavemaking apparatus at Froude's parsonage at Dartington. To properly model the ship's behavior, Bell needed to know its center of gravity. The estimation of the

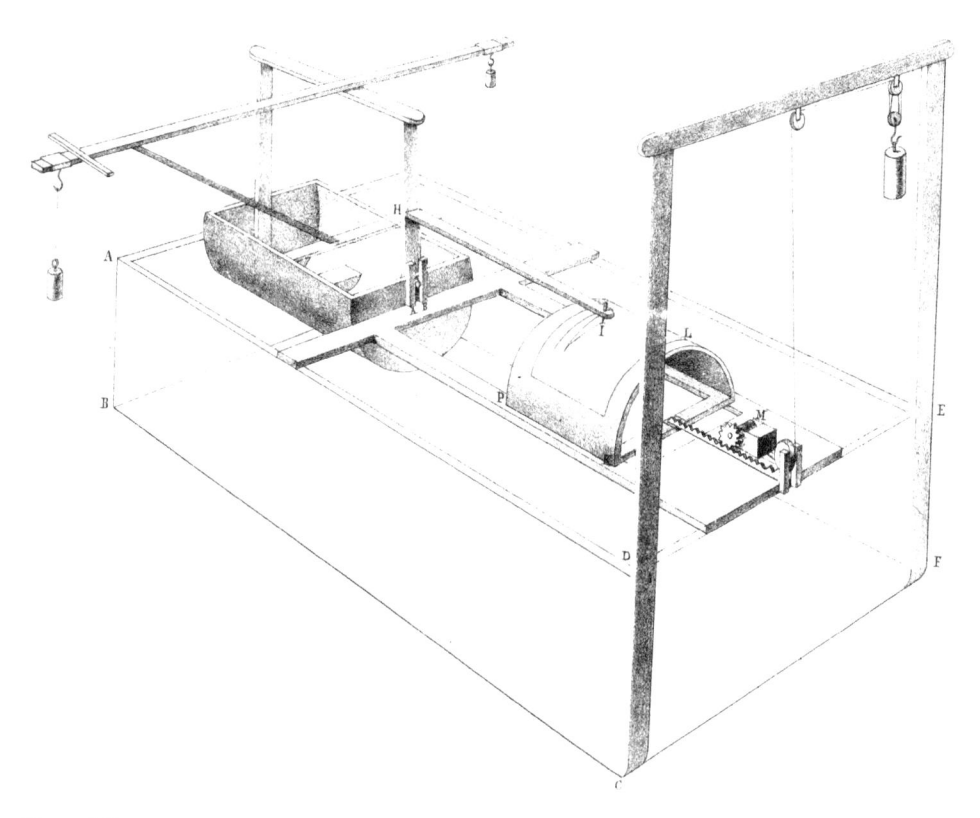

**Figure 3.1**
Moseley's experimental apparatus for measuring rolling, 1850.
Credit: JSTOR.

center of gravity is essentially a long, tedious, error-prone process of calculating both the weight and the exact position of each of the thousands of elements of the ship—hull plates, deck plates, girders, and so on—then squaring, multiplying, and summing each of them to get the ship's moment of inertia, finally dividing by the total weight to obtain the center of gravity of the overall ship. The ship's builder, John Scott Russell, had understandably demurred when Brunel asked for this—the calculations would have taken valuable weeks and months by Russell and his staff, now racing to complete the ship—so Brunel simply tasked William Bell to do it instead.[7]

By the summer of 1857 Froude and Bell had greatly progressed in their studies, going well beyond their original remit of simply looking at whether Brunel's ship would stick in the mud and instead trying to develop a general theory of "the work done by the ship in rolling," as Froude told Brunel in January of that year. Froude's most important

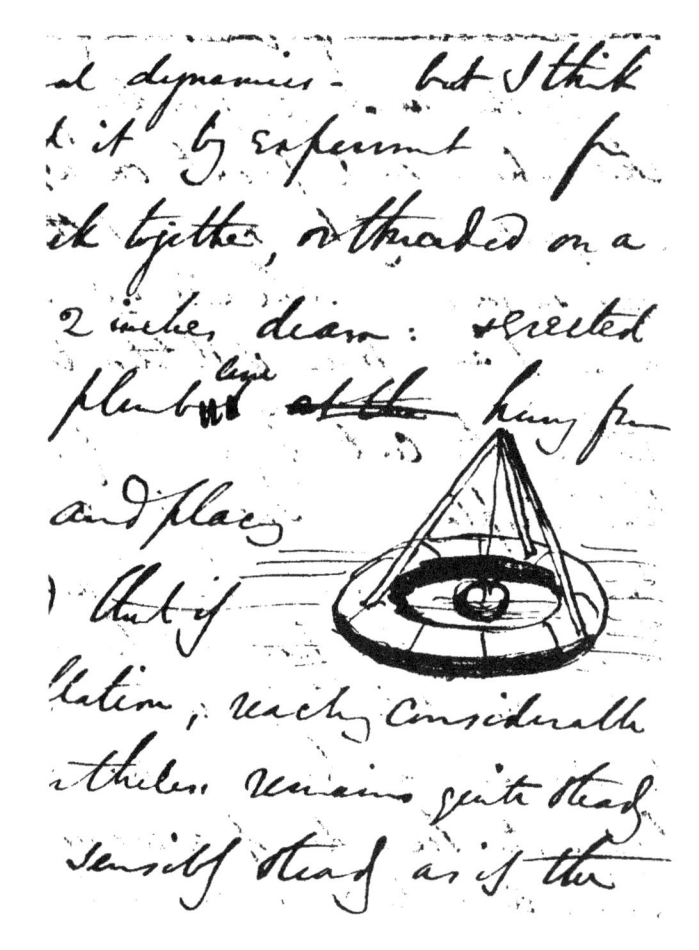

**Figure 3.2**
Froude's first experiments on rolling, 1857. Froude's sketch of his ring float with plumb bob (this page). Ring float in waves (opposite page).
Credits: Brunel University Special Collections (this page); author's collection (opposite page).

insight into the dynamics of ship motions came from a simple experiment using a four-inch-diameter cork ring with a plumb bob suspended over it (figure 3.2), which he tested first in waves generated in the little tank on the parsonage grounds and later in the rolling surf, presumably at the nearby beach in Paignton. Froude noted that the plumb bob would always point to the wave surface and not vertically down, leading him to the conclusion that "the forces by which a body floats on the surface of a wave … [are] at right angles to the surface of the wave," as he excitedly told Brunel on 16 August.[8] Froude's insight from this was that the "fundamental law which governs

FIG. 2.—EXPERIMENT OF FLOAT IN WAVES.

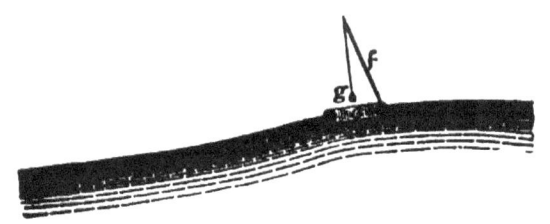

*e.* Float made of a cork ring.
*f.* Short mast in the float carrying.
*g.* A small plumb-bob which hangs in conformity with the apparent direction of gravity, and thus at right angles to the surface of the water.

FIG. 3.—EXPERIMENT OF FLOAT IN BREAKING WAVE.

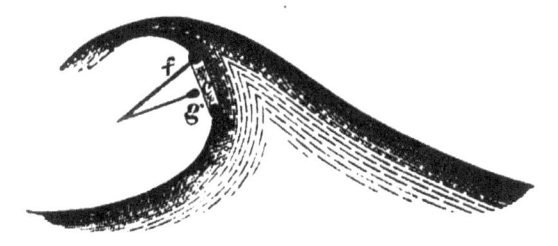

*e.* Float made of a cork ring.
*f.* Short mast in the float, carrying
*g.* A small plumb-bob which hangs in conformity with the apparent direction of gravity, and thus at right angles to the surface of the water.

the motion of a ship on waves" is that the ship floating in waves is subject to the same forces as the mass of fluid particles it displaces. Since any particle at the wave surface has its momentary axis of equilibrium—local vertical—at right angles to the surface of the wave, the ship will also have its momentary axis of equilibrium in the same direction. Thus, stability of a ship is a "measure of her effort to become normal to the waves in undulating water," as he reported in 1861 to the Institution of Naval Architects.[9]

Here, Froude was recapitulating the most commonly accepted theory at the time for ocean waves, first elucidated in 1802 by the Bohemian scientist František Josef Gerstner. Gerstner's model of oscillating deepwater gravity waves assumed that each of the fluid particles in the wave followed a circular path, where the radius decreased with depth. The resulting equations gave the waves a trochoidal profile, with the surface of the water forming a cycloid—this was the "wave of replacement" that John Scott Russell assumed for his wave-line theory.[10] The greater the wave slope (the angle of the wave surface), the greater the disturbing moment of that wave. Froude's resulting equation of motion for rolling in waves equated (using modern notation) the righting moment (on the equation's left) of the ship with the wave-disturbing moment (on the equation's right):[11]

$$I\frac{d^2\varphi}{dt^2} + W\overline{GM}\varphi = W\overline{GM}(2\pi\xi_a/L_w)\sin\omega_w t,$$

where $I$ is the ship's inertia, $\varphi$ is the ship's heel angle, $W$ the displacement, $\overline{GM}$ the metacentric height, $\xi_a$ the wave amplitude, $L_w$ the wave length, and $\omega_w$ the wave frequency. The implication for this equation was that the metacentric height alone governed the resulting motion, both period and amplitude, which he confirmed with numerical results by forcibly rolling the small models of waves in his little basin using the wavemaker. This equation confirmed previous observations, that if the wave period came close to the natural frequency of roll for the ship, a synchronous oscillation would occur and greatly increase the maximum roll angles. Finally, Froude and Bell used their model of *Great Eastern* to establish the roll decay equation, by "hoving it down" to 45 degrees of heel and then counting the oscillations until it stopped noticeably seesawing.

Froude's work generated considerable discussion at the Institution of Naval Architects, particularly from John Scott Russell, who continued to believe that a ship's form, and not its metacentric height, controlled its roll behavior. But by this time, Froude's father had died, so he was forced to leave the parsonage in Dartington and temporarily move to Paignton. For several years, he was otherwise occupied with family affairs and overseeing the construction of his new home, called Chelston Cross, in nearby Torquay. During this period, he only sporadically revisited the question of rolling, and few other scientists paid much attention to the subject; even the 1869 British Association for the Advancement of Science (BAAS) report "State of Existing Knowledge on the Stability, Propulsion, and Sea-Going Qualities of Ships" (of which more later) reported little advancement on rolling beyond the works of Moseley and Froude.[12]

That all changed in September 1870 with the capsize and sinking of the ironclad HMS *Captain* and the loss of almost its entire crew. The ship, which was the brainchild of a naval artillery officer named Cowper Phipps Coles, was approved by the Admiralty even over criticism by its own naval constructors. The saga of *Captain*, as discussed here and in subsequent chapters, was perhaps the most significant event in the development of the naval architecture profession in the late nineteenth century; it had far-reaching effects concerning the nature of technical authority and the role of science in ship design that went well beyond technical arguments over the ship's lack of stability.[13]

Coles had developed the idea of a rotating gun turret during the Crimean War, at about the same time as did John Ericsson, but it was Ericsson's turret aboard USS *Monitor* that was the first to see action, in 1861 during the U.S. Civil War. For several years afterward the merits of turret versus broadside armament were debated in the navy, Parliament, and professional and public journals. In 1865 the British chief constructor Edward J. Reed, who had just replaced HMS *Warrior*'s designer Isaac Watts, began to design Britain's first purpose-built turreted warship, HMS *Monarch*. The turrets would be mounted on the upper deck and were somewhat encumbered by the masts and sails. Coles—who still claimed parental rights over the idea of the turret—argued vociferously in countless letters to politicians, magazines, and newspapers that "a sea going turret-ship" should have its guns low to the water to provide a more stable firing platform. Although Reed objected to the planned freeboard of just eight feet (2.4 meters) as not being "satisfactory for a sea-going cruising ship," the Admiralty bowed to the political pressure in favor of Coles and approved the plans for *Captain*, even as *Monarch* was still under construction.

The Laird shipyard in Birkenhead was selected to build the ship, but because of the way the contracts were written, no single entity—Coles, Laird, or Reed—had full authority over the entire design. From the time the keel was laid in 1867 until completion in 1870, the ship's weight grew so much that its freeboard decreased to just six and a half feet (two meters). Soon after the ship was commissioned in April 1870, Reed resigned in protest over his treatment by the Admiralty, which ignored his advice against the ship's construction. Instead, Reed's former assistant (and brother-in-law) Nathaniel Barnaby took over as chief naval architect. Barnaby tasked one of his staff, Frederick K. Barnes, to evaluate *Captain*'s stability. Barnes used the data from Laird's inclining experiment to calculate (for what appears to have been the first time for any ship) the $\overline{GZ}$ curve that had been developed in 1798 by Vial du Clairbois and Atwood to measure the full range of stability over all angles of heel. Barnes determined that while the metacentric height was adequate to carry sail in calm water (it was about 0.7 meter), the ship's stability began to diminish at a heel angle of 20 degrees, compared with HMS *Monarch*'s angle of maximum stability of 40 degrees (figure 3.3). Both Moseley and Froude had already shown that a ship's momentum in roll would cause it to tilt far more than the steady angle of heel from the same wind force, so Barnes noted that if the ship were subjected to a strong gust of wind combined with a rolling sea, "there would be some danger of the ship foundering." But by the time Barnes had come to this conclusion on 29 August 1870, the ship—with Coles aboard—had already departed Gibraltar with its squadron. Shortly after midnight on 7 September, off Galicia, Spain, in roaring winds and heavy seas, HMS *Captain* foundered and sank with 480 men lost, including Coles. More men died that night, on that one ship, than the entire British fleet had lost at the Battle of Trafalgar.

Such an enormous loss understandably caused a national uproar, but amid the court-martials and finger-pointing that accompany any such disaster, there also began

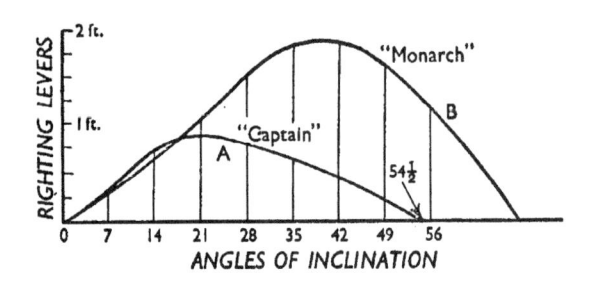

**Figure 3.3**
$\overline{GZ}$ curves of HMS *Captain* versus HMS *Monarch*.
Credit: Author's collection.

a concerted effort to scientifically understand and improve the behavior of ships at sea. In January 1871 the British Admiralty established a Committee on Designs for Ships of War to rigorously examine previous warship designs and help establish future policy. Although led by a career diplomat and top-heavy with naval officers, the committee also included prominent British scientists and engineers such as William John Macquorn Rankine, William Thomson (soon to be Lord Kelvin), and William Froude. The committee tasked Froude to verify the rolling characteristics of several ships, including the sloop HMS *Greyhound* and a new turret ship then under construction, HMS *Devastation*, the first British capital warship to be built without sails.[14]

Froude was well prepared to continue the rolling investigations from where he had left off. He had by then finished building his manor home Chelston Cross, which in addition to a grand ballroom and central heating, had workshops and what looked like an attached chapel but actually housed a small in-ground basin for testing ship models. Froude had developed in his workshops a mechanism for automatically recording roll motions, from which he could plot and mathematically analyze the curves of motion to determine the model's performance. In March and April 1871 he used his basin to test two models of *Devastation*, three meters and six meters long. In addition to determining the roll motions and decay, he also tested each model with and without bilge keels—longitudinal strakes jutting out from the hull—to confirm that they were indeed effective at dampening roll motions. On the basis of Froude's results, the committee recommended in 1872 that the freeboard of *Devastation* should be raised and that the Admiralty should adopt bilge keels and take measures to ensure adequate stability at large angles of heel.[15]

But Froude understood that model tests alone would not convince the public and the Parliament, still reeling from the *Captain* disaster, of the safety of the *Devastation*.

In 1873 he approached the Admiralty, arguing that only experiments using the actual ship would engender public trust in that type of vessel. Privately, of course, he wanted to carry out full-scale trials to verify his theory and model tests (he also knew, from personal correspondence with the French constructor Louis-Émile Bertin, that the French navy was already conducting such trials).[16] Despite the misgivings of some naval officers, Froude was given permission to board *Devastation* with his apparatus for measuring roll at sea (figure 3.4). The device used two pendulums—one shorter and lighter, which would follow the ship's rolling motion, the other longer and heavier, which would stay reasonably vertical—to record the roll motion on a drum driven by clockwork.

The first trials were carried out in April in Gibraltar harbor, where the rolling was induced by the crew running back and forth across the deck. Later that month, and then again in September, the ship went out in search of waves to record roll motions and verify the seaworthiness of the vessel. *Devastation*'s officers were pleased with the vessel's performance in heavy seas—the additional freeboard kept the ship drier—while Froude said that "the resulting oscillations [of the ship], reduced to a diagram and analyzed … were precisely of the character which the theory prescribes." The success of both the ship and of Froude's methods helped the British navy regain the trust it had lost. Around 1880 HMS *Devastation* was fitted with bilge keels, just as Froude had recommended, and the ship went on to have a long and successful career.[17]

**Predicting Speed and Power**

William Froude may have made his reputation by developing the means to predict ship motions, but it was his methods for predicting hull resistance and powering that made him world famous. As I describe at the beginning of *Ships and Science*, I was driven to this project of chronicling the history of naval architecture in part because William Froude is almost the only name known to modern naval architects, despite decades of progress both before and after his time.[18] Yet Froude did not arrive at his now well-known method of using small-scale models for estimating speed and power simply as an extension of his studies on ship rolling. His efforts occurred just as Britain's ascent as the workshop of the world created the economic impetus for ever-greater speed and efficiency of shipping and communications, which created the environment that allowed Froude to flourish in his research.

By the 1860s, when Froude was first developing his ideas for predicting speed and power, steamships accounted for only 10 percent of the world's shipping tunnage but already carried much of the high-value cargo like mail and passengers. The economic forces driving the steamship trade were captured in a magisterial three-volume study

**Figure 3.4**
Froude's instrument for automatically recording the rolling of ships, 1873.
Credit: Princeton University.

by Eugène Flachat titled *Navigation à vapeur transocéanienne* (Transocean steam navigation), published in 1866 for the Compagnie Générale Transatlantique, a syndicate of railroad magnates that, like Brunel's Great Western companies, was promoting the idea of extending its rail lines across the Atlantic Ocean. Flachat's analysis of the economics of steamships compared with sailing vessels was very much in the mold of the French technocracy of the era, which had already made a science of quantifying costs and benefits for public works. For example, he determined that while sailing ships had a 1 in 200 chance of sinking, the rate for steamships was 1 in 1,000, or one-fifth. Steamships also vastly improved the speed of communications across the oceans; even before the telegraph was introduced, steamships had cut the transmission time of information around the globe by three-quarters in just the space of 40 years.[19]

Flachat's fixation on costs found its focus in estimating fuel efficiency—the word "coal" (*charbon*) appeared almost 200 times—and he devoted considerable space reviewing theoretical developments in marine engineering and naval architecture in the preceding century. Marine engineering at that time was advancing more quickly than naval architecture, for the reason that naval architecture was focused only on ships, while marine steam engines were but one small subset of the much larger array of steam applications. From pumping water to running factories to driving railroads, far more resources from many different industrial sectors were poured into the improvement of steam power; the compound steam engine, as noted, was first developed on land before being applied on the oceans. At the same time, British scientists like Rankine were developing the first theoretical insights into the new science of thermodynamics and applying them to steam machinery, and across the Atlantic, the U.S. Navy's engineer in chief Benjamin Isherwood was conducting the first systematic experiments of marine steam engines.[20]

Although the science of naval architecture had also made significant advances since the turn of the century, it still lagged behind marine engineering. As explained in chapter 1, Newton's original concept of resistance as due to fluid shock was proved false by the experiments of Charles Bossut and gradually replaced with the concept of pressure as the principal mechanism. The shuttering of the French Academy of Sciences in 1793 had marked the end of large, government-backed research into naval architecture, and that vacuum was only sporadically filled by research carried out privately, like Mark Beaufoy's self-funded experiments in hydrodynamics, or supported by entities such as the BAAS, which backed Russell's wave-line research. For a while, these experiments provided some measure of prediction; Robert Fulton, in his 1809 steamboat patent application, used Beaufoy's resistance data to determine the engine power needed to propel his boats at a specific speed.[21] But as ships grew larger and faster, these results became increasingly outdated. John Scott Russell's wave-line theory had singularly failed to provide a means of predicting speed and power, and even by the time Flachat had published his work, no other theories had appeared on the horizon that adequately explained fluid resistance in order to make such predictions.

Before adequate theories of ship resistance could be developed, engineers and scientists needed standardized means of measuring speed and power. Ship speed through water had been estimated for centuries using chip logs payed out over the side or stern; the word "knot" derives from the number of regularly spaced rope knots that passed a point in a given time interval. Later, mechanical devices such as patent logs (also called taffrail logs) automatically recorded speed using a spinning device in the water attached to a dial indicator aboard the ship. But these mechanisms were notoriously inaccurate, so the idea of measuring speed along a known distance arose almost as soon as the first steam engines were on the water. As noted in chapter 2, in 1790 John Fitch set up the first measured mile—sets of flagpoles set exactly one mile apart—along the Philadelphia waterfront to measure the speed of his oar-driven steamboat. In 1818, Britain's first measured mile was set up by Boulton and Watt along the Thames. Within a few years, shipbuilding contracts for steam packets began calling for a specified trial speed (speed made good along a measured mile); Laird's 1837 contract for the iron steamer *Rainbow* required 12 knots, with bonuses added for greater speed. For some time, measured mile courses were established on an ad hoc basis, but in 1866 the British Admiralty established a permanent measured mile in the Firth of Clyde that gave uniform results for all shipbuilders, a practice repeated around Britain and across the globe.[22] Nevertheless, many pilots preferred to test their ship's speed in steamboat races, which were particularly notorious in the United States for resulting in catastrophes like the *Lucy Walker* boiler explosion on the Ohio River in 1844 and *Henry Clay*'s shipboard fire and grounding on the Hudson River in 1852.

The standardized measurement of "horsepower," the term James Watt invented in 1783 to market his steam engine, came about at roughly the same time as the measured mile. In 1796, Watt and his assistant developed the indicator diagram, which plotted piston volume versus pressure during the stroke cycle. The indicated horsepower (IHP) was derived from these diagrams and showed how much power the engine theoretically produced, but its value did not account for friction or other losses, nor was its definition or usage universally agreed upon. In 1822 the French mathematician and engineer Gaspard de Prony developed the first dynamometer to directly measure the actual power output of an engine. The friction brake, or Prony brake, as it became known, used a belt and sheave system to measure the torque and speed of the engine's output shaft and thus yield the brake horsepower (BHP). Other dynamometers were soon developed

# The Quest for Accuracy

using broadly similar principles; for example, in 1842 Auguste Taurines, a professor at the artillery school in Brest, invented a dynamometer that used springs instead of a belt to account for the torque. It measured engine output in BHP in several trials of screw corvettes. But marine steam engines were rarely tested using these large and cumbersome dynamometers, so that IHP became the most commonly used measure of power until the turn of the twentieth century, when the torsion meter (which measures a drive shaft's twist, which is transformed to transmitted power in order to indicate shaft horsepower, or SHP) became a regularly installed instrument on propeller shafts.[23]

Shipbuilders, increasingly under contractual obligations to meet speed requirements, soon availed themselves of measured speed and power to develop methods of predicting both. Beaufoy's experimental data was proving inadequate for the purpose; the translation of his various elements of hull resistance into horsepower was fraught with inconsistencies and guesswork, and as one shipbuilder noted in 1830, "[It was] too tedious for practical use."[24] Instead, engineers and naval architects began creating and using formulas based on data measured from existing vessels to estimate the speed and power for new ships. These formulas all followed the same basic pattern: a coefficient linking factors of speed, IHP, and some measure of displacement or midship sectional area that were derived from existing ships and then extrapolated to new designs by multiplying this coefficient by the factors for the proposed vessel.

Boulton and Watt was the first company to conduct a series of comparative trials of its steamships, starting in 1828, to establish a coefficient of performance ($C_P$) for predicting the power requirements for newer ships. Its formula was straightforward, a function of velocity and midship sectional area $A_M$:

$C_P = (V^3 \cdot A_M) / \text{IHP}$.

In 1833 a British consulting engineer, John Farey Jr., who had learned of Beaufoy's work and found it wanting, described a more complicated coefficient to represent resistance ($C_R$) that also accounted for ship length $L$ and maximum breadth $B$:

$C_R = V^3 [(\sqrt{A_M} \cdot L) / B + A_M] / (1000 \cdot \text{IHP})$.

It is not clear that these formulas had widespread use; it is more likely that most shipbuilders avoided calculations of speed and midship sections and instead stuck to using a simple ratio of between 2.5 and 4 IHP per registered tun to size their engines, as one anonymous Manchester shipbuilder noted in an 1841 letter published in a widely read technical journal.[25]

That state of affairs unexpectedly changed in 1853 with the publication of *The Capability of Steam Ships* by Charles Atherton. Atherton was then chief engineer at the Woolwich Dockyard, having already served in a number of positions as a civil and marine engineer. Atherton was also an active member of the BAAS and, as part of its emerging campaign to collect "statistics of improvement in our mercantile marine," was uniquely placed to bridge the knowledge gap between naval and merchant shipbuilding. Atherton's goal was to "systematize the science of steamship construction" with comprehensive ship and engine data. As part of the Admiralty establishment, he was able to gather the trial data collected by the British navy on steamships like HMS *Rattler* and use them to predict speed, power, coal consumption, and so on, for new designs. To provide a common basis for estimating the "locomotive performance" of different-sized vessels, Atherton created an "index number" based on existing ships. Instead of using midship sectional area, Atherton chose to use displacement ($\Delta$) because it was a more commonly known value during design, but he raised it to the 2/3 power to approximate the midship sectional area used in previous coefficients. This index number soon was known as the Admiralty coefficient and became widely used (then as today) in the very early stages of ship design for estimating speed and installed horsepower:

Admiralty coefficient = $(V^3 \cdot \Delta^{2/3})$ / IHP.

For example, the British assistant constructors who designed the steam warships HMS *Temeraire* and HMS *Iris* in 1875 averaged the coefficients of several earlier vessels to determine the appropriate Admiralty coefficient to estimate the required installed horsepower. As for Atherton's quest to collect and use performance data to improve mercantile shipping, it eventually foundered under the political maneuverings of the BAAS.[26] Yet as is discussed shortly, those efforts eventually led to Froude's novel approach to the problem of predicting speed and power.

Although the Admiralty coefficient was adopted by constructors and shipbuilders as a more scientific method for estimating speed and power than had been previously available, it was in fact little more than a repackaging of Newton's long-debunked impact theory of fluid resistance.[27] Moreover, as the constructors and engineers who used the method would note, the actual coefficients were widely scattered even for similar ships and were increasingly unreliable at the higher speeds that steam plants could now deliver.

The Scottish shipbuilder James R. Napier, who as noted in chapter 1 had previously used the wave-line theory without much success, was not only unhappy with the wave line but also found the Admiralty coefficient singularly unsuited for the power predictions he needed to make in order to meet contract requirements for the Russian packet ship *Admiral* he was building. In 1857 Napier asked William Rankine, then a professor at Glasgow University, to suggest an alternative method. Rankine communicated his formula privately to Napier, who used it to predict *Admiral*'s power and speed, which met with a success Napier later called "unprecedented." Rankine did not publicize his formula until 1861, when he delivered it first in outline form to the BAAS, then a year later in a detailed explanation to the Royal Society. Rankine assumed, as had Russell and Beaufoy before him, that a ship's resistance was due to three factors—bow pressure, stern pressure, and friction. But in Rankine's estimation, only the friction component was of any significance, and it was a function of fluid density, hull roughness, and the geometry of the hull. The formula he had delivered to Napier and now made public was

$$\text{IHP} = f\rho V^3 GL\,[1 + (\pi^2 B^2) / L^2],$$

where $f$ is the friction coefficient, $\rho$ the density of water, and $G$ the mean girth around the ship; the argument in square brackets is the "augmented surface" of the hull that accounts for the effect of curvature. Rankine sought to further understand the effects of curvature on friction by examining the flow lines around a ship and in particular "the resistance [of the hull] due to frictional eddies." In a paper delivered to the Royal Society in 1863 (and extended in 1871, one of his last works before his untimely death the following year), he noted that frictional resistance depended on the smoothness and fairness of the hull (i.e., no discontinuities or rapid change of shape) and wondered if Russell's wave line prevented not only waves but also frictional eddies. Rankine accordingly developed a potential flow model of a ship's hull to calculate this resistance. He chose a hull shape in which the streamlines (figure 3.5) had minimal variations in velocity from their neighboring streamlines, which indeed turned out to be very much like Russell's wave-line hull form. But Rankine's streamlines provided little in the way of real predictive power, and his resistance equation, though having proved useful to James Napier, was too complex in practice to be of use to other nineteenth-century shipbuilders. However, Rankine's potential flow model for a body in a fluid, in which the surface was visualized as a group of streamlines flowing through a distribution of point-like foci (later called sources and sinks, where fluid streams emanate from and are drawn in to, respectively), would in the twentieth century become a powerful mathematical tool for analyzing ideal fluids—without mass or friction—as a means of solving real-world problems.[28]

Across the Channel in France, complexity was not a problem for a generation of naval *officiers savants* who had faced the mathematically daunting entrance exams for the École Polytechnique and the École d'application du Génie Maritime (of which more in chapter 5) and who had been steeped in the eighteenth-century hydrodynamic works of Pierre Bouguer, Charles Bossut, and Pierre Du Buat. During the 1830s and 1840s, the most famous instructor at the École d'Application in Lorient was Frédéric Reech, a brilliant Alsatian engineer who combined the studies of thermodynamics and naval architecture to select "hull forms that offer least resistance" to improve the efficiency of steam power. Reech at first proposed, following in the footsteps of

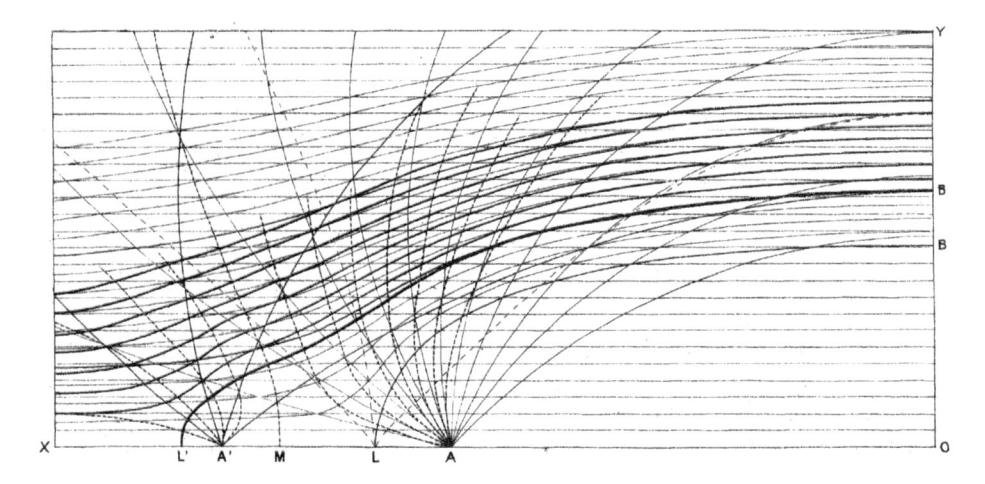

**Figure 3.5**
Rankine's streamlines around a ship shape.
Credit: JSTOR.

Bossut and Du Buat, that resistance varied as the area of midship section. Henri Dupuy de Lôme, who had been one of Reech's more gifted students, tested that hypothesis with a series of full-scale trials by pulling one vessel behind another and measuring the towing force with a dynamometer installed aboard. He conducted the first trials on small paddle wheel vessels in Toulon harbor beginning in 1841. The French navy then built the screw *avisos* (dispatch boats) *Pingouin* and *Pélican* to experiment with various propeller configurations, which in 1847 were used to conduct measured towing trials in Bordeaux. These tests and trials gave Dupuy de Lôme enough data to develop the machinery and propeller configuration for his next project, the 90-gun screw steamer *Napoléon* and to determine that the coefficient of form, $K$, equivalent to the Admiralty coefficient, took the following form:[29]

$K = V^3 A_M / \text{IHP}.$

As with the Admiralty coefficient, French shipbuilder and constructors were soon dissatisfied with the coefficient of form $K$. Further trials with *Napoléon* convinced Dupuy de Lôme that resistance was due not just to the midship section but also to other factors such as friction and hull curvature, and in 1852 he developed a new formula for ship resistance that anticipated Rankine's "augmented surfaces":

$R = K A_M (V^2 + 0.145 V^3) + K' A'^3_M \sqrt{V},$

where $K$ was a new coefficient based on the curvature of the hull form, and $K'$ was a coefficient of friction based on hull roughness. Several years later, one of France's most celebrated *officiers savants*, Siméon Bourgois, started with Dupuy de Lôme's theories and experiments to develop a complex book-length synthesis of ship resistance and powering, *Memoire sur la resistance de l'eau* (1857), containing equations that accounted for hull form, surface roughness, and even air resistance. This work, however, would turn out to be the last gasp of a purely mathematical way of predicting speed and power, because the idea of making such predictions using scale models was already beginning to take hold.[30]

The previous scale-model experiments by Bossut, Beaufoy, and Russell were all done in attempts to find general laws of hydrodynamics, but in no case was there an attempt to use small-scale models to directly predict full-scale ship speed and power; all the predictive equations thus far were based on comparing full-scale ships of roughly equal sizes. Frédéric Reech was the first to propose using small-scale models for prediction at full scale, although his ideas took time to evolve. As noted earlier, Reech had at first stuck to the old idea of calculating ship resistance on the basis of the midship area, but he also allowed that power likely scaled as a function of length, though admitting that "the complete development of the subject could not be pushed very far here." In 1848 he came upon an essay by the mathematician Joseph Bertrand, who had studied the works of Galileo and Newton on the subject of "mechanical similitude"—that is, representing the forces acting on bodies of different sizes. Bertrand argued that to correctly represent friction and traction forces acting on, say, a full-scale locomotive with length $L$ and speed $V$, one could use a smaller scale model of length $l$ operating at a speed proportional to $\sqrt{V}$.

Reech further developed Bertrand's theorem four years later, at the very end of his 1852 textbook *Cours de mécanique* (Course of mechanics) destined for the École Polytechnique. Using steamships instead of locomotives, he stated that speeds between models will scale as a proportion of the square root of length, and inertial and gravitational forces scale proportionally to the cube of length. This theorem, Reech claimed, would allow one "to experiment [with a steam vessel or boat] on a model of a smaller scale" and make the science of hydrodynamics "much simpler and more powerful" than the present specialized formulas for ship resistance, which he considered "of absolutely no use" in a mechanics course—even the formulas of his former student Dupuy de Lôme, who conversely considered it "wrong to compare the resistance of different ships by means of experiments made on models to reduced scales."[31]

Reech's theorem was literally an afterthought in his textbook, and there is no evidence that anything was done with it in France. But across the Channel, Reech's theorem was noted with interest by Charles Merrifield, a mathematician and principal of the newly opened School of Naval Architecture in London. He was a correspondent of Eugène Flachat and had read about Reech's theorem in *Navigation à vapeur*. At that time

(1868) Merrifield was lobbying the BAAS to fund a solution to what he termed "The Necessity for Further Experimental Knowledge Respecting the Propulsion of Ships," on the grounds that the "law of the resistance of ships" was still "inexact," and needed experimental observations "as a proper basis for theory." The previous attempts by Charles Atherton to collect data on real ship performance, as noted earlier, had foundered despite a decadelong struggle by five separate BAAS committees to do so. The latest efforts sponsored by the BAAS in 1864 involved testing two 1.5-meter models on a lake in Blackheath, but these gave inconclusive results that were never applied to full-scale ships. In answer to Merrifield's petition, the BAAS agreed to form yet *another* committee, which included Rankine, Froude, and Francis Galton (today known as a pioneer in statistics and genetics), to "report on the state of existing knowledge on the stability, propulsion, and sea-going qualities of ships."[32]

Merrifield's committee presented its findings to the BAAS at the annual meeting in August 1869. The report reprised the various theories of ship resistance going back to Euler, Bossut, and Beaufoy, highlighted the more recent equations developed by Dupuy de Lôme and Bourgois, and summarized Russell's wave line and Rankine's augmented surfaces. It also summarized the research up to that point concerning rolling and motions in waves. Merrifield recommended, as had all BAAS committees beforehand, that the Admiralty should conduct measured trials of full-scale ships with "considerable size and fine form," which could yield an adequate theory of resistance and prediction of performance. But when discussing small-scale models as an alternative to full-scale trials, Merrifield was both cursory in his description and purblind in his conclusions:

> It was supposed that models would most aptly represent ships at the same speed both for the ship and the model. Experiments ... seem to show a dissatisfaction with the results of small models, and some time later, M. Reech pointed out that models of different sizes intended for comparison should be made to move at velocities varying as the square roots of their lineal dimensions. In this case the actual resistances would vary as the cube of the lineal dimensions. This would follow from the theory of the resistance of submerged bodies, on the supposition that the resistance varies as the square of the speed.

Merrifield had correctly identified that the problem with previous small-scale model tests was that they were not conducted at the proper speed, and he had acknowledged that Reech provided the clue to make model-scale testing successful by obeying the square-root law; he then singularly failed to follow through on this observation, opting once again for full-scale trials. All the committee members unconditionally signed off on the report, except one—William Froude.[33]

Merrifield and other members of the committee rejected the use of small-scale models, but Froude would have none of it. To Merrifield's report he appended a six-page

"Explanations" that outlined the research he had been conducting on small-scale models since 1865. By then Froude had become dubious of Russell's wave-line theory, so he decided to compare it with a more rounded form derived, as he stated, "by eye from water birds." He made two sets of models, respectively named *Raven* (wave-line) and *Swan* (blunt-ended), each having versions one, two, and three meters long, that he towed around Dartmouth Harbor using a steam-powered launch fitted with a dynamometer (figure 3.6). The results showed that the blunt-ended *Swan* models had less resistance at higher speeds than the wave-line *Raven*.

Of greater interest to Froude in his "Explanations" was that the principal component of resistance, the "generation of surface waves," scaled predictably among the three sizes of model as the square root of the length of the model (i.e., $V/\sqrt{L}$ was identical for all), and he deduced that this theory of similitude should hold true all the way to full-sized ships. This finding gave him the confidence to state "that experiments of models of rational size ... can be relied on as truly representing the ships of which they are the models." It is likely that Froude developed this theory quite independently without knowing of Reech's previous work, although Reech himself always spoke of Froude "in terms of appreciation" for proving his hypothesis. Nevertheless, the resulting Froude's law of similitude is today referred to in France as *la loi Reech-Froude de similitude*.[34]

**Figure 3.6**
Froude's original models of *Raven* and *Swan*.
Credit: Science Museum, London.

Froude's "Explanations" might have been overlooked as yet another academic quarrel between scientists were it not for his very last sentence: "Unless the reliability of small-scale experiments is emphatically disproved, it is useless to spend vast sums of money upon full-size trials." Froude's statement that his concept would save "vast sums of money" came at precisely the right time in history, for it was exactly what the newly elected government of William Gladstone wanted to hear. Gladstone's Liberal Party had swept into office in 1868 promising balanced budgets, low taxes, and free trade. Controlling navy costs was one of Gladstone's priorities, and Hugh Childers, his first lord of the Admiralty, was committed to reducing outlays, reorganizing the bureaucracy, and cutting manpower. Suspicious of his own technical experts, especially Edward Reed, Childers had authorized the commissioning of HMS *Captain* and lost his own son aboard when it foundered off Spain.

Froude had been advocating the use of small-scale models to the Admiralty even before Gladstone came to power. He had been working with Henry Brunel, son of Isambard, on the testing of *Raven* and *Swan*, and in late 1868 they put together a proposal to Edward Reed for the Admiralty to sponsor the construction of a model-testing tank on leased property adjacent to Froude's home at Chelston Cross, for a sum of £2,000 (about $3 million today), and Froude would run the tank free of charge. Froude emphasized to Reed that "the results of small scale experiments are capable of a direct and accurate application to full size ships ... at once exhaustive, rapid and economical [in] contrast to full size trials." After Childers settled into his office the following year, Froude also sent him missives emphasizing the savings in coal consumption that would accrue from more efficient hull forms. In short, although Froude knew his testing tank would further the cause of naval science, he sold it to the Admiralty as an economy measure. For perhaps the last time before his resignation, Reed was in agreement with Childers; Froude's proposal for small-scale testing was preferable to the full-scale trials proposed by the Merrifield Commission. On 9 February 1870, the head civil servant of the Admiralty, Vernon Lushington, sent two letters: To Froude he said, "After a full consideration of your proposal ... [the Admiralty] is pleased to approve of the same." To Merrifield he said, "[The Admiralty] is unable to give a general assent to conduct experiments on Her Majesty's ships ... but [is] pleased to sanction experiments by Mr. Froude." Merrifield accepted this decision with "good grace," and Froude immediately got to work building his new facility.[35]

Froude signed the contract for building the new tank on 29 April 1870. The site was a narrow field in Torquay owned by Froude's brother-in-law, diagonally across Seaway Lane from the back of Chelston Cross. Almost everything about the tank was new to hydrodynamics. In his proposal to Reed, Froude carefully laid out his rationale for its features and operating procedures and followed them closely in the actual construction and operation. Because so many of them are today considered standard features in experimental model basins around the world, it is worth describing them here in some detail.[36]

Previous experiments carried out by the likes of Beaufoy and Russell were done in open water, with models towed by falling weights. Froude recognized that weather, currents, and other phenomena not only affected accuracy (for example, towed models tended to yaw) but also cut down on the time available to run experiments, so the tank at Torquay was enclosed in a roofed timber building that also contained a drawing room, offices, and workshops (figure 3.7). Instead of building the models from wood—an expensive and time-consuming process—they were milled from dense paraffin wax using an ingenious model-cutting machine of Froude's design (wax models are widely used in towing tanks even today). Instead of falling weights, the former railway

**Figure 3.7**
Froude's Torquay model tank, 1872.
Credit: David K. Brown.

engineer devised a truck (today called a carriage) that ran on rails and was powered by a steam engine. The truck was attached to the model with a lightweight truss lever system (today referred to as grasshoppers) that transmitted towing forces to the onboard dynamometer and also prevented the model from yawing during the experimental runs. The basin itself was dug using canal-building methods and lined with asphalt to prevent leakage. Its measurements—85 meters long, 11 meters wide, and 3 meters deep—ensured that the 3-meter-long models had enough room to accelerate and decelerate and that the tests would not be affected by the proximity of the models to the basin sides or bottom. Froude's test procedures allowed sufficient time between test runs—usually 20 to 30 minutes—"to enable the water to restore itself completely"—that is, to let the model-generated waves die out.

Construction of the tank took just over a year and was finished by June 1871, during which time its Admiralty patron Edward Reed had resigned and been replaced by Nathaniel Barnaby, who also held Froude in high esteem. But the work of running towed model tests had been delayed by Froude's appointment to the Admiralty's Committee on Designs for Ships of War in January of that year, on the heels of the loss of HMS *Captain* the previous fall. Even as Froude was carrying out small-scale rolling tests on models of HMS *Devastation* in the basin at his home, he recommended to the committee that rolling trials be conducted on HMS *Greyhound*, a sloop that had been recently reduced to harbor duty in Portsmouth, as full-scale verification for the model tests he planned for later that year. The committee agreed to Froude's proposal, and even before he carried out full-scale rolling experiments on the ship, he and Henry Brunel arranged for measured towing trials to be conducted on the ship in August and September in the Solent.

The steam corvette HMS *Active* was put into service as the towing ship, with a 14-meter wooden outrigger fitted starboard. The smaller *Greyhound* was towed 60 meters aft by a hawser connected to a dynamometer on the bow, with a chip log to measure speed (figure 3.8). The weather was generally fine during the six-week period, allowing resistance measurements to be taken at a variety of displacements and trim angles. In October, Froude returned to his now-ready tank and began fabricating the 1/16 scale *Greyhound* model, which he tested from December 1871 through May 1872 (figure 3.9). With both model and full-scale resistance data in hand, Froude now turned his attention to the most critical part of his experimental plan, the estimation of frictional resistance.

In his original 1868 proposal to Reed, Froude outlined his methodology for conducting model tests. "The resistance experienced by any given model [or ship] consists of two parts," he claimed, "ordinary," or frictional, resistance, and "excess" resistance due to wave generation. Froude understood that these two parts of resistance did not obey

*Plan of H.M.S. "Active" towing the "Greyhound."*

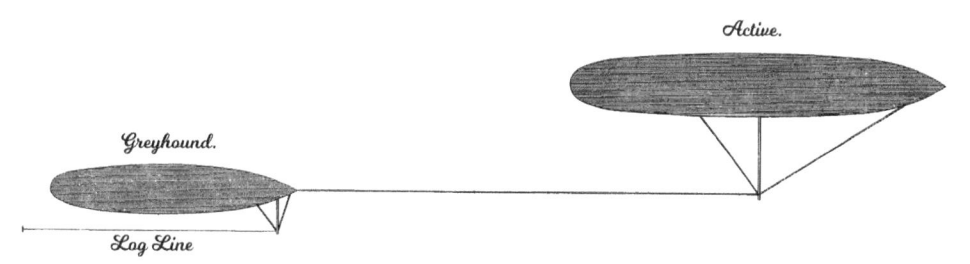

**Figure 3.8**
Full-scale measured towing trials of HMS *Greyhound*, Portsmouth, August–September 1871. Credit: RINA.

the same scaling laws from model to full size, and would have to be examined separately. To estimate the total resistance of a full-scale ship, he would first need to determine a value for the "ordinary" resistance based upon the frictional resistance of an equivalent-area flat plate moving underwater at the same speed, for both model-scale and full-sized hulls; to that he would add the "excess" resistance due to wavemaking, as determined by model tests and scaled up proportional to the formula $V/\sqrt{L}$. Froude, who as a civil engineer had studied the effects of friction in Torquay's main water pipes, considered the whole issue of frictional resistance as insufficiently studied and full of "misconceptions."[37] He may have been correct about the misconceptions, but they remained not from lack of effort, for fluid friction had been the subject of extensive theory and experimentation for well over a century.

The aforementioned experiments by Charles Bossut in the 1770s in which he claimed that "the whole friction against the sides" would be vanishingly small compared with bow resistance was soon refuted by other experiments conducted by the naval officers Antoine-Jean-Marie Thévenard and Nicolas Romme and especially the civil engineer Pierre Du Buat, who identified skin friction as the principal element of resistance. Beaufoy, as noted, subsequently towed thin planks to determine laws of frictional resistance. In 1800, the French engineer Charles Augustin de Coulomb, who was already famous for having developed a comprehensive theory of static and dynamic friction of sliding surfaces, turned his attention to fluid friction. He determined that fluid resistance varied linearly with velocity at low speeds because of internal friction (between fluid molecules) but as the square of velocity at higher speeds because of friction with a surface. Other scientists such as Claude-Louis Navier in France and George Gabriel Stokes in Britain independently extended Coulomb's observations to develop a set of differential

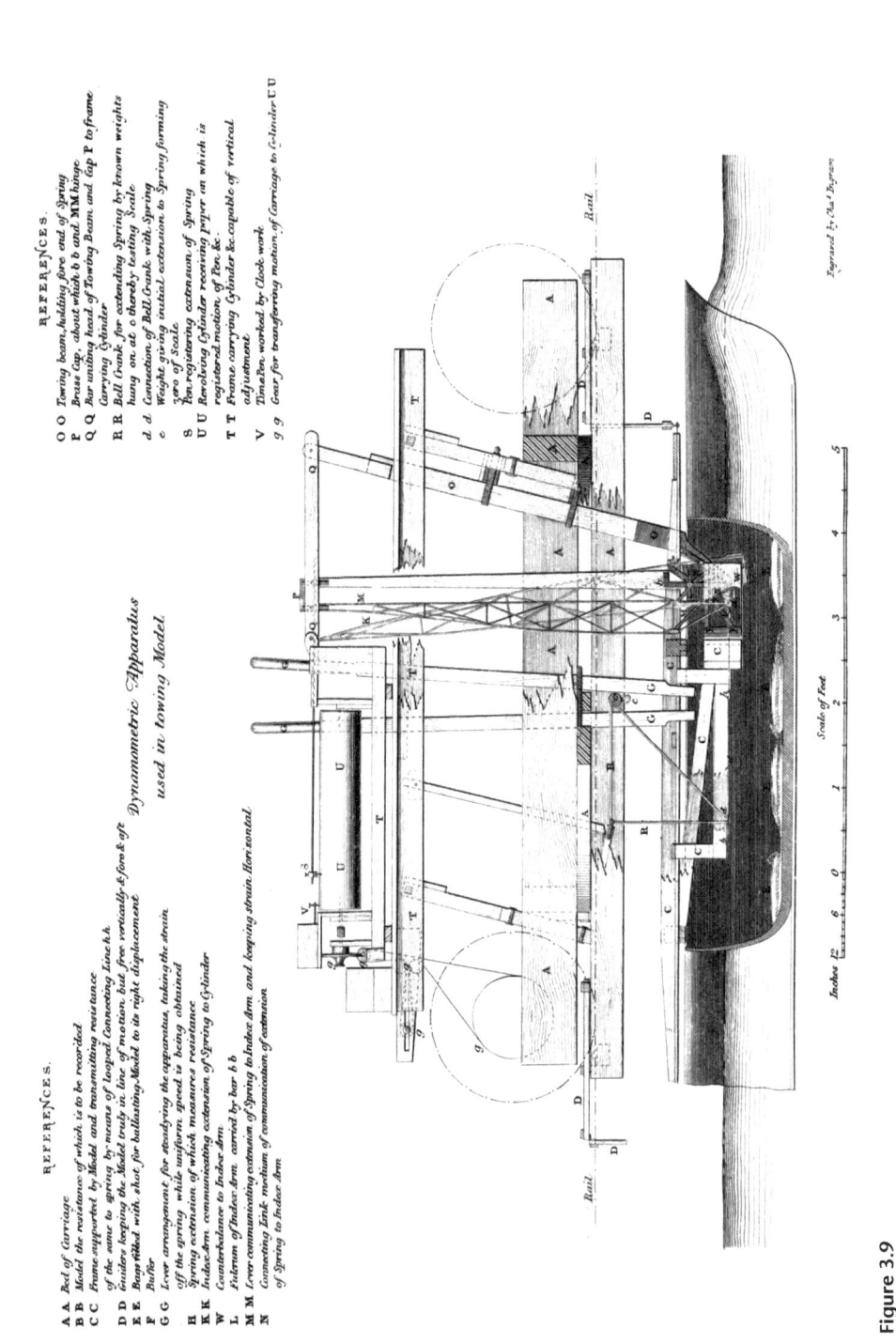

**Figure 3.9**
Model test rig for HMS *Greyhound*, Torquay, December 1871–May 1872.
Credit: RINA.

equations (today called the Navier-Stokes equations) that describe viscous fluid flow in terms of shear stresses (due to surface friction and internal friction), pressure, and energy. But at the time that Froude was exploring the problem of fluid friction, these equations were still regarded by scientists more as mathematical curiosities than useful physical tools.[38]

From April to November 1872, Froude tested many combinations of planks, using 4 different lengths and 11 different coatings, to determine frictional resistance factors. Taking a leaf from Beaufoy's experiments, he attached the dynamometer to planks held vertically and entirely underwater, with the leading-edge support shaped so as to permit a streamlined flow over the rest of the plank. On the basis of the results of these hundreds of experimental runs, he developed an empirical formula for frictional resistance ($R_f$) that owed nothing to Coulomb, Navier, or Stokes:

$$R_f = fSV^n,$$

where $S$ is the immersed surface area, $V$ is speed, and $f$ the frictional coefficient that varied with surface roughness and board length—lowest for long, smoothly varnished planks and highest for short planks covered in rough sand. He found the speed exponent $n$ to be nominally 2 but as low as 1.83 for long, varnished planks.

By the beginning of 1873 Froude was armed with the complete set of data—the full-scale trials of *Greyhound*, the model test results, and the plank friction data—so that he could validate his theory of geometric similitude. As shown in the upper plate of figure 3.10, he started with the overall resistance of the model (solid line A), then deducted the frictional resistance (dashed line B), which he calculated as that of a plank equal to the immersed hull area of the model. The difference between A and B, he stated, is "the resistance of the ship without surface-friction"—that is, the "excess" resistance—and was the value to which the $V^2$ scaling law applied. In the final step, he calculated the full-scale value of the "excess" resistance, added it to the fictional resistance for the full-scale ship (again modeled as a plank of equal area to the hull, dotted line C). The resulting predictions are shown in the lower plate of figure 3.10, labeled "Actual Resistance of Ship compared with that above deduced from Model." Froude's original assumption was that *Greyhound*'s coppered hull (upper solid line B) was equivalent in roughness to a smoothly varnished plank, but the predicted resistance (lower dotted line A) was considerably lower. But when he substituted a friction formula between smooth varnish and rougher calico cotton, the resistance predicted from the model test (line C) lay almost exactly atop the actual full-scale trial data. Froude was now confident that these results validated his model prediction theory, even down to the similarity of wavemaking at model scale and full scale: "With the *Greyhound* model, the resemblance to the

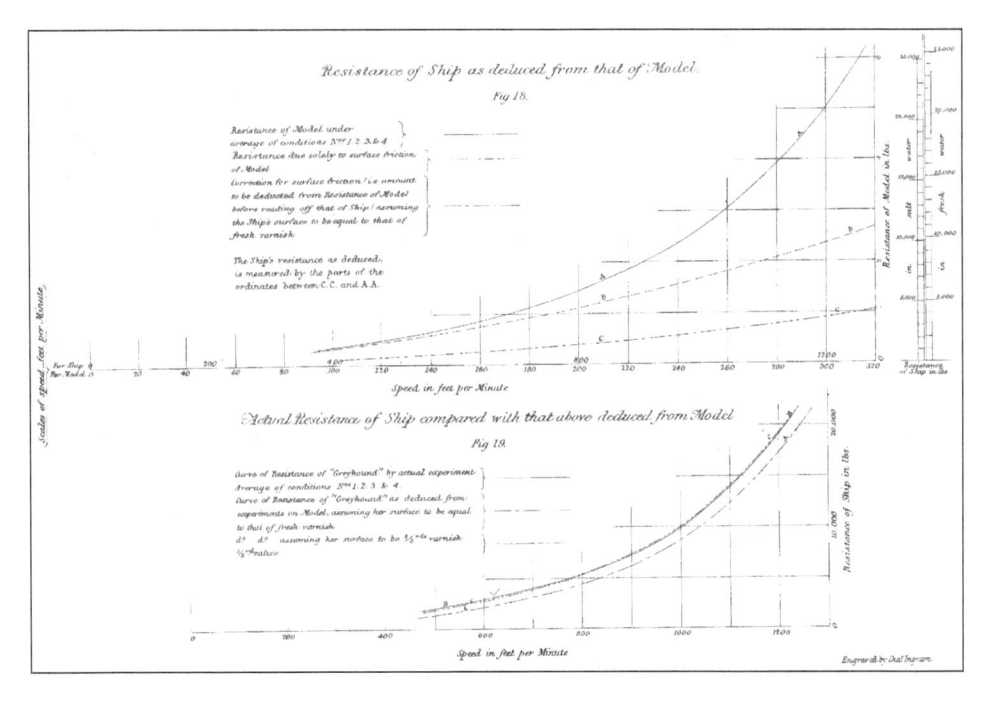

**Figure 3.10**
HMS *Greyhound* full-scale trial versus model test data.
Credit: RINA.

waves developed by the ship at corresponding speeds was most striking, even to the peculiar features of the surge at the bow," he noted. In March 1873 Froude reported these results to the Admiralty Committee on Designs for Ships of War, and a year later read them before the Institute of Naval Architects.[39]

Froude's *Greyhound* results were well accepted by the Admiralty committee, and Chief Constructor Barnaby continued to place his faith in Froude's work. But these results, combined with Froude's successful rolling tests on HMS *Devastation*, sparked anxiety among naval officers who believed their sea legs afforded them, and not mathematicians like Froude, the authority over the merits of a design—a sentiment that Edmund Fishbourne, a senior naval officer with substantial technical knowledge, facetiously described: "Because I have been to sea and have got practical experience … I do not need to understand the theory of naval architecture." Fishbourne then dismissed this sentiment and instead argued that modern naval officers needed to combine practical experience with mathematical ability lest they cede their authority to engineers and scientists.[40] Fittingly, just as Fishbourne was voicing these ideas, Gilbert and Sullivan were penning *The Pirates of Penzance*, in which Stanley, the "model of a modern Major-General," boasted that he was "very well acquainted, too, with matters mathematical."

William Froude had little time to see Gilbert and Sullivan, or operas of any kind, after the *Greyhound* tests, for he was bombarded with requests from the British Admiralty and the navies of France, Russia, and Brazil to conduct model tests that had suddenly become an essential part of the warship design process. Although Froude abjured any salary, the Admiralty's annual stipend of £1,550—made absent any formal contract—paid for his staff of a dozen men. These included Frank Purvis, who would go on to chair the department of naval architecture at the University of Tokyo, Henry Brunel, who provided general (though unsalaried) assistance, and junior constructors seconded from the Admiralty such as Phillip Watts, who would go on to become the director of naval construction. By far the most important staffer was Robert Edmund Froude, William's fourth child and the only one to follow in his father's footsteps. He frequently ran the Torquay facility when his father was called to attend one of the many committees and commissions under the Admiralty and the BAAS.[41]

In addition to assisting in warship designs, Froude was also carrying out a methodical program of experimentation to extend the fundamental understanding of naval architecture theory. He examined the principles of wave interference by testing models with various lengths of parallel middle body (figure 3.11), which showed that the wave patterns generated at the bow and stern would create interference patterns that could augment or decrease the energy drawn off by the wave system, giving ship designers a tool to help control overall resistance. He examined eddy-making resistance, the increase in viscous resistance over simple flat-plate friction resistance, which would later become the subject of extensive research into turbulent flow. He looked at propeller design and efficiency, sinkage and trim, and even planing hulls. But this period of extraordinary achievements would last only five years.

In July 1878 Froude's wife of 40 years, Catherine, died of tuberculosis. William was of course distraught, so his friend Commodore Frederick Richards invited Froude to accompany him to the gentler climes of South Africa as Richards took over the Cape of Good Hope naval station at Simon's Town. While in London on the way to board ship at Chatham, Froude finally took in a Gilbert and Sullivan opera, *HMS Pinafore*, courtesy of Henry Brunel. Richards and Froude sailed in December and arrived at Simon's Town a month later, where Froude lodged at the Admiralty House. It was summer there, and Froude profited by observing the flight of birds and speculating how it could be applied to a flying machine. In April, he was preparing to return to Britain when he contracted dysentery. By 1 May his physician reported that his condition had improved, but within hours he took a turn for the worse and died on 4 May 1879, just shy of age 69.

**Figure 3.11**
Wave system for a ship with long, parallel middle body.
Credit: RINA.

He was buried with full military honors two days later at the Seaforth Cemetery, which was later marked by a stone cross that was a duplicate of the one at Catherine Froude's gravesite (figure 3.12).[42]

Despite the upheaval caused by his father's death, Robert Edmund continued his work for the Admiralty at an unslackened pace. By then the fame of the Torquay facility was widespread and Robert Edmund often hosted visitors from other navies. One such visitor in May 1879 was Bruno Joannes Tideman, the chief naval constructor for the Royal Netherlands Navy, based in Amsterdam (figure 3.13). Tideman was already personally acquainted with British scientists like Rankine and had corresponded with William Froude in 1875 about the model-cutting machine used at Torquay. The following year Tideman took over a section of the Amsterdam *rijkswerf* (dockyard) in Kattenburg, roofed it over and began carrying out his own experiments on paraffin-wax ship models, starting with that of his newly built cruiser *Atjeh*. Lack of funds had forced Tideman to make do with much cruder techniques and instruments than Froude had used; his towing apparatus was powered by four dockyard workers instead of a steam engine, and he could measure only speed directly and had to estimate towing force. Despite these handicaps—among other problems, the open-air towing basin was frozen

**Figure 3.12**
Author and son at William Froude's grave, Seaforth Cemetery, Simon's Town, South Africa. Credit: Bill Rice and author.

during the winter months—Tideman managed to complete a number of experiments with useful results for predicting speed and power.

Of greater interest was that as early as 1877 Tideman was presenting his data in *non-dimensional* notation, which meant that the data for small-scale experiments and full-sized trials could be shown on the same graph. For example, William Froude had always used the term $V/\sqrt{L}$ for his law of similitude, although the units cannot be directly translated from small to large scale. But by including the force of gravity $g$ in the equation,

**Figure 3.13**
Bruno Joannes Tideman.
Credit: Bruno Tideman.

the corresponding proportion—today called the Froude number $F_n$, although William Froude himself never used it or referred to it—is nondimensional and can be used at both model and full scale:

$$F_n = \frac{V}{\sqrt{gL}} = \frac{\text{m/sec}}{\sqrt{(\text{m/sec}^2)(\text{m})}} = \frac{\text{m/sec}}{\text{m/sec}} = 1.$$

Osborne Reynolds, another British scientist working on fluid frictional resistance at the same time as Froude, also employed a nondimensional number, today called the

Reynolds number, $R_n$, as the constant of proportionality by this equation (where ν is the kinematic viscosity of the fluid in m²/sec):

$$R_n = \frac{VL}{\nu}.$$

When Tideman visited Robert Edmund Froude in May 1879—neither one knowing that the elder Froude was already interred in Simon's Town—it is quite likely that the Dutch constructor introduced the British scientist to the idea of nondimensional notation, for a short time later Robert Edmund began representing his own data to the Admiralty in nondimensional terms, which he referred to as the "constant system of notation," or circle notation (e.g., Ⓒ = total resistance divided by displacement), and which became standard practice worldwide.[43]

Robert Edmond quickly filled his father's role as the Admiralty's technical authority; Barnaby's successor William H. White stated, "[I would often] telegraph Mr. Froude that I am coming down to see him when some problem going beyond my experience has to be solved ... and in half an hour [Froude] places me in a position of security." But the requirements of the British navy and shipbuilders were already outstripping the ability of the Torquay facility to fulfill them. In 1883 the Scottish shipbuilding firm William Denny and Brothers, anticipating the need for the "application of science" in shipbuilding, constructed its own Ship Model Experiment Tank in Dumbarton almost exactly on the plan of the Torquay tank and hired one of Froude's former assistants, Frank Purvis, to run it (figure 3.14). Meanwhile, the original Torquay tank was already well past its planned two-year lifespan and was visibly deteriorating. In 1882 Robert Edmund sketched out a plan for a new, larger facility to be built at the old Gunboat Yard at Haslar near Portsmouth. The Torquay tank continued operations while the Haslar facility was built. On Tuesday, 5 January 1886, the 46,190th and final test at the Torquay tank was completed, and the staff moved to Haslar the following month to continue their work. The old building and land were sold off to an estate agent. The Denny and Haslar tanks were the first of a line of hundreds of new model basin facilities that would be built around the world over the course of the next century, differing little in concept from the ideas first laid down by William Froude.[44]

## Propulsion and Maneuvering Theory

Like the research efforts aimed at calculating a ship's rolling motions, speed, and power, the quest to accurately predict its propulsion and maneuvering characteristics began with the collection of empirical and experimental data. Unlike rolling and speed and power, however, the theoretical developments for understanding how propellers worked and

**Figure 3.14**
Denny Ship Model Experiment Tank today in Dumbarton.
Credit: Author's photo.

how ships maneuvered would lag far behind data collection until twentieth-century investigations into aircraft aerodynamics provided the tools needed to link theory and practice.

This lag is somewhat surprising, given that the first efforts in propulsion and maneuvering theory, by eighteenth-century luminaries such as Leonhard Euler and Pierre Bouguer, predated some of their groundbreaking research into rolling and resistance. Euler, who as noted in chapter 2 had developed the first idea for a rotating propeller in response to a 1753 French Academy of Sciences prize, also made the first attempt at developing a theory for its action, modeling the blades as flat surfaces that developed thrust by means of Newton's impact theory. Bouguer, shortly thereafter, wrote his monumental *De la Manoeuvre des vaisseaux* (On the maneuvering of vessels), which used Newton's theories and experimental verification to explain the interaction of rudders and sails, including estimating time for maneuvers, the most efficient angle of the rudder, and the placement of masts and sails for greatest speed.[45]

The introduction of steam power accelerated the development of propulsion technologies faster than theoretical developments could keep up. The paddle wheel, introduced at the end of the eighteenth century, underwent continuous improvements and advances for several decades without any scientific theory to back them up. Again, this was surprising given that, at the same time, scientists and engineers such as John Smeaton in Britain and Jacques-André Mallet in Switzerland were calculating the power and efficiency of vertical waterwheels for industrial applications but ignoring any parallels to shipboard paddle wheels. The French engineer Marc Seguin made a desultory attempt in 1828 to apply the theory of fluids to paddle wheels in order to improve his new steamboat service on the Rhône River. His theory, derived from experiments conducted 50 years earlier by Pierre Du Buat, used the difference in height between the water levels at the front and back of a paddle wheel blade to derive its propulsive force. However, his memoir went almost completely unnoticed and would soon be overshadowed by a literal tug-of-war between paddle wheels and propellers.[46]

Chapter 2 documented the progress of the steam screw propeller, starting with Josef Ressel's original concept in 1827, its unauthorized adoption by Francis Pettit Smith in 1838, and Brunel's decision in 1840 to test Smith's screw steamer *Archimedes* for his new transatlantic ship *Great Britain*. Almost all experimentation during this time was aimed at simply proving that the propeller concept worked, but Brunel, as always, went one step further by insisting upon accurate measurements of its performance. When he hired *Archimedes* to test the configurations of eight different propellers, he also had the engines fitted with a modern engine indicator to record horsepower for comparative purposes. The results were contained in a detailed technical report to the Great Western Steamship Company, and the report convinced the company to fit *Great Britain* with a propeller instead of paddle wheels. The British Admiralty Board, which had been keenly following the development of the screw propeller, requested a copy of Brunel's report. This, along with their own trials of *Archimedes*, convinced the board that the screw propeller was the way ahead. The question now was how to move forward.

In March 1841, the Admiralty Board approved a plan to build a screw steamer for conducting comparative trials with a paddle wheel ship. Impressed with Brunel's thoroughness and also conscious that an impartial authority was needed to quell any public doubts, the board asked him to oversee the trials. Brunel quickly agreed, for the trials afforded him the opportunity to test various propeller configurations for his own *Great Britain*, largely at Admiralty expense. After a year of preliminary engineering work on engine and hull criteria, the screw steamship HMS *Rattler* was laid down in April 1842 and completed 18 months later. At Brunel's urging, the ship was equipped with engines from his close colleagues at Maudslay, Sons, and Field and fitted with a large aperture aft to accommodate a wide variety of screw propellers for testing. In October 1843 *Rattler* began a 14-month-long series of trials, 32 in all, that tested six different propeller configurations of two, three, and four blades, with various modifications to three of the configurations. Brunel insisted that the engines be fitted with permanent horsepower indicators and that each of the trials, which were conducted along a measured mile course in the Thames, be repeated several times in both directions to account for tides and current. By 1844 it became apparent that the most efficient propeller was a two-bladed model designed by Smith (figure 3.15) that could be hoisted out of the well while under sail. Toward the end of the trials, *Rattler* was also fitted with a dynamometer to directly measure the power delivered to the propeller.

**Figure 3.15**
Final two-bladed Smith propeller of HMS *Rattler*, 1844.
Credit: Author's photo.

These trials gave Brunel the information he needed to select a propeller for *Great Britain*, and the British navy, convinced by the early results of the utility of the propeller, had already ordered four screw frigates even before the trials were finished. Still, the Admiralty opted for a very public demonstration of the propeller versus the paddle wheel by staging a series of races and towing trials, surreptitiously rigged in favor of the propeller. In March 1845 *Rattler* was joined in Yarmouth by the paddle wheel sloop HMS *Alecto*, somewhat smaller than *Rattler* in power and tunnage, and the two spent several days racing side by side in the North Sea, with *Rattler* generally outperforming *Alecto*. Finally, on 3 April 1845, came the now-famous tug-of-war. With the two ships lashed stern to stern, *Alecto* started its engines and pulled *Rattler* backward. Then *Rattler*'s more powerful engines came to life and towed *Alecto* backward, its paddle wheels thrashing the water, over the course of several miles at 2.8 knots. Even if these races and tugs-of-war were more of a publicity stunt than an actual engineering experiment, the onboard engine indicator and dynamometer did provide useful data on engine and propeller performance, showing, for example, that the gears, shafting, and auxiliary machinery contributed to a substantial 15–25 percent loss between the engine and the propeller.[47]

The *Rattler-Alecto* spectacle was merely the most famous of a series of propeller trials carried out during the 1840s. In Britain, the steam sloops *Niger* and *Basilisk* held a similar tug-of-war in 1849. In France, the steam packet *Napoléon* was tested in 1842 with eight different models of screw propeller, the results carefully recorded using Taurines's

**Figure 3.16**
*Rattler* and *Alecto* tug-of-war, 3 April 1845.
Credit: National Maritime Museum, Greenwich, London.

newly invented dynamometer. Another steam packet, *Pélican*, was the subject of a series of experiments from 1847 to 1849 by Siméon Bourgois and Charles-Henri Moll. Although these trials provided empirical data on hull design, transmission losses, and efficiency, the only direct measure of performance of the propeller itself was the "slip," the ratio between the theoretical advance of the screw during a complete turn (the pitch) and the actual distance traveled. Engineers were still hampered by a lack of understanding of how screw propellers actually functioned, and without that knowledge it was difficult to further the state of the art except by slow trial and error.

The first steps to understand the theory of the propeller were being carried out even as the trials of *Rattler* and *Napoléon* were underway. In 1842, Auguste Taurines published an article, "Théorie de la vis d'Archimède" (Theory of the Archimedes screw) in *Annales maritimes et colonials*. Taurines imagined the screw as a sort of helical paddle wheel contained within a cylinder or cone, and he modeled the propulsive forces as if the screw were pushing the water perpendicular to its surface along the length of the helix. Three years later Siméon Bourgois published his own theory of the propeller, which built upon the work of Taurines but added the component of frictional resistance to the propulsive action. Although both works were mathematically precise in their equations to characterize blade geometry, they were both founded upon the long-disproved Newtonian theory of fluid shock to describe the dynamics, and they both ended up as blind alleys in the search for a theory of the screw propeller.[48]

Rankine and William Froude recognized this failure when several decades later they took up the problem of propeller action; Rankine pointed out that the "[Newtonian] theory leads to results very much inferior to the actual performance of screws," and Froude noted that "the old law ... was entirely in error" and that only recently had the subject "received a sound theoretical basis." Rankine, in 1865, was the first to develop a physical concept, which came to be known as momentum theory. Froude's concept, published in 1878, was called the blade element theory. But these two theories modeled the propeller completely differently and were apparently incompatible with each other, and Rankine and Froude both died before they could find a way to bridge the divide. Instead, the task of developing a workable theory of the propeller would be mostly carried out by Robert E. Froude and primarily reported in the pages of the *Transactions of the Institution of Naval Architects*.[49]

Momentum theory, roughly speaking, modeled the effect of the propeller on the surrounding water. The propeller was idealized as an infinitely thin actuator disk with an infinite number of blades, through which an ideal fluid—the slipstream—is accelerated. The action of the disk is presumed to cause an instantaneous jump in the pressure of the fluid, increasing the momentum of the slipstream aft of the disk and thus providing forward thrust. Rankine originally assumed that the propeller did not produce any rotation in the fluid, a simplification later corrected by Robert E. Froude and the British mathematician Alfred G. Greenhill. The advantage of this model was that it provided a reasonable approximation of the efficiency of the overall propeller and allowed ship designers to select the appropriate propeller diameter for a given horsepower and speed. The problem with using an ideal disk in an ideal fluid is that the model did not provide any means of calculating thrust and torque, nor did it give any hint as to how to design the actual propeller—number of blades, rotation speed, blade sections, and so on.

Blade element theory, by contrast, modeled the effect of the water upon the propeller blades themselves. William Froude proposed that each element of a propeller blade, from root to tip, be modeled as if it were an infinitely thin ring section. The resultant forces on each element—from water inflow and rotation—were then summed as an integral across the entire blade to provide the total thrust and torque. Although this was a sound basis for physical analysis, there was no provision to account for friction, nonuniform loading, or other factors, so it appeared as if the propeller were 100 percent efficient (in reality, typical ship propellers are 60–80 percent efficient), thus the model would overpredict the performance.

Neither momentum theory nor blade element theory gave a complete picture of how propellers operate, so their results were both incorrect and incompatible. It would not be until the advent of aircraft aerodynamics in the twentieth century that a paradigm shift in understanding lifting surfaces—wings and propellers—would occur. As discussed in chapter 6, circulation theory would provide the missing piece to make accurate theoretical models of propeller action. The British polymath Frederick W. Lanchester tried to introduce circulation theory to the world of naval architecture in a 1915 *Transactions* paper correcting the Rankine-Froude momentum theory. But this, and similar ideas offered by German and Russian scientists, fell on deaf ears. Only in the 1930s did a complete theoretical model of ship propellers emerge that could provide accurate predictions of their performance.[50]

Instead, William Froude's empirical methods of model testing were extended by his son Robert Edmund into the realm of propeller design. He divided the problem of propeller performance into two parts: "the efficiency of screws working by themselves in undisturbed water" (i.e., open-water performance) and the interaction between the screw and the hull. As he reported to the Institution of Naval Architects in 1883 and 1886, the Torquay facility was equipped with a specially built dynamometer that could directly measure the power delivered to the model propeller to assess both open-water performance and hull interactions. For the open-water results, Robert Edmund Froude reported on a small series of experiments varying the number of blades, pitch, diameter, and rotational speed, producing charts of comparative performance for the different configurations. Such results enabled naval architects and constructors to quickly home in on a design solution without resorting to extensive calculations. These propeller series would not be substantially improved on for almost half a century, until more extensive series were developed by naval model basins in the United States, Britain, and the Netherlands.

The younger Froude also paid close attention to propellers operating much differently behind a ship than in open water, owing to water flowing into the propeller being substantially affected by its passage around the ship. To test this, his original testing arrangement had the propeller motor and dynamometer on a separate platform connected to the ship model; later configurations placed them within the model itself. Robert Edmund defined several elements to measure these effects. The first element he termed "wake fraction," the difference between ship speed and the speed of water flowing into the propeller, which can be higher or lower than the ship speed depending upon the stern configuration. The second, which he alternately dubbed "augment of resistance" and "thrust deduction" (we use the latter term today), describes the effect of the propeller reducing pressure on the hull itself, which in turn diminishes the overall thrust delivered to the ship. The third element, relative rotational efficiency, accounts for the nonuniform flow into the propeller. This process of testing propellers, first in open-water conditions and then in self-propulsion tests behind the hull, is used today to predict the full-scale performance of ships, with methods that have changed little since Robert Edmund's day. Just as the elder Froude's procedures became the standard for all future ship model testing, so too did the younger Froude's methodology become the basis for propeller model testing until today.[51]

Perhaps surprisingly, it was the very development of the propeller that gave rise to the need for maneuvering theory, because maneuver became the preferred tactic of weaker navies, like those of France and Italy, in the face of the overwhelming size and firepower of British warships. As noted in chapter 2, in 1842 the French navy sent a young lieutenant named Nicolas-Hippolyte Labrousse to Britain to witness the propeller trials of *Archimedes*. Convinced that coupling the screw propeller with an iron ram at the bow could turn French warships into great ramming ships like the ancient Greek triremes, Labrousse circulated a widely read memorandum that argued for bolstering the French fleet's inferior position with respect to the larger and more heavily gunned British fleet through the absolute combat of ramming. At first, rams were fitted to Dupuy de Lôme's ironclads like *Solférino* and *Magenta* (1861), with the idea that the ships' heavy guns would disable the enemy and the ram would deliver the coup de grâce. But by 1865, the French navy was constructing smaller, high-speed coastal defense ships like *Taureau*, which as the name implies, would use the ram as the principal weapon against the larger, more cumbersome British warships. Other nations soon followed suit; Italy's *Affondatore* was built along the same principles, and even Britain tried out the concept with its *Hotspur*.[52]

As an outgrowth of these smaller, high-speed rams, the idea of using high-speed torpedo boats against larger British ships gained capital from the 1870s to the 1890s under the naval strategic concept known as the Jeune École (Young School). After its stunning defeat in 1871 in the Franco-Prussian War, France had to rethink its entire naval strategy in light of its weakened economy. French naval theorists, concerned about the looming British threat, argued that its navy should not attempt to go toe-to-toe with the larger British navy but rather should resort to *guerre de course* (commerce raiding) to cut off the vulnerable British supply lines and focus on coastal defense. Using many small, inexpensive, high-speed torpedo boats against large British battleships, went the argument, would defeat them just as surely as "microbes were victorious over giants," an idea in vogue thanks to Louis Pasteur's recent discoveries in the germ theory of disease. The French navy, followed by the navies of Italy, Germany, and Austria-Hungary, built dozens of torpedo boats, which Britain countered by building fast torpedo-boat destroyers.[53]

At the heart of ramming and torpedo boat tactics was the notion that to the most agile belonged the victory. Siméon Bourgois was the first to develop the theoretical underpinning of the *cercle mort* (dead circle), which any modern fighter pilot would recognize as the ability to turn inside your adversary to gain the tactical advantage; for example, in a contest of rams, equally matched opponents would find themselves in a head-to-head situation with no clear winner (figure 3.17). To gain the upper hand, went the thinking, it was necessary to have a complete mathematical model of the movements of friend and foe alike. Similar considerations held true when trying to line up for a torpedo attack on a maneuvering enemy. Bourgois, along with other French and Italian *officiers savants*, wrote numerous calculus-heavy articles on the subject for the French *Revue Maritime et Coloniale* and its Italian counterpart *Rivista Marittima*, both journals widely read for their technical content as well their essays on tactics and strategy. These articles constituted the first attempts at a theory of maneuvering under steam, calculating the ship's pivot point in a turn, advance, drift and transfer, tactical diameter, directional stability, rudder angles, and angles of heel in a turn. But these equations were only kinematic in nature; they did not describe the forces involved in steering and turns, which limited their utility in design.[54]

The need for more maneuverable warships also dictated improvements to the rudder and steering system. During the age of sail, a ship's rudder was typically a flap attached

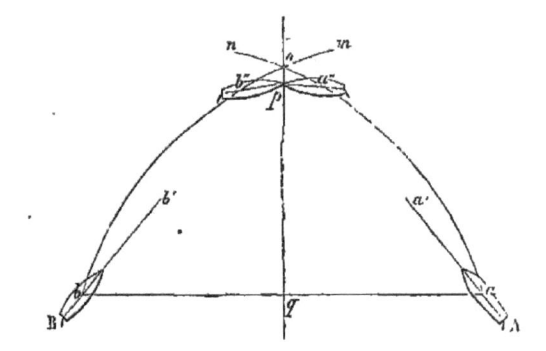

**Figure 3.17**
Maneuvering while ramming, 1869.
Credit: Bibliothèque Nationale de France/Gallica.

to the sternpost, which was used in conjunction with the sails to turn the ship. As early as 1790, Charles Stanhope—one of the founders of the Society for the Improvement of Naval Architecture—recognized that this unbalanced rudder was hard to turn because the hinge was right at the leading edge. He developed a concept for a balanced rudder in which the hinge point was nearer the quarter-point of the rudder, allowing for easier steering. The balanced rudder (figure 3.18) was at first slowly adopted by the steamship community—Brunel used one for *Great Britain*—but with the advent of ramming, its use became widespread in steam warships to allow tighter turns. At the same time, the manual steering wheel and tiller were being replaced by steam servomotors to more quickly turn the rudder, while inventors like John McFarlane Gray and Joseph Farcot developed automatic feedback mechanisms so the rudder would respond more precisely to the helm. To accurately estimate the power of the motor needed for turning, in 1873 the French naval constructor Joseph Joëssel measured the resistance of flat plates in the current of the Loire River at Indret, developing simple equations to predict rudder torque that even today are called the Joëssel method.[55]

While these developments in maneuvering theory and rudder control constituted a major step forward, even naval architects of the time admitted that they did not have a fundamental understanding of the underlying forces and dynamics involved. Compared with hull resistance and propeller design, steering and maneuverability got short shrift when it came to experiments using scale models; only one perfunctory effort on behalf of the BAAS was made by Osborne Reynolds in 1875, using steam and clockwork models to measure the ability of ships to avoid collisions. As with propellers, it would not be until the advent of aircraft aerodynamics in the twentieth century that

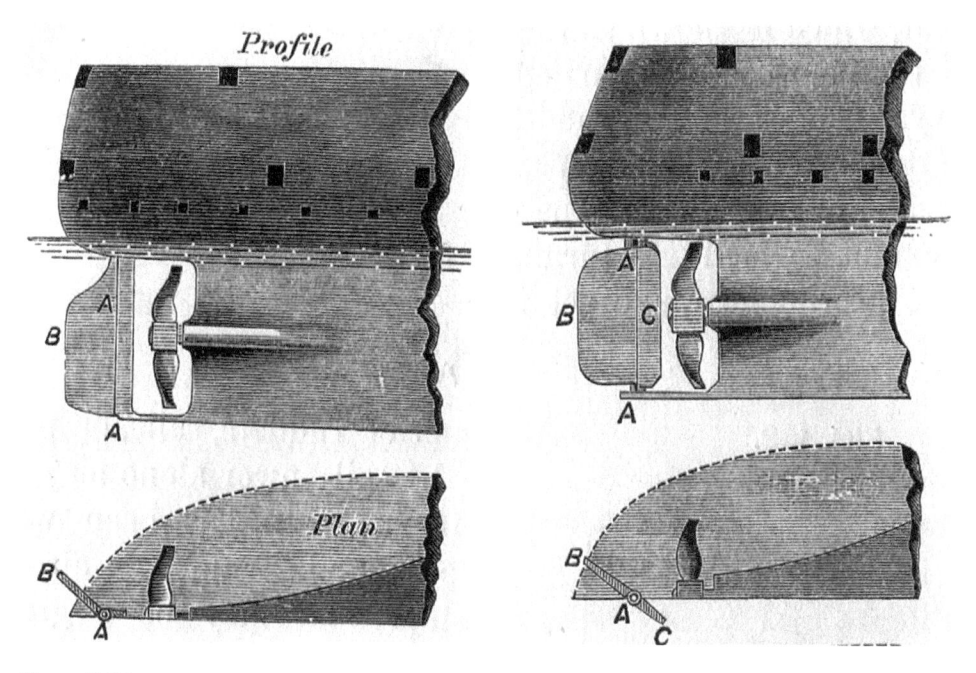

**Figure 3.18**
Unbalanced rudder (left) and balanced rudder (right).
Credit: Author's collection.

scale-model tests would reveal the actions of the hull and the rudder to be comparable to airfoils with lift and drag. On the basis of those aircraft experiments, as described in chapter 6, a complete six-degrees-of-freedom model, defining both the kinematics and the dynamics of ship maneuvering and control, would be developed.[56]

## Iron Shipbuilding and the Theory of Structures

The quest to understand how structures behave under loads predated by two centuries the widespread use of iron. Galileo, in his 1639 treatise *Two New Sciences*, examined the behavior of wooden beams, arguing that the strength of the beams was proportional to their cross-sectional area and that they would fracture when the stress—the load divided by the cross-sectional area—exceeded the material's "absolute resistance" (akin to ultimate tensile strength). This model, however, did not account for elasticity, or bending along the length of the beam. By 1686, the French scientist Edmé Mariotte had developed a theory that took bending into account and modeled the beam more accurately by considering it as containing fibers that stretched or compressed according to the load.[57]

Even at this early stage, scientists were thinking about the strength of ships and their components. Galileo, for example, had first considered the strength of oars while working for the Venice Arsenal. In 1697, the French mathematician Paul Hoste, in the first attempted synthesis of naval architecture, *Théorie de la construction des vaisseaux* (Theory of the construction of vessels), applied Galileo's studies to the breaking of ship's hulls, but because he inexplicably failed to account for Mariotte's more recent work, his theories failed to provide any useful results.[58] It was Pierre Bouguer, in his seminal 1746 *Traité du navire* (Treatise of the ship), who first provided a useful theory of the strength of ships by describing how to calculate the hull bending loads due to the opposing forces of weight and buoyancy. Bouguer assumed that the ship weight distribution was uniform over its length and that the buoyancy force took the form of a parabola (figure 3.19). Integrating the two functions, he arrived at the formula for a ship floating in still water:

$$\text{Hull bending moment} = \frac{39}{520} \cdot L \cdot \Delta.$$

This marked the first appearance of what became known as the bending moment coefficient, used by naval architects even today to provide a rough estimate of bending loads in hull structure design. The Spanish constructor Jorge Juan extended Bouguer's

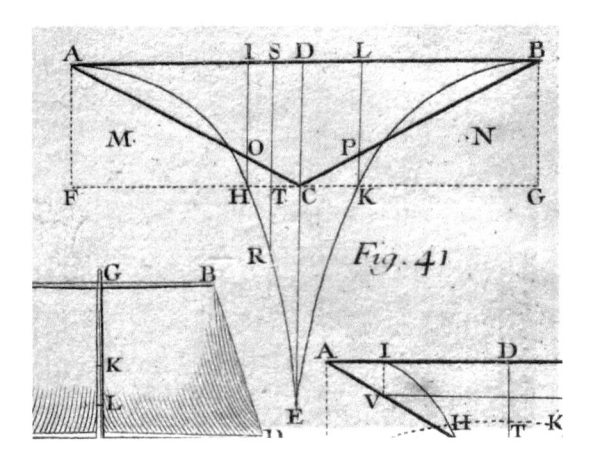

**Figure 3.19**
Bouguer's explanation of the hull bending moment, 1746.
Credit: Author's collection.

analyses, using the nascent theories of stress in structural beams to determine the correct thickness of timber to resist loads.[59]

While Bouguer and Jorge Juan were developing the foundations of ship theory, other scientists were extending the application of rational mechanics to structures and materials. Leonhard Euler in particular advanced the state of knowledge of structural theory, showing that the deflection of a beam under an applied load is proportional to the product of the elasticity of the material $E$ times the second moment of inertia of the structure $I$. For many decades it was unclear whether $EI$ was a single property of a structure (as Euler believed) or could be separated into component parts. Even in 1807, when Thomas Young first described the modulus of elasticity $E$ (today referred to as Young's modulus), he did not clearly differentiate between $E$ and $I$. It was not until the 1826 work of Claude-Louis Navier (of Navier-Stokes fame) that the two properties were definitively separated. At the same time, Navier and Thomas Tredgold independently popularized the concept of the neutral axis, the plane through a beam or plate where no tension or compression occurs during bending. By the mid-1820s, therefore, the major elements of beam bending theory—elastic modulus, second moment of inertia, tension, compression, and neutral axis—were in place.[60]

Even before a complete theory of beam bending was in place, civil engineers were attempting to calculate the size of beams needed to carry specified loads. When Charles Bage built his flax mill at Leeds in 1803, he used the theories of Galileo and Mariotte to size the cast-iron supporting beams and verified his equations by test loading a full-scale beam until it broke. Thomas Tredgold's beam theories became more widespread starting in the 1820s, when the architect John Nash used them to size the cast-iron beams of Buckingham Palace. Structural load calculations were well advanced by the 1840s and were evident in the long-span iron roofs that were built to cover the shipbuilding ways at Chatham dockyard. Even so, structural theory at the time was still viewed with some suspicion. Engineers routinely constructed and load tested scale models of new bridges; Samuel Brown, for example, built a 30-meter trial bridge to test its behavior under loads, before constructing the much longer Union Suspension Bridge that links England and Scotland over the River Tweed. At the same time, engineers incorporated a healthy factor of safety in their design loads to account for uncertainties in both calculations and their underlying assumptions. The American engineer James Finley routinely multiplied the weight of the roadway by five or six to size the structural components. This practice became standardized in 1849, when the British Royal Commission on the Application of Iron to Railway Structures concluded that a factor of safety of six would be appropriate for bridge girders.[61]

William Fairbairn and Isambard Kingdom Brunel were perhaps Britain's most respected engineers, in part because of their unceasing quest for accuracy using both theory and experiment. As described in the prologue, Fairbairn conducted a series of scale-model tests on tubular beams from 1845 to 1846 to properly design the tubular Britannia Bridge over the Menai Straits. Brunel followed Fairbairn's lead in 1849 when he conducted large-scale experiments on bridge girders, before constructing the Chepstow Bridge. Their extensive experience with structural iron gave them the confidence to undertake large-scale shipbuilding projects such as *Great Eastern*, because both men envisioned the hull girder as behaving much as a bridge does when subject to external loads.[62] But whereas Brunel and Fairbairn could define and calculate static bridge loads—the weights of the bridge, rails, locomotives, and so on—they quickly ran into the problem that would plague naval architects and constructors for the next two centuries: it is almost impossible to accurately calculate the dynamic three-dimensional loads on a ship moving in a seaway or to create meaningful scale models for experimentation to develop the static forces needed to size the hull structures.

Ship constructors, many coming from the civil engineering world, understood that the hull design for longer iron ships would be dominated by bending loads, just as bridge structures were. At first, they simply applied bridge-building assumptions to the design of ship structures. Brunel, for example, decided that *Great Eastern* would be constructed like Fairbairn's Britannia Bridge over the Menai Strait, specifying that it must be "all iron, double bottom, and sides up to the water line, with ribs longitudinal like the Britannia tube" (see the prologue). His concept of the ship as a bridge extended to his calculations for longitudinal strength. Evidently concerned after *Great Britain*'s stranding in Ireland, he carefully designed the ship to survive a similar grounding. He envisioned the ship as suspended between two rocks and modeled the hull girder as a beam simply supported at two points, precisely the sort of calculations he and other engineers were by then using to determine the strength requirements for bridge girders supported by piers at each end.[63]

In 1863, a paper by the Liverpool shipbuilder John Vernon used this same assumption of suspending a ship between or on top of rocks to calculate the stress on the hull girder (figure 3.20). Vernon determined the stress in the upper deck to be about 17 tons per square inch (230 megapascals), about the same as the ultimate tensile strength of wrought iron; in other words, in that condition the hull would fracture. The paper caught the attention of William Rankine, who had already been working with shipbuilders like James R. Napier on the subject. In 1864 he reported to the BAAS that, in Vernon's assumption, "the strains [he meant stresses] which would be thus produced are far more severe than any which have to be borne by a ship afloat." Instead, he proposed placing the ship "afloat ... supported amidships upon a wave crest and dry at the ends," and after making some assumptions about the distribution of weights and buoyancy, he arrived at the following formula for a ship on a wave:

$$\text{Hull bending moment} = \frac{1}{20} \cdot L \cdot \Delta.$$

Whether by intent or by coincidence, this wave bending moment coefficient of 1/20 was exactly 50 percent greater than the still-water bending moment calculated by Pierre Bouguer a century earlier. In Rankine's calculation, the highest stress in the hull structure would be just 4 tons per square inch (60 megapascals), giving a factor of safety of six, equivalent to best practice for bridge design.[64]

In 1865 and 1866, William Fairbairn built on Vernon's and Rankine's work to outline a method for determining structural stresses using a quasi-static analysis, based on the static method used to analyze bridge girders. He proposed that the ship hull should be represented as if it were momentarily balanced on two types of waves: first, between the crests of the waves at the bow and the stern (figure 3.21, top), similar to what Vernon had proposed for balancing upon two rocks, and second, balanced upon a wave amidships (figure 3.21, bottom), as Rankine had proposed. The first condition would give the greatest sagging moment, compressing the upper deck and tensioning the keel; the second condition would yield the greatest hogging moment, putting the upper deck in tension and the keel in compression. By analyzing the hull as an equivalent girder under both hogging and sagging loads, its scantlings (the dimensions of the structural members) could be determined.[65]

Although Fairbairn had established the basic principle of balancing a ship on a wave, he did not provide any details on how to do so. Nevertheless, the British navy's chief constructor Edward J. Reed saw the importance of the method, and with one of his assistants, William H. White, he developed a graphical method of determining bending moments and loads for naval ships.[66] In 1877 White provided a detailed explanation of the process in his textbook *A Manual of Naval Architecture*.[67] First, the hull is divided into transverse sections—Reed and White chose 20 "slices"—and the weight and buoyancy of each section are determined and then summed to find the resultant shearing force for each section. Integrating the shearing force curve along the length of the ship provides the still-water bending moment. While this was the same as the method first proposed over a century earlier by Alexandre Gobert and Pierre Bouguer (see chapter 1), White took it a step further by next balancing the ship on a quasi-static wave.

White then examined the nature of the waves to be used for this part of the calculation; on the basis of the then-current ideas of ocean waves and the published

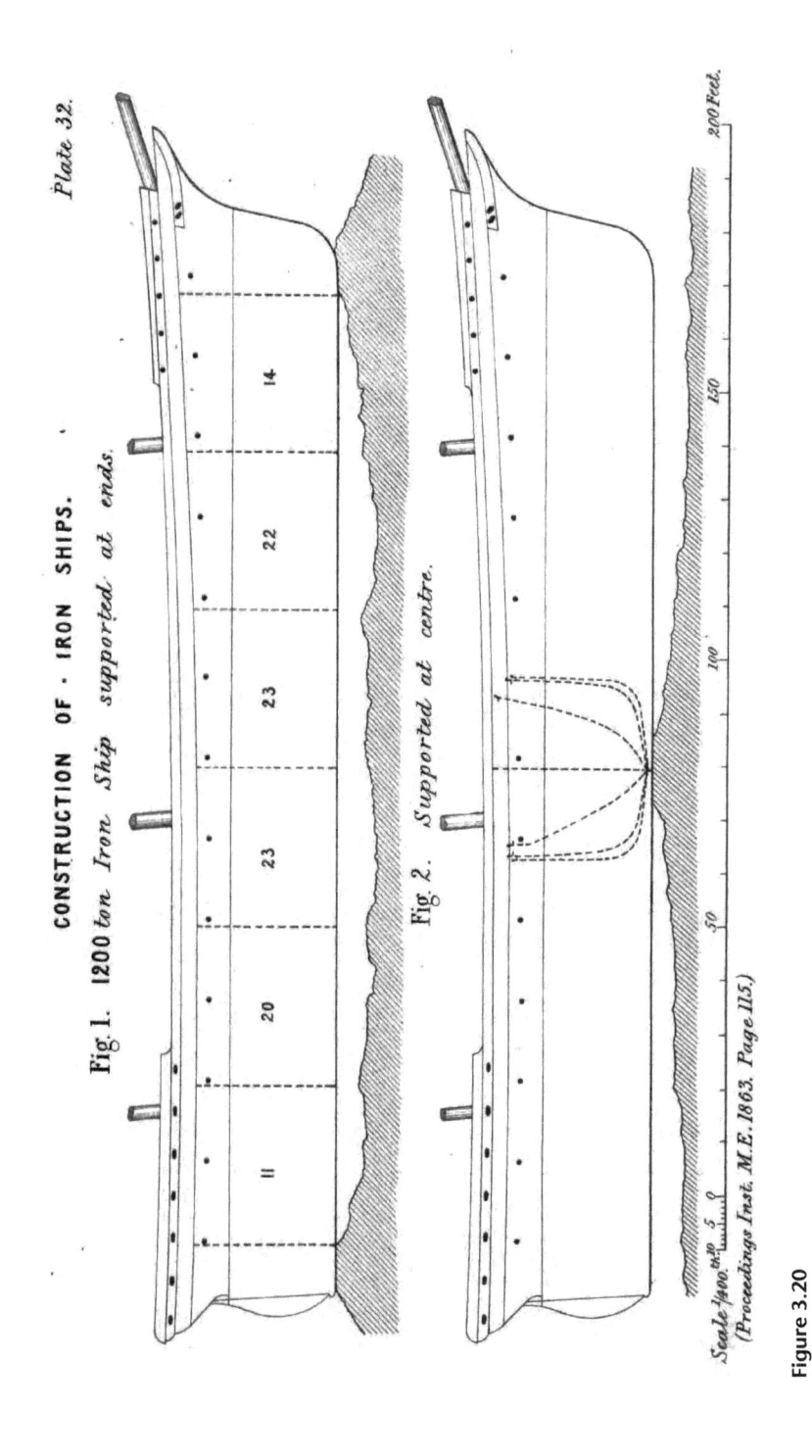

**Figure 3.20**
Ship supported on rocks, sag (top) and hog (bottom), 1863.
Credit: University of Minnesota.

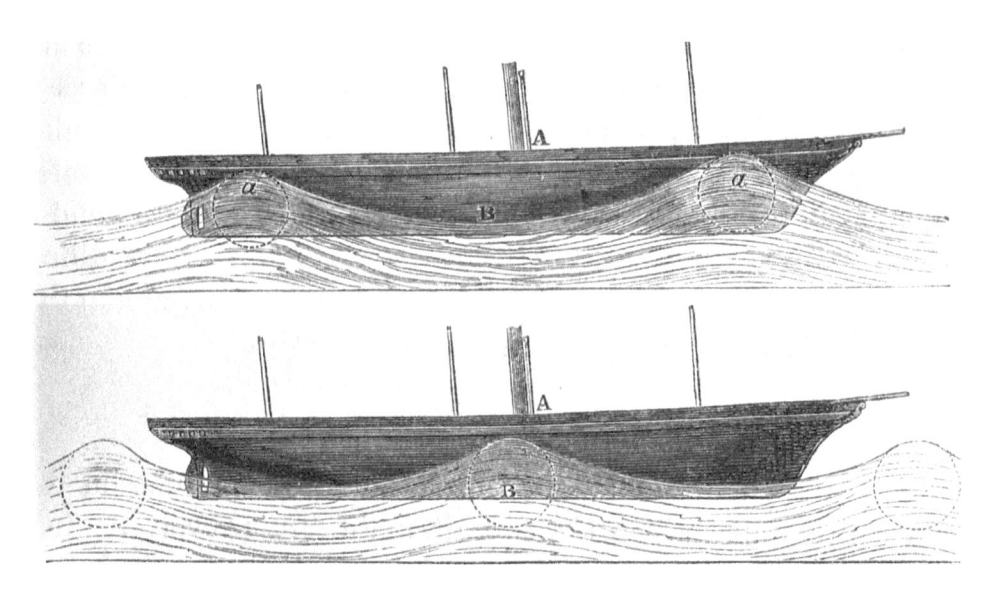

**Figure 3.21**
Ship balanced on waves, sag (top) and hog (bottom), 1865.
Credit: Author's collection.

observations of the French naval officers Armand Paris and Charles Antoine, he determined that the most representative wave would be a trochoid, with a length equal to the length of the ship, and a height 1/20th the wavelength. Next, the paper ship was set upon that imaginary wave (figure 3.22; in reality, the wave would be drawn on tracing paper and placed over the drawing of the ship's profile) and then trimmed (ship draft adjusted fore and aft) until the proper overall buoyancy was reached. This was a time-consuming, trial-and-error process of first estimating the proper hull draft and trim angle, using Bonjean curves to calculate them directly, comparing them to see if the estimate equaled the actual, and then readjusting the estimates to start over.[68] A more precise analysis might incorporate the Smith effect (named for the British naval constructor William E. Smith, who first described it), which takes into account the dynamic pressure within the wave itself; this so-called Smith correction can increase or decrease the wave bending moment by more than 20 percent.[69]

For all this apparent precision in calculation, which caused future generations of student and professional naval architects to spend countless hours hunched over their drawing boards and slide rules, the actual concept of balancing a ship on wave was, in fact, a theory built on a foundation of sand. It was, as explained, directly borrowed from the design methods of bridge builders, for the simple reason that builders of

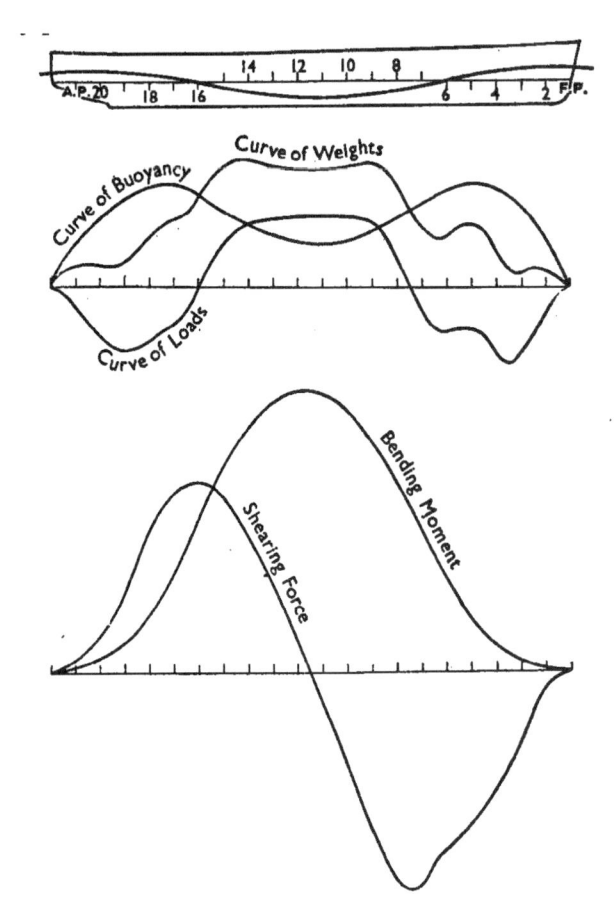

**Figure 3.22**
Ship balanced on wave in sag—curves of weights, loads, and bending moments.
Credit: Author's collection.

bridges and ships were often the same people. However, even scientifically grounded engineers like Fairbairn and Brunel, who were able to verify bridge-building theories using scale models and experimental testing of loads, were never able to do the same for the balance-on-a-wave theory, because it was almost impossible at the time to measure structural loads at sea or adequately test them at small scale. Indeed, compared with the other domains of naval architecture discussed in this chapter—ship motion, speed and power, propulsion, and maneuvering—the development of ship structure theory was notably devoid of any real collection of empirical and experimental data. As one twentieth-century author noted, "For many years naval architects were faced with an unusual set of circumstances ... unlike other branches of engineering, information was lacking as to the exact magnitude of the loads imposed on the structure when in service."[70]

Some attempts were made to verify the balance-on-a-wave assumption, the most famous of which were the HMS *Wolf* trials from 1902 to 1903. Spurred by the loss of the destroyer HMS *Cobra*, which broke apart at sea in 1901—among other things, the accident cast doubts on the still-newtechnology of steel construction—the British Admiralty commissioned John H. Biles of the University of Glasgow to perform a series of tests using strain gauges upon another destroyer, HMS *Wolf*, to verify the basic assumptions of hull loading. The ship was placed upon blocks in a Portsmouth dry dock to test static bending, first loaded at amidships and then at the ends, according to the concepts of Fairbairn and Reed. The ship was then taken to sea in search of bad weather, and strains were measured there. The results showed that the actual hull stresses were close to those corresponding to Reed's assumptions on static and wave bending.[71] For much of the naval architecture community, that one data point seemed to confirm the balance-on-a-wave theory, which continued as the accepted model through most of the twentieth century.

The visualization of a ship hull as a girder under bending loads did bring some additional benefit to ship structural design. The first ironclad, HMS *Warrior*, had been built in 1860 largely following a transverse framing system, which (as for wooden ships) situated most of the load-bearing structure in the transverse, or athwartships, direction. In 1863, Edward Reed, assisted by Nathaniel Barnaby, designed a new type of ironclad, HMS *Bellerophon*, with what they referred to as a "bracket frame system." This was in reality a longitudinal framing system inspired by John Scott Russell's *Great Eastern* (which in turn was inspired by Fairbairn's Britannia Bridge). Compared with *Warrior* built just three years earlier, *Bellerophon*'s longitudinal framing made the overall hull girder stiffer and more efficient, allowing ships of ever-greater length to be designed and built.[72]

The longitudinal framing system for ships' hulls and the balance-on-a-wave model for hull bending became the principal achievements of ship structure theory during the nineteenth and early twentieth centuries, although elements such as the shape and size of the model wave would evolve over the years. Naval architects would also adopt many civil engineering practices , such as the use of the moment-distribution method developed in the 1930s, to assess the strength of other ship structures.[73] Toward the end of the twentieth century, the rise of electronic computers (discussed in chapter 7) would usher in more advanced structural design methods such as deterministic evaluation of loads, finite element analyses, and direct real-time measurements of forces in a seaway.

# 4  The Demand for Standardization

Predictability was the motivation for adopting the technologies of steam, iron, and steel in shipbuilding; naval architecture theory and experimentation were the mechanisms to achieve accuracy in predicting their performance; and standardization was the reason that practicing constructors and shipbuilders integrated naval architecture theory and experimentation into ship design and construction. Standardization comes in different forms, which accomplish different economic purposes. Standards that establish minimum levels of safety and quality are useful when buyers are not, or cannot be, fully informed of every aspect of a product—for example, when a shipowner must decide on competing bids from different shipyards. If shipowners cannot readily distinguish higher-quality vessels from lower-quality ones, high-quality shipyards find it difficult to sustain their price premium and are undercut by low-quality sellers. Thus, minimum industry standards establish a floor for comparison and also reduce risk for the shipowner. Standardized products—sometimes called variety-reducing standards—limit the number of variations and levels of quality and safety, thus enabling economies of scale for sourcing and production of material by the shipbuilder. These standards also reduce risk for shipowner and shipbuilder alike, as they create a critical mass for emerging markets to have confidence in developing new technologies, such as when the shipping industry settled on compound steam engines as the industry norm.[1]

During the nineteenth and twentieth centuries, the rationale for standardization had changed little since the days of sailing ships, when navies as well as state-owned firms like the Dutch East India Company had established uniform ship types and designs to reduce costs, streamline spare parts, and ease tactical planning. Naval architecture, as I explain in my previous book *Ships and Science*, played a central role in this standardization. As far back as 1678, the French minister of the navy Jean-Baptiste Colbert demanded that his constructors "establish a theory on the subject of ship construction ... so that all the designs are correct down to the foot and inch." From that point on, ship theory was developed and applied as a means for French naval administrators to centrally control the way its constructors designed and built ships, as opposed to allowing different constructors in each port or dockyard to design and build vessels according to their own interpretation. This process of standardization was firmly fixed by France's 1765 naval ordnance, which established the calculations for stability, speed, sail plan, and so on, that had to be carried out for all vessels. First Spain, then the Netherlands, Denmark, and other countries began imitating this development, so that by the end of the eighteenth century, ship theory had become one of the primary means by which European nations ensured a standardized, centralized system of organization and control over naval ship design and construction.[2]

At the beginning of the industrial age, therefore, most navies already had established firm control over ship design and shipbuilding, had vertically integrated their engineering and construction processes, and had established their own standards and practices. This had not been the case for the commercial shipbuilding industries, which during the era of wood and sail had remained largely fragmented. This all changed with the introduction of steam, iron, and steel from 1800 until the end of the nineteenth century, when the worldwide volume of trade grew fivefold, while the ships themselves became ever more expensive to buy, finance, and maintain. To justify their cost, shipowners demanded tighter scheduling and improved performance, while their insurance underwriters wanted surety for the safe delivery of cargoes.[3] This in turn required shipbuilders to develop more sophisticated and cost-effective means of controlling the design and construction process. At first, standardization was imposed at the very top levels of the shipping industry through the rise of classification societies, which had developed industry-wide norms by the middle of the century, and later through the involvement of governmental and then international bodies. By the 1870s, the high cost of fabricating iron and steel steamships, coupled with increasing demands by owners for reliable ships that met specific performance goals, drove commercial shipyards to consolidate their operations in vertically integrated organizations in which engineering theory and naval architecture played a key role in the management process. The design standards that emerged from this integration and consolidation of commercial shipyards often differed dramatically from those that most navies had already instituted decades earlier, resulting in significant differences between the naval architectural practices between naval and commercial vessels that would persist through the twentieth century.

**Classification Societies and International Regulations**

The dangers of seafaring have been recognized since ancient times. Roman law prohibited navigation during the stormy winter months, a practice that persisted in other

# The Demand for Standardization

parts of Europe through the Middle Ages. The first known regulations concerning ship design were the Venetian Maritime Statutes of 1255, which established load waterlines according to the age of the ship, using an iron cross fixed to the hull. Similar laws were soon instituted in Pisa, Barcelona, and Genoa. As transoceanic discoveries drove shipbuilders to construct larger seafaring vessels, regular surveys to ascertain the condition of the ships were instituted by the Hanseatic League, city-states like Genoa, and nation-states like France. But as the rapidly evolving technologies of steam and iron began to dominate ocean trade in the nineteenth century, shipowners and shipbuilders were increasingly worried that governments could impose overly restrictive regulations that would place them at a disadvantage in an already fiercely competitive international market. Instead, attempts to achieve greater safety and reliability evolved organically within the commercial marketplace itself, in the form of classification societies.[4]

Yet classification societies such as Lloyd's Register, Bureau Veritas, and the American Bureau of Shipping were initially established not as guarantors of safety but rather as clearinghouses for information. The first of these, Lloyd's Register, had its roots in Edward Lloyd's coffee house in the City of London, established around 1690. Like many coffee houses of the day, it served as the place of business for groups of merchants who rarely had offices of their own. Lloyd's shop was near the customs house and the old offices of the British navy, so it naturally attracted shipowners and underwriters to its spacious but sparely furnished ground-floor coffee room. Lloyd distinguished himself from other coffee house owners by regularly collecting and publishing shipping intelligence and setting up auctions and sales of merchandise. The name Lloyd's lived on after his death in 1713 and became associated with three separate but related products: Lloyd's List, a weekly—later daily—publication of ships and shipping news; Lloyd's Corporation, often called Lloyd's of London, the syndicate of underwriters that insures ships, shipping, and later all manner of business (including, most famously, the actress Betty Grable's legs); and Lloyd's Register, which was established in 1760 as an extension of Lloyd's List, providing additional information to underwriters and merchants on the condition of ship hulls and equipment. This was the beginning of the classification society and would be repeated in Europe, the Americas, and around the world.[5]

For its first 70 years, the surveyors of Lloyd's Register evaluated and recorded the condition of existing vessel hulls and equipment—A, E, I, O, and U for hulls in decreasing order of condition, and 1, 2, 3, and 4 for equipment (so A1 meant highest class). By the turn of the nineteenth century, the different methods being employed in the registers for shipowners (recorded in the Red Book) and underwriters (Green Book), exacerbated by the preferential treatment afforded Thames-built ships, created considerable friction within the shipping industry. This situation was not resolved until 1834, when Lloyd's Register was rechartered as a single entity for classing both British and foreign vessels and drew up new rules for the classification of ships, which gave specific details on acceptable frame spacing, thickness of planking, fastenings, and workmanship based on ship tunnage. These rules now reached a new audience other than shipowners and underwriters, as shipbuilders for the first time were given guidance on how to deliver a ship built to class.

A new class of surveyors was also born, the shipwright surveyor, whose role was to inspect new vessels while building. Lloyd's frequently poached constructors from British navy dockyards to fill this role, as they had the scientific training and engineering experience needed for this role. Among the most famous of these was Augustin Creuze, who was hired as a principal surveyor in 1844 after service in the Portsmouth dockyard and authorship of numerous texts on naval architecture (discussed in chapter 5) and who created the Lloyd's Register library. Another was Bernard Waymouth, hired in 1853 from the Deptford dockyard, who not only designed—with Lloyd's permission—the record-setting composite clipper *Thermopylae* but was also instrumental in creating rules for iron and then steel ships that reflected advances in naval architecture and structural theory. Lloyd's first rules for iron ships, issued in 1855, had largely followed the practice in wooden ship rules of specifying scantlings based solely on tunnage. By the late 1860s, shipbuilders felt that these rules were outmoded and resulted in ships that were too heavy and too expensive. At the same time, theoretical work carried out by Rankine and Fairbairn showed how stresses varied along the hull as a result of wave loading—greater at midships, lesser toward the ends—so Waymouth proposed a new set of rules for iron scantlings "determined not by [the vessel's] tunnage, but by their dimensions ... and proportion of breadth to length, which by better distribution of the material in the structure, admitted of considerable reductions in the scantlings previously insisted upon." After considerable discussion and political maneuvering, Waymouth's new rules for iron ships were adopted in 1870, followed in 1888 by similar rules for steel ships.[6]

During the nineteenth century, the Lloyd's name became synonymous with underwriters and ship registries, as other bureaus in Europe and the Americas adopted the name even when they had no connection with the original London coffee-house: Österreichischer (Austrian) Lloyd in 1833, American Lloyd's in 1857, and Germanischer Lloyd in 1867. But Lloyd's by no means held the monopoly on name recognition. In 1828 the company that became known as Bureau Veritas (after the goddess of Truth) was formed in Antwerp to "furnish detailed descriptions of the qualities of vessels" in and around the North Sea, a direct answer to Lloyd's, which registered only British hulls. In 1830 it moved to Paris, home to many of the marine underwriters it served. The move was spurred in part by the Belgian Revolution but also because Bureau Veritas had by then expanded well beyond the North Sea and was quickly filling the vacuum for classing European vessels; within two years it had registered almost one-third the number of ships under Lloyd's. Like Lloyd's, it also developed rules for the classification of ships, issuing its first set of rules for wooden ships in 1851, iron ships in 1858, and steel ships in 1880. But even though Bureau Veritas positioned itself to be the classification society for all Europe, Norwegian shipowners and underwriters felt their rates to be usurious, so in 1864 they formed a national classification society, defiantly named Det norske Veritas, to provide "reliable and uniform classification and taxation of Norwegian ships," publishing its first rules for wooden ship classification in 1867 and for iron ships in 1871. The society generally followed Lloyd's rules in terms of technical content to remain competitive.[7]

Other classification societies also emerged to compete with the market domination of Lloyd's and Bureau Veritas, such as Registro Italiano Navale in 1861, the British Corporation Register of Shipping in 1890, and the above-named Lloyd's in Austria, Germany, and the United States. Another entity, the American Shipmasters Association, was established in 1862 to certify the competency of mariners, but after the end of the Civil War, fearing that Lloyd's Register was unjustly downgrading American ships, it began surveying and classing U.S. vessels. Its first wooden ship rules—using Lloyd's Register designations such as A1 for first quality—came out in 1870, followed by iron and steel rules in 1877 and 1890, respectively. By this time, the U.S. government had largely taken over the role of certifying mariners, so in 1899 the society changed its name to the American Bureau of Shipping and focused its efforts on ship classification.[8]

The U.S. government's role in certifying mariners was part of a larger trend, starting in the mid-nineteenth century, of maritime nations exerting increasing regulatory authority over safety at sea. The British Merchant Shipping Act of 1854, the first major piece of maritime legislation in the industrial age, included a comprehensive section on "building, equipment, and inspection of all sea-going vessels with a view to safety and the prevention of accidents." The act required that "every steamship built of iron ... shall be divided by substantial watertight bulkheads," which would limit flooding in case of catastrophic collision or grounding. Even though the rationale was sound, industry pressure forced Parliament to repeal it in 1862 on the grounds that "it was better to leave builders and designers unfettered in providing extra strength and security." The assumption was that shipbuilders would always build safer ships than regulations required, when of course the opposite was true; the loss of the overloaded passenger ship SS *London* during a storm in 1866 was partly blamed on lax regulations.[9]

The term "coffin ship" became popular as a result of such accidents, which were all too common at this time—almost 2,000 vessels were wrecked or damaged each year around the British Isles, and one out of every five British mariners perished at sea. The cause of these seamen was taken up by Samuel Plimsoll, a Liberal Party member of Parliament from the landlocked city of Derby who had previously defended the rights of coal miners. Convinced that overloading was the principal cause of these ocean tragedies, Plimsoll spent several years gathering information on the subject. He wrote an influential pamphlet, *Our Seaman* (1873), which publicized his findings, and presented bills to Parliament requiring ships to have their safe-load waterlines established before setting sail. He met stiff resistance from Liberal and Conservative parliamentarians alike, but by 1876 many of these politicians—notably Prime Minister Benjamin Disraeli—had changed their minds. A new Merchant Shipping Act was passed that year that required vessels to have their maximum load waterline—which immediately became known as the Plimsoll mark or line—indicated on the hull. Even so, many shipowners resisted this demand; Alfred Holt, owner of the Blue Funnel Line, one of the most respected fleets in the industry, objected to being told how to run his ships and called the act "hasty and most arbitrary."[10]

But the Merchant Shipping Act of 1876 had neither substance nor teeth. Load lines could be placed wherever the owner wanted; one Cardiff shipowner reportedly placed his Plimsoll mark on the funnel. The Board of Trade—the parliamentary body that oversaw shipping and seamen—recognized that the proper placement of load lines would have to consider factors such as required freeboard (height of deck above water) and reserve buoyancy, a measure of the enclosed volume above the waterline that determined how far the ship could be flooded before it foundered. Lacking the expertise to provide guidance or establish rules, the board turned to Lloyd's Register for assistance. Lloyd's Register's chief surveyor Benjamin Martell had already been preparing tables of load lines based on calculations of reserve buoyancy, which by 1882 were published as a voluntary set of rules that the Board of Trade also adopted as guidance.

The following year the board began considering the overall problem of loading and stability after damage and established a series of committees to examine these subjects—a load line committee, a committee on lifesaving and lifeboats (chaired by the president of the White Star Line, Thomas H. Ismay), and a bulkhead committee led by the shipbuilder Edward Harland. The three committees were interlinked; as the Life Saving Committee noted, proper placement of bulkheads and load lines was necessary for a "ship to remain afloat for some length of time after an accident has occurred" so that lifeboats could be properly deployed. Once again, Lloyd's was ahead of the game. The utility of watertight bulkheads to limit flooding after damage had been well understood in Britain ever since Samuel Bentham fitted them on his sloops *Arrow* and *Dart* in 1795, but they were also expensive and limited a ship's cargo capacity. In the constant tug-of-war between safety and economy, Lloyd's came down on the side of safety by requiring a specific number of bulkheads based on ship length. The Admiralty representative to the Board of Trade, James Dunn, commissioned the construction of "fair-sized" models of merchant ships, floated in large tanks, to examine the effect of bulkheads on flooding and stability after damage (figure 4.1). The idea of using models to test damaged stability was then in the air, as the French constructor Émile Bertin had already used small-scale models to verify the concept of "cellular" construction to limit flooding after artillery or torpedo damage for his proposed cruiser *Sfax*. In Dunn's tests and in subsequent analyses, the acceptable level of flooding was set by a margin line three inches (eight centimeters) below the main deck. The acceptable subdivision—distance between bulkheads—was calculated by developing floodable length curves (which employed Antoine Bonjean's methods for calculating displacement) so that a single accident would not sink the vessel.[11]

In 1890 a new Merchant Shipping Act codified these findings into law. Among other things, it established that passenger ships longer than 130 meters must be able to survive flooding of at least two compartments, while cargo ships over 90 meters had to survive one-compartment damage. Load lines were now required and indicated by the Plimsoll mark—a circle bisected by a line, later adopted by the London Underground as a symbol of surety—with mandatory lines fixed for conditions in summer, winter, and the most onerous, winter in the North Atlantic (figure 4.2). Another law set the minimum number of lifeboats and lifejackets required for passenger ships on the basis of tunnage. Finally, for the first time the Board of Trade gave authority to three classification societies—Lloyd's, Bureau Veritas, and the newly formed British Corporation Register of Shipping—to assign load lines for British merchant vessels. The same load line authority was soon granted by other nations to their classification societies, including Germanischer Lloyd, American Bureau of Shipping, and Registro Italiano Navale. Even today, the initials of the classification society appear on either side of the Plimsoll mark to indicate which authority established the load line.

The Board of Trade's decision that Lloyd's would carry out load line assignments, and the adoption of that practice by other nations, was a recognition that classification societies had expanded from the purely commercial role of clearing information for merchants and underwriters and now encompassed the governmental responsibility of ensuring that standards were complied with to safeguard life and property at sea. This was in part due to the recognition by the British government, and others as well, that ships that were classed and regularly inspected by a classification society had a casualty rate less than half that of vessels that were not classed.[12] At the turn of the twentieth century, it certainly appeared that the shipping world, through a judicious

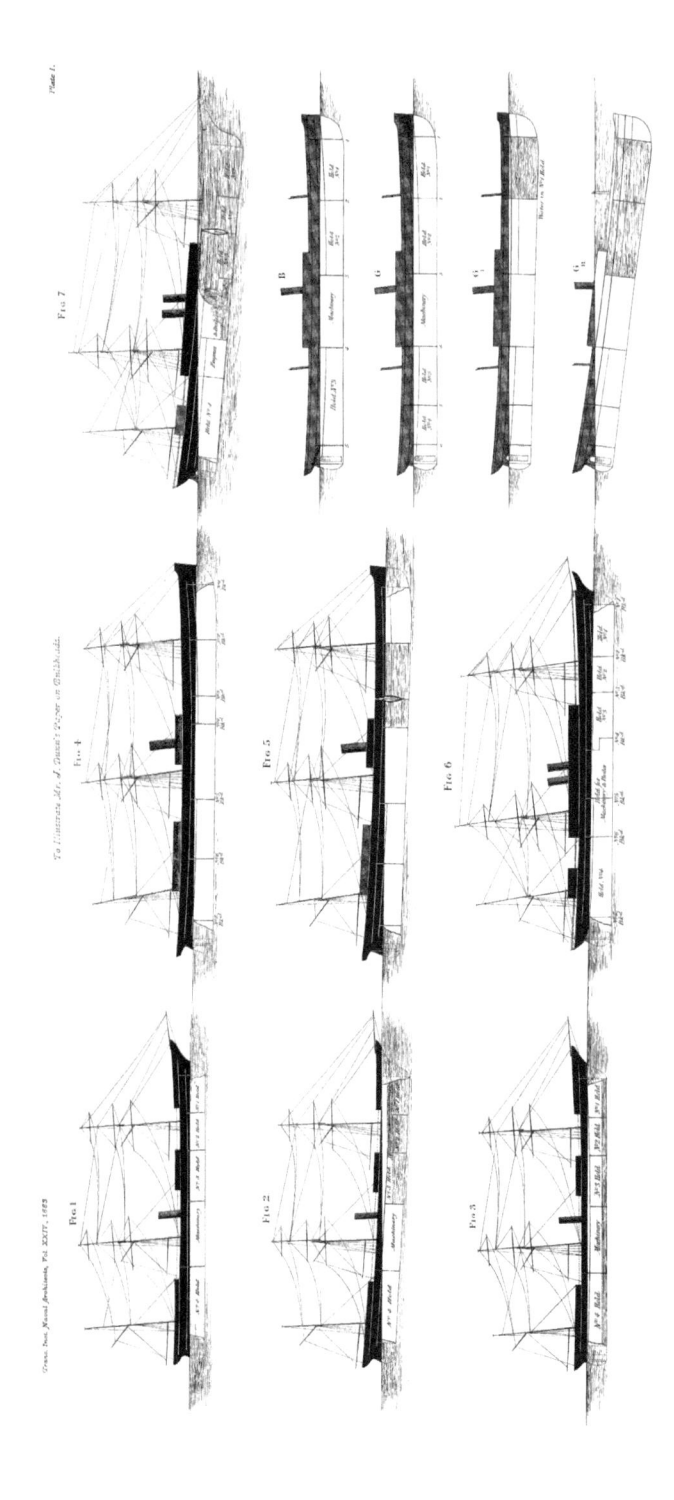

**Figure 4.1**
James Dunn's model tests of damaged stability, 1883.
Credit: RINA.

The Demand for Standardization

**Figure 4.2**
Plimsoll mark on a modern container ship, and as adapted by the London Underground.
Credits: Salim R. Rezaie (Plimsoll mark); author's photo (London Underground).

combination of governmental regulations and classification rules, had adeptly established the set of standards for these technological marvels of steam, iron, and steel that would keep commerce running smoothly around the globe.

Much of that commerce ran across the North Atlantic, carried by freighters and passenger liners. In the early years of the twentieth century a four-funnel passenger liner, designed for speed and highly subdivided for safety in case of damage, was steaming full ahead off the Grand Banks near Newfoundland when an iceberg suddenly appeared dead ahead through mist. The helmsman put the ship hard to port and reversed the engines, while the officers closed all the watertight doors, but by then it was too late—the ship struck the iceberg a glancing blow, dumping ice on the decks and damaging the bow plating. This was not RMS *Titanic* but rather SS *Kronprinz Wilhelm*, the date 8 July 1907 not 14 April 1912, and the ship continued steaming to Hoboken, New Jersey, with the damage fully under control. This scenario had played out many times in the past; from 1874 to 1912, some 53 iron or steel ships had collided with icebergs and survived, with only two ships abandoned after collision and with no loss of life. Just three days before *Titanic* had its fateful encounter, the French liner *Niagara* traveling westbound also struck an iceberg and continued to New York. Given this record of steel-hulled impunity, neither *Titanic*'s builders nor its crew had any reason to fear for its safety. The ocean was vast, icebergs were not considered a threat, and collisions between ships were rare except in crowded roadsteads. That was why lifeboats were allocated on the basis of tunnage, not on manning; their purpose was seen as shuttling people from a sinking ship to other ships waiting nearby, as opposed to saving the entire complement of passengers and crew far at sea.[13]

*Titanic* was ordered by White Star Line in 1908 and built between 1909 and 1912 at the Harland and Wolff shipyard in Belfast beside its elder sister and running mate *Olympic*.[14] It was common at the time for North Atlantic passenger ships to be built in pairs, with one going westbound while its running mate was headed eastbound. These were the largest ocean liners in the world at the time and among the most luxurious, though certainly not the fastest—a distinction denoted by the Blue Riband for the fastest transatlantic crossing, held at the time by the Cunard liner RMS *Mauretania*. In an ironic twist, neither *Olympic* nor *Titanic* were classed with Lloyd's Register or the British Corporation, even though the names Harland and Ismay were on the committees that led to these classification societies becoming an integral part of safety at sea. Nevertheless, the shipyard and shipowners—now led by William Pirrie and Bruce Ismay, respectively—claimed that the combination of double bottom, 15 watertight bulkheads, and electrically operated watertight doors rendered the ships "considerably in excess of the requirements" of Lloyd's. The ships' reputation for safety was cemented in 1911 when *Olympic* was struck by the British cruiser *Hawke* off the Isle of Wight. *Hawke*'s ram bow tore into *Olympic*, flooding several compartments, but its master, Captain Edward Smith, managed to bring the ship back to port for repairs. Absolved of any blame for the incident, he soon was given command of *Titanic*. Captain Smith was so impressed by *Olympic*'s ability to withstand the ramming that he declared that "the *Olympic* is unsinkable, and the *Titanic* will be the same when she is put in commission."[15]

The ramming of *Olympic* may have colored Smith's perception of *Titanic*'s invulnerability just enough that on its maiden voyage to New York, he kept the engine order telegraph at full ahead, making 22 knots, even though he knew they were in an ice field. A few minutes before midnight the lookouts reported an "iceberg right ahead" and the first officer ordered the ship to veer to port to avoid it. The ship grazed the iceberg on its starboard side, the impact buckling the hull plates and popping rivets from the bow to the first two boiler rooms. With the seams between plates now opened several centimeters, water gushed in to flood a total six watertight subdivisions. *Titanic*'s naval architect Thomas Andrews, a nephew of Pirrie, was aboard and knew the ship could withstand four subdivisions flooded, but not six. He also knew there were not enough lifeboats for all on board, as this scenario had never been contemplated. Remarkably, *Titanic* stayed upright even as it was sinking by the head—most sinking ships lose transverse stability early and roll over—so that many of the passengers and crew had time to abandon ship before the strains on the hull caused it to shear in two and take the final plunge, about two hours after the collision. Even so, two-thirds of the people aboard died, including Captain Smith and Thomas Andrews. Bruce Ismay, also aboard, made it to one of the collapsible lifeboats and survived, only to face two boards of inquiry—the first one by the U.S. Senate and convened the day after SS *Carpathia* landed in New York with *Titanic* survivors and the second one for the British Board of Trade held a month later.

Both boards of inquiry made numerous recommendations that were quickly enacted by the British and American governments, such as mandating that all ships monitor radio distress frequencies and establishing an ice patrol. Both governments also required sufficient lifeboats to be carried for everyone aboard. This last requirement had at least one tragic and unforeseen consequence, as most existing ships were not designed to carry the weight of the additional lifeboats needed. This was the case for the Great Lakes steamer *Eastland*, which having had its lifeboat capacity doubled in early July 1915, capsized at its pier on the Chicago River later that month the very first time it embarked passengers, most of whom were Western Electric employees going to a company picnic. A total of 844 people died, still the largest loss of life on the Great Lakes. A later inquiry found that the extra lifeboats and lifejackets mounted high in the ship brought its metacentric height at full load from an already small 10 centimeters to effectively zero, or even negative.[16]

The ineffectiveness of creating and enforcing strictly national regulations had been recognized as far back as 1609, when the Dutch jurist Hugo Grotius published *Mare Liberum* (Freedom of the seas), noting that oceans were the "common property of all" and not subject to any one nation's jurisdiction. Maritime trade was highly competitive, which as noted, led the shipping industry to resist national regulations and instead create its own self-governing system of classification. The loss of *Titanic* was a political and cultural shock across the entire globe, on a scale that has not yet been equaled, and it brought into focus the need for a collective set of safety standards and regulations. The argument for a minimum set of universally applicable standards, then as now, was to prevent a race to the bottom of less scrupulous nations to obtain a competitive edge, at the expense of lives and matériel.[17]

Just over a year after the British and American boards of inquiry published their findings on *Titanic*, 16 maritime nations took the first tangible steps to internationalize the regulation of safety at sea. In November 1913 at the invitation of the British government, delegates from France, the United States, Russia, Australia, and other states met at Whitehall in London for two months to hammer out an international agreement that covered radio, navigation, construction practices (including bulkhead and load lines), and lifesaving devices. The Safety of Life at Sea (SOLAS) convention was signed by all the participating nations in January 1914 and was to take effect the following year. The outbreak of World War I in July 1914 suspended ratification by most countries. Although it did not go into effect, its provisions were adopted by many nations after the end of the war.

The first SOLAS convention set the stage for subsequent initiatives in internationalizing safety at sea in the early twentieth century. Britain was still the largest maritime power and dominated the proceedings of the first few SOLAS conventions. The second convention, with 18 nations, was held in London in 1929 and updated the issues from the first convention plus fire safety. Its provisions went into effect in 1933. The third SOLAS took place after World War II, in 1948, also in London. That same year in Geneva, the newly created United Nations established a specialized agency for maritime safety, the International Maritime Consultative Organization (IMCO), which updated SOLAS in 1960 and again in 1974. By that time, Britain had fallen a long way from its preeminence in shipping and shipbuilding, so those conventions and subsequent programs were really between peer and near-peer nations. In 1982, IMCO was renamed the International Maritime Organization (IMO), which today maintains the international regulatory framework for safety and for environmental protection and maritime security. Yet the IMO does almost nothing on its own; all the actual work of developing and enforcing regulations is carried out by the 172 member states, through their national bodies such as the U.S. Coast Guard and Britain's Maritime and Coastguard Agency, as well as through the dozen classification societies and other nongovernmental and intergovernmental bodies that make up IMO's membership.[18]

These national bodies and classification societies also carried out work independently from the international context, the most famous of which were the series of investigations during World War II into the fracturing of welded steel ships. Riveting had been the preferred method of joining steel plates through the nineteenth and early twentieth centuries, but it was an expensive and time-consuming process. The development of welding metals using electrodes and oxyacetylene torches had evolved gradually since the 1880s. Though it was recognized to produce lighter but stronger and more durable structures than riveting and also required less manpower (one welder could join seams that would require three to five riveters), welding was used only sparingly in shipbuilding. By the 1930s, welding machinery powerful enough to reliably join heavy steel plates came into common use for bridges and skyscrapers. At the same time, a series of post–World War I naval treaties in 1922, 1930, and 1936 limited the displacement of capital ships (generally battleships and aircraft carriers), spurring naval shipbuilders to fully adopt welding to keep structural weight to a minimum. With the outbreak of World War II, shipbuilders like Kaiser and Bechtel in the United States, and Cammell Laird and Vickers-Armstrong in Britain, were under immense pressure to quickly build thousands of cargo ships (*Empire* and *Liberty* ships) and oil tankers (T-2) and devoted enormous resources to training large numbers of welders to get the ships into the water in a hurry.

The night of 23 January 1943 at the Kaiser Shipyard in Portland, Oregon, the T-2 tanker SS *Schenectady* ruptured with an explosive boom, just days after its first sea trials. At daybreak, the extent of damage was clearly visible; resting in calm water at the pier, the entire ship had cracked in half (figure 4.3). This would become the best-known example of a spate of catastrophic failures that beleaguered hundreds of T-2 and *Liberty* ships even as they were needed to keep the massive logistical train running between the United States and war-ravaged Europe. The Maritime Commission, a recently created government regulatory body, convened a board of investigation composed of technical and scientific experts from the U.S. Navy, Coast Guard, and American Bureau of Shipping to investigate the cause of the failures and record solutions. At first the board focused on the hurried training of welders, and then it looked at residual stresses that were locked in after construction. By the middle of 1943, it became apparent that the fractures were due to a combination of flawed design practices such as sharp hatch corners, which initiated cracks, combined with unsuitable steel alloys that became brittle at low temperatures. The immediate fixes were straightforward—rounded hatch corners, strategically placed reinforcements, crack arrestors at the sheer strake (where the shell connects to the deck), and improved metallurgy. The longer-term fix was to establish in 1946 a permanent Ship Structure Committee, with member agencies from the American and Canadian governments as well as the American Bureau of Shipping, which even today carries out large-scale experiments and research on both naval and commercial ship structures.[19]

**Figure 4.3**
T-2 tanker SS *Schenectady* after breaking in two, 1943.
Credit: Ship Structure Committee.

In Britain, a similar board, the Admiralty Ship Welding Committee, was also formed in 1943 and continued long after the war's end. Even though the problems of fracture were not as pervasive in British yards, a consortium of researchers from government, academia, and industry (including Lloyd's Register) investigated the root causes of welding failure, including full-scale trials on tankers in still water and at sea to determine hogging and sagging stresses and deformations. The Admiralty group also lent their expertise to the Americans; it was a British researcher from Cambridge University, Constance Tipper, who first identified the problem of low-temperature brittleness in the steel of *Liberty* ships and developed what is now known as the Tipper test for determining brittleness in steel. Despite the efforts expended during the war to improve welding, most British shipbuilders doggedly continued with riveting all the way through the 1960s, steadily losing economic and technological ground to other nations, both in Europe and in Asia, that had already staked their shipbuilding futures on welded steel vessels.[20]

## Vertical Integration of Commercial Shipyards

The decline of British shipbuilding after World War II marked the end of a century-long global domination by British shipyards in building the world's merchant tunnage. This domination had been made possible by the steady integration and expansion of its commercial shipyards, begun as a response to the capital-intensive technologies of steam, iron, and steel and the increasing demands by shipowners for ships that met specific performance standards. The rise of classification societies and national regulations alone were not enough to meet these demands, as they focused on a narrow range of safety issues. When shipowners specified speed, cost, and material quality for increasingly complex and costly vessels, the shipyards themselves had to develop the management techniques, scientific and engineering capabilities, and corporate structures to achieve those demands on a regular, systematic basis. Invariably, these techniques and structures involved the vertical integration of production—organizing all parts of shipbuilding, from steel production through boiler manufacture and hull construction, all under one roof—and the creation of a strong middle-management structure that relied not on the invisible hand of the marketplace but rather on the "visible hand" of the corporate manager, as the business historian Alfred Chandler so eloquently put it.[21]

Naval administrations had already developed their own vertically integrated dockyards and hierarchical management structures during the age of sail a century before, with ship science and engineering theory emerging as a principal means for exerting standardization and control. The industrial age saw a wholesale expansion of these dockyards and management hierarchies to organize and control the integration of materials, engines, weapons, and systems into the construction and maintenance of complex warships. In France, the Lorient dockyard was the first to be outfitted for iron production, with Brest, Rochefort, Toulon, and Cherbourg following soon after. The extensive collection of British naval dockyards—Chatham on the Thames, Portsmouth and Plymouth on the south coast, Pembroke in Wales—expanded to maintain Britain's strategic capability of being numerically superior to the next two navies combined. The Amsterdam *rijkswerf* in Kattenburg was modernized in 1867 to build ironclads. Even the American navy, fettered at its beginning by disagreements over its national role and strategic mission, built its first dockyards in Boston, Norfolk, and New York in the 1830s and 1840s, which were expanded to meet wartime contingencies during the Civil War.[22]

The vertical integration of commercial shipyards in Europe and America closely followed the pattern of national economies during the Industrial Revolution as they transitioned away from agriculture and crafts and toward machine-based manufacture. In the early part of the nineteenth century, most industrial firms were family owned or partnerships and could be characterized as single-unit enterprises, employing a small number of men and creating one or just a few products. The hierarchy could best be described as binary; owners and workers. Owners were generally responsible for design, production, quality, receivable accounts, and marketing, all at the same time. Workers produced the products. Most of the processes were developed by careful trial and error over extended periods, resulting in effective (but often inflexible) rules of thumb for guidance.

The period between 1870 and 1910 saw the merging of the Industrial Revolution with two other revolutions, managerial and scientific, which together created a new type of industrial enterprise. The managerial revolution led to the vertically integrated, hierarchical firm that brought many units under its control. This was marked by the growth of specialized divisions within the firm—for example, separate branches for accounting, engineering, and production. Each of these divisions was led by middle managers, a new breed of worker who was neither an owner nor a laborer but a salaried professional responsible for the people, material, and processes in his division. The coordination of the activities of these divisions, and of the entire company, was carried out by upper-level managers, who themselves often rose up from the middle ranks.

In conjunction with this was a revolution in the development and application of scientific knowledge. The modern, hierarchical enterprise demanded increasingly standardized processes; and as the customer and supplier networks grew, the creation of standardized products grew as well. Industrial standardization was based upon scientific standardization:, first, specify and predict the characteristics and performance of the technology while still in the design stage; second, precisely control the production processes; and third, accurately measure the results. Middle- and upper-level managers of engineering firms were increasingly men who did not come up from the shop floor but rather had a solid academic background in addition to practical training. In short, the period from 1870 to 1910 marked a major transition of industries, in Britain, Europe, and the Americas from small-scale family-owned and partnership firms to the modern, vertically integrated business enterprise.

One of the first to undergo this sweeping change was the railroad industry. From around the 1850s, railroads across the globe evolved into modern, vertically integrated organizations, with salaried middle managers in charge of internal coordination of the various parts of the firm—rolling stock, fixed stock, coal, and so on—arrayed across a wide area. By the 1870s, railroad firms had developed something of a modern engineering organization to oversee the design and testing of components, creation of standards and specifications for rail shapes, composition of alloys, and other technologies. During this period, many firms, such as the Great Western Railway in Britain and the Baldwin Locomotive Works in the United States, established drafting rooms and laboratories overseen by engineers, who were increasingly university educated and who belonged to professional societies. Locomotive design became more and more informed by theoretical developments in metallurgy and thermodynamics.

During the period 1870–1910, the vertical integration of organizations, and professionalization of its management, quickly spread to other manufacturing industries, which often borrowed talent from the railroad industry. The American steel manufacturer Carnegie Company, for example, hired former railroad men who brought into the firm the engineering and management practices that made railroads so successful. Former railroad engineers working for architectural firms gave the newfangled skyscrapers their lightweight steel skeletons, based heavily on railroad truss bridges. Many other industries were also becoming vertically integrated and were increasingly employing scientists and engineers to oversee the internal workings of their plants. For example, chemical-based industries such as petroleum distilleries, alkali producers such as glassmakers, and even beer breweries began using ever more sophisticated chemical processes and laboratory analyses to improve production flow and quality, as well as to develop new products.

The integration and professionalization of the maritime industries in this era was "particularly complex," notes economic historian John Hutchins, during which time both shipping companies and shipyards progressed from an "unorganized, competitive system" to a "highly organized, rationalized and concentrated type of organization."[23] The process of incorporating scientifically based naval architecture into the design of ships was a critical part of this transformation in industries across Britain, Europe, and the Americas. The gradual adoption of scientific naval architecture into the management structure of these commercial shipyards can be traced in their business records, including wage and salary books, organization charts, and shipbuilding contract files.[24]

A microcosm of this development can be seen in the six largest and most innovative shipyards on the "two Clydes": William Denny and Brothers, Robert Napier and Sons, and J. & G. Thomson on the River Clyde near Glasgow and William Cramp and Sons, John Roach and Sons, and Harlan and Hollingsworth on the "American Clyde," the Delaware River near Philadelphia. The selection of these shipyards is not haphazard. By the 1890s, British shipbuilders accounted for 75 percent of the world's ship construction, and the tunnage it built for foreign owners alone almost equaled the rest of the world's shipbuilding combined. Two-thirds of British-built ships were launched on the River Clyde. The United States was in distant second place, building 10 percent of the world's tunnage, of which half was built along the Delaware River. Together the two nations had built seven out of every eight ships plying the world's oceans and waterways.

**Figure 4.4**
Denny shipyard on the River Clyde.
Credit: Scottish Maritime Museum.

The River Clyde and its Firth were once the location of over 300 shipbuilding firms and saw the construction of over 25,000 ships over three centuries. Innovations in steam, iron, and steel shipbuilding developed more quickly in the Clyde region than in many other parts of Britain, due both to its proximity to Atlantic trade routes and to the network of universities and engineering industries in the region. The term "Clyde-built" became synonymous with quality, in part due to the precise engineering and workmanship of these shipyards.

William Denny and Brothers in Dumbarton (figure 4.4) was established in 1844 and by the turn of the twentieth century was regarded as one of the most technologically advanced yards in the world. Denny often led the way among commercial shipbuilders in the use of scientific naval architecture. As early as 1869, the shipyard was computing the displacement of its ships, a rare practice at the time for commercial shipbuilders. Stability calculations first appeared in ships' plans in 1880. And as noted in chapter 3, in 1883 Denny was the first shipyard to erect a Ship Model Experiment Tank based on the plan of Froude's Torquay tank.

Denny began evolving its management structure in a vertically integrated organization with clear hierarchies and specialization of functions. The design offices were separated into the technical department (steelwork), arrangements department, and scientific department. In particular, the appearance of the scientific department charged with "calculations as to weights, capacities, displacement, stability, speed, trim, etc.," demonstrated the rise of the new middle-manager class; a worker with both the scientific and the practical experience to oversee the production process and develop innovative design approaches. This development coincided with the specialization and vertical integration of the production side of the shipyard, which was producing and installing high-efficiency triple-expansion engines and newly developed electrical generating and distribution systems. Denny's attention to detail led the company to develop a complete set of rules that standardized each part of the design and production process, from specific calculations that had to be completed for each ship to the flow of materials in the production yard.

The increasing importance of naval architecture in Denny's corporate hierarchy can also be seen in the position descriptions described in the yard's salary books. For example, Charles Henry Johnson, who began as a draughtsman in 1870, saw a change in title from chief draughtsman to chief designer in 1882, reflecting changes in the company structure as well as the role of what is regarded as a modern day naval architect. His assistant, Frank P. Purvis, was, at the same time, also promoted from draughtsman to assistant designer and soon became the superintendent of the experiment tank. Both their salaries nearly doubled in the space of two years. Similarly, from 1880 to 1882, the firm hired two draughtsman (a term embodying both drawing and calculations) for its scientific department, James Thomson and Richard Mumford. Another employee, William Gray, started at Denny in 1884 with a five-year apprenticeship at the experiment tank. In 1892 he became head of the scientific department and in 1896 was promoted to "Head Draughtsman of the Scientific Department." These records indicate a substantial change in both design and managerial practices: that the rule-of-thumb building practices (which required only drawing skills) were being superseded by a revolutionary scientific approach demanding both designers and calculators and that these roles were being staffed by well-salaried, technically skilled professionals to deal with the increasing complexity of the organization and its products.

Robert Napier and Sons opened in Govan in 1841. As noted in earlier chapters, Napier worked with eminent scientists such as William J. M. Rankine and was the training ground for some of the most famous naval architects and marine engineers, including John Elder and John and George Thomson, before being taken over by the William Beardmore Company in 1900. The surviving business records do not include details of the shipyard organization. They do, however, show that the shipyard, under the influence of the scientifically minded partner James R. Napier, was quite advanced in using scientific naval architecture. As early as the 1860s, he was carrying out longitudinal and transverse strength calculations for iron paddle wheel steamers. By the 1870s, he was making increasingly sophisticated calculations for ship launch, including the vertical and longitudinal travel of the center of gravity. From 1872 to 1876, Napier was corresponding with William Froude on the subject of ship rolling.

The shipyard of J. & G. Thomson, founded in 1871 by pioneering brothers John and George Thomson after their employment with Napier, was later absorbed into the John Brown Company. Within the first decade, naval architecture began playing a key role in Thomson's shipbuilding management. The first wage books up to 1880 showed a business with highly specialized divisions: foreman, drawing office, "girls" (to do tracing), "counting office" (accounting), and other intermediate roles. By March 1881 there was a new science department to carry out calculations (similar to Denny's scientific department) and a restructuring of the drawing offices to reflect greater specialization and control. At the same time, the Thomson shipyard was also rationalizing and concentrating its production, creating a set of specialized, vertically integrated facilities (boiler works, engine works, and fitting-out basin) in a single site.

In 1880, Thomson hired John Harvard Biles, a graduate of the Royal Naval College at Greenwich, as its first naval architect, reflecting the growing influence of scientific ship design. His assistant William David Archer was one of the highest paid members of the shipyard, which suggests that not only had he been there a long time but that his position within the company was one of a high level of authority and responsibility. After Biles's departure from Thomson in 1891, Archer was elevated to the position of naval architect.

Prior to 1880, little technical information was included in ships' plans or notebooks. Starting in 1880, coinciding with John Biles's arrival, the general calculation notebooks begin exhibiting specifications and calculations that demonstrate a sophisticated degree of scientific naval architecture:

- General calculations for the passenger ship *Servia* (1881), which include longitudinal and transverse metacenters, moment of buoyancy, added weights, and an inclining experiment. Additionally, launching drafts are discussed, compared, and calculated, referring to previously built vessels *Arab* (1879) and *Trojan* (1880).
- Launching calculations for the passenger ship *America* (1883), showing stability parameters such as the draft and trim, freeboard, load lines, center of gravity, and the metacentric height.

# The Demand for Standardization 167

- Resistance and thrust calculations for paddle wheels and propellers—for example, for *Trojan* (1880).

By contrast to British iron and steel shipbuilding, American iron and steel shipbuilding lagged several decades behind, as noted in chapter 2. Before the end of the Civil War, most shipyards were small family-owned firms building wooden vessels. Facilities were generally quite simple; a slipway on a river or protected bay and small woodworking shops sometimes fitted with steam-driven saws. A foreman oversaw the small crew (generally between 8 and 25 men), consisting of carpenters, caulkers, joiners, riggers, and rope makers, to construct a medium-sized vessel. The introduction of iron hulls into American shipbuilding during the 1840s and 1850s did not automatically result in a restructuring of the shipyards. In many cases the new iron shipbuilders remained small in size and followed the same organizational hierarchy as older wood shipbuilders.

The changeover from small-scale shipyards to large-scale, vertically integrated shipbuilding firms began in the 1870s and was in full swing a decade later. By the 1880s, the shipbuilding business had become concentrated along the Delaware River, in Philadelphia, Chester, and Wilmington. The "American Clyde" was already a major hub of locomotive building, and the availability of materials like iron and coal, railroad and transport infrastructure, and skilled labor and management gave rise to the most important metal-and-steam shipbuilding center in the nation.

William Cramp and Sons (figure 4.5) was founded as builder of wooden ships in 1825 and was unusual in that, unlike most wood yards, it successfully made the transition to iron and steam during the Civil War. After war's end, it became aligned with the railroad industry, at a time when many railroad magnates were extending their domains into shipping and vice versa. This concept was adopted by the Pennsylvania Railroad, which pioneered the intermodal railroad-shipping industry by funding a fleet of iron steamships, starting with SS *Pennsylvania* in 1872. In this way they could transport passengers cross-country by rail and then across the Atlantic by ship. That the railroad opted to build the ship at Cramp was due to the shipyard having already begun creating a vertically integrated system similar to that found in locomotive builders and having already sent its chief engineers to British Clyde shipyards to study their shipbuilding and engineering techniques.

William Cramp noted, "The growth of complexity in modern ships have entailed upon the naval architect and constructor demands and difficulties never dreamed of in earlier days…. The staff required to design and construct [a modern ship], and the complexity of its organisation, has augmented almost infinitely." To improve efficiency, Cramp initiated a systematic design and production schema so that "the form of every plate must be sketched before it is ordered." Thus, by 1880, draftsmen had moved

**Figure 4.5**
Cramp shipyard on the Delaware River.
Credit: Independence Seaport Museum, Philadelphia.

beyond simply drawing hull lines to creating highly complex shell expansion plans, which define a precise two-dimensional shape for a plate that will be bent and curved to fit a three-dimensional hull. By 1883, simple curves of form for displacement were becoming commonplace. By 1890, full stability curves including metacentric height were being developed, launching calculations began appearing in 1898, and by 1901 hull structural stress calculations began to appear.[25]

John Roach and Sons was founded in 1864 as an offshoot of an iron and engine manufacturer. Like Cramp, Roach saw that the technology, equipment, and expertise required for large, integrated industrial shipyards did not exist in the United States, so in the 1870s he toured the British Clyde to see how it was done. Upon his return, he used what he had learned to create a vertically integrated plant "in which [he] could build ships from the ore up," including the production of plate, frames, boilers, and piping. During the 1880s, Roach enlarged his workforce by 50 percent, to over 1,500 men, consolidated functions into specific departments under direct control of supervisors, eliminated outmoded craft specialties like blacksmiths, and greatly increased the skilled trades like boilermaker. He also formed a strong middle-management team of specialists in finance, labor management, and engineering, including a department of 12 naval architecture draftsmen to prepare ships' plans. These draftsmen were considered valuable employees, with their daily wage of $3.19 (about $60 today) being twice that of a ship's carpenter.[26]

The drive to vertical integration paid off. Between 1871 and 1885 Roach was the largest shipbuilding company in the United States. The navy turned to Roach in 1883 to build its first modern, all-steel ABCD warships (*Atlanta*, *Boston*, and *Chicago* cruisers and *Dolphin* dispatch ship), as his yard was the only one that could meet the navy's exacting standards for steel quality and production at an acceptable cost. Because Roach had already made extensive contacts on the Clyde, he was able to work with the U.S. Navy to facilitate the purchase from Britain of thousands of plans and drawings for armored cruisers and compound steam engines, which formed the basis for the ABCD ships. These vessels, which signaled the birth of America's new steel navy, came at a personal cost for Roach, who became embroiled in scandal and was forced into bankruptcy.

Harlan and Hollingsworth was originally founded in 1837 as a railroad car manufacturer but soon moved into shipbuilding, specializing in destroyers, ferries, and coastal steamers. The engineering and naval architecture capabilities of the shipyard grew in step with the increasing complexity of the company and the ships it built, especially between 1880 and 1900. Harlan's drafting department, which had been staffed by company-trained draftsmen, was by the 1880s renamed the naval architecture department and was employing university-educated engineers.

It is instructive to note the difference in education and training of a father and son employed at Harlan. Thomas Jackson came to Harlan in 1843 as a machinist, having been previously apprenticed in a wool mill. He rose up through the ranks and by 1856 was head of the drafting department, where he became "an authority upon many branches of marine architecture" by dint of his experience. His son Edward Jackson, by contrast, went to Cornell University and received a bachelor of science degree in 1875 before coming to the firm as a draftsman. This development coincided with increasingly prescriptive shipbuilding contracts and specifications taken on by Harlan, which included requirements for specific plate and frame sizes and hull speed. All these requirements could be met only by individuals trained in naval architecture and could be supported only by a corporation organized to carry those engineering predictions through from paper to steel.

The progress of integrating theoretical naval architecture as an institutional part of the ship design and shipbuilding management process is shown in figure 4.6. This process, perhaps unsurprisingly, followed the same evolutionary path taken by naval administrations and dockyards a century earlier when they integrated and centralized

|  | 1870s | 1880s | 1890s | 1900s |
|---|---|---|---|---|
| **Denny (UK)** | Draughtsmen | Chief designer<br>Scientific draughtsmen<br>Experiment tank | Scientific department |  |
| **Napier (UK)** | Strength calculations<br>Launch calculations<br>Froude correspondence |  |  |  |
| **Thomson (UK)** |  | Naval architects<br>Stability calculations |  |  |
| **Cramp (US)** |  | Curves of form<br>Shell expansion | Stability calculations<br>Launch calculations | Hull stress calculations |
| **Roach (US)** | Draftsmen | Naval architecture draftsmen |  |  |
| **Harlan (US)** | Draftsmen | Engineering-trained draftsmen<br>Detailed specifications |  |  |

**Figure 4.6**
Milestones in development of naval architecture on the two Clydes.
Credit: Author.

controls over their increasingly complex naval fleets.[27] In the case of these commercial shipyards on the two Clydes, the process unfolded primarily between 1870 and 1910, beginning with industry leaders like William Denny and John Roach and quickly spreading among major shipbuilders in the United States and Britain as they grew even larger and more complex; the New York Shipbuilding Company in Camden, New Jersey, for example, was specially built in 1899 in the model of an efficient manufacturer of ships.

The rise of vertical integration was not limited to shipyards in Britain and the United States. European yards underwent a similar evolution during the same period, though in many cases there were political as well as economic factors that contributed to this development. This era was marked by a burst of nation building across the Continent. The Austro-Hungarian Empire was formed in 1867, while the unification of Italy was completed in 1871, the same year that marked the union of Germany and the defeat of France in the Franco-Prussian War. The governments of those nations subsequently embarked on accelerated programs of industrialization as a means of recovering from the economic and social upheavals, and shipbuilding was at the center of these efforts.[28]

In most of those cases, vertically integrated iron and steel shipyards were built from the ground up, either upon the site of defunct wood shipyards or in greenfield sites. In Germany, the shipyard of Meyer Werft in Papenberg had been building wooden sailing ships since 1795, but by the late 1860s it was foundering as a business. The unification of Germany spurred a sudden flurry of investment capital, and Joseph Meyer took advantage of it by building iron foundries and boiler and engine works atop the remains of the old wood boatyard. At first building small, iron-hulled coastal steamboats, he instituted a strict division of labor among workers and brought in skilled engineers from other industries. Although vertical integration was built in from the start, Meyer only slowly expanded the yard's capabilities to large steel-hulled oceangoing vessels after the end of World War I. By contrast, in 1877 Hermann Blohm and Ernst Voss, both of whom had spent the previous three years working in Clyde shipyards, teamed up to erect a new integrated yard on a former cow pasture in Hamburg, which was dedicated right from the start to building large steel oceangoing vessels. Within 10 years, Blohm und Voss was ready to produce small steel warships, and by 1898 it was designing and building 12,000-tonne battleships.[29]

The creation of the Austro-Hungarian Empire also spurred the industrialization of the shipyards in its main port of Trieste, which already had a long history of technical innovation—the Panfilli shipyard built one of the first screw steamboats, *Civetta*, in 1829. Gaspare and Giuseppe Tonello, who came from Panfilli, tried to transition their San Marco shipyard from wood to iron, but shortly after Italian unification the yard closed. Meanwhile, the Stabilimento Tecnico Triestino led by Austrian industrialist

Georg Strudthoff, began as an engine builder and quickly grew. After unification it added dry docks and foundries to its San Rocco yard and built two of the navy's first four steel warships. By 1896 a consortium of investors led by the Rothschild family allowed Stabilimento Tecnico Triestino to buy the old San Marco yard and significantly scale up and integrate its machinery, construction, and repair facilities, becoming the sole provider of large capital ships for the Austro-Hungarian Empire. Meanwhile, the unification of Italy led to the vertical integration of Ansaldo, a company established as a railroad builder in Genoa in 1853, to include not just shipyards but also steel mills, foundries, and engine and electrical manufacture. By the early twentieth century Ansaldo was building transatlantic liners as well as capital ships for the Italian, Spanish, Japanese, and South American navies. In France, a major postwar construction subsidy in 1881 revived the dormant shipbuilding industry in Saint-Nazaire "like an electric current." With investments in foundries, cranes, rail systems, and new dry docks in shipyards like Penhoët, employment tripled and some of the world's most luxurious transatlantic liners began coming off the ways into the Loire River.[30]

By the beginning of the twentieth century, a 40-year transformation had changed the face of the shipbuilding industry from a collection of small craft shops to a few industrialized, vertically integrated plants that could produce large, complex vessels. But the drive for increased control by management did not stop with vertical integration, which was primarily focused on facilities and processes. The challenge to exert more management control over the workforce itself was met by a variety of methods and initiatives aimed at creating a predictable outcome to their labors. The most famous of these initiatives was the scientific management system made popular by the American engineer Frederick W. Taylor. Taylor had begun his pursuit of time-motion studies in the 1870s. His hallmark was to examine each step of the production process to determine the most efficient use of the worker, design, and placement of tooling and the flow of materials. From 1894 to 1895 Taylor was contracted by William Cramp and Sons to improve its steel production, but his final recommendations for new facilities were costly compared with rather dubious future savings and were largely ignored. One of Taylor's acolytes, a U.S. Navy officer named Holden A. Evans, instituted similar changes at the Mare Island Navy Yard from 1906 to 1908 but was unsuccessful at convincing senior navy leadership to adopt the program wholesale throughout other dockyards.[31] Instead, a system of professionalization of the workforce, as discussed in chapter 5, became the mechanism for ensuring that the engineers and workers in these vertically integrated shipyards would be able to design and build complex ships to the predictable standards demanded by an increasingly globalized shipping industry.

## Naval and Commercial Standards and Practices

By the turn of the twentieth century the methods for calculating most major elements of naval architecture—speed and power, structures and stability—had been developed and were in wide use. But the standards for acceptable criteria (e.g., structural strength, stability after damage) and the engineering practices used to make those calculations differed greatly between commercial and naval vessels and indeed between one navy and another.

The Danish-American naval architect William Hovgaard, who from 1887 to 1930 was active in merchant and naval ship construction on both sides of the Atlantic, observed the reasons for the differences between these standards and practices:

> The merchant shipbuilder is able to determine the general construction of the hull as soon as he has settled the type, size, and principal dimensions of the ship. These rules and tables, which are very detailed and complete, are framed by the so-called Classification Societies on the basis of the enormous mass of experience and information which these institutions have at their disposal. The object of these societies, Lloyd [Register], [Bureau] Veritas, British Corporation, Germanischer Lloyd, and others, is to establish and maintain certified standards of construction. When a ship is built according to their rules, the shipowners, the shippers, and the underwriters have a guarantee that the construction is satisfactory.... This does not prevent other naval architects from proposing new departures in ship construction, but, ordinarily, they must be approved by the Classification Societies before they can be adopted. In fact, this procedure has been followed more frequently during recent years than was formerly the case, after the Societies have adopted a more liberal and progressive policy.
>
> In warship design no rules or tables exist to guide the designer, except such as may from time to time be adopted by any navy for certain details. In general, the rapid and great changes that take place in warship design render it unprofitable or useless to frame rules or to compile tables so complete and detailed as those that are used in merchant shipbuilding. The experience which can be obtained with new features, even in the largest navies, is far more limited than that which is at the disposal of the Classification Societies, and warships [are] much more complex than merchant ships. For these reasons, the problem before the designer of warships is much more difficult than that which the designer of merchant vessels has to solve.... It was often difficult to discover why certain features were adopted and in some cases why they differed in different navies. The explanation is in general that once a certain mode of construction has been introduced in one of the leading navies and found satisfactory, it becomes a standard.[32]

The most noticeable differences between commercial and naval standards and practices, and indeed those among navies themselves, are seen first in comparisons of structural design practices and second in stability requirements. The structural design of commercial ships has diverged from and converged with naval ships throughout history. Greek and Roman triremes were built to be lightweight but with a heavily reinforced keel, a distinct divergence from heavier merchant amphora ships of the era. The

Middle Ages saw a convergence of merchants and warship design, since few nations had standing navies and most warfare was conducted from merchant ships carrying light guns. In the 1500s and 1600s the advent of heavy guns and gunports led to another divergence, since the newly developed ship of the line required reinforced decks and hulls to absorb the weight and recoil of the guns. In the eighteenth century, the rise of the East India trading companies led to a convergence of structural design philosophies, as many East Indiamen were fitted with heavy cannon and often built in the same shipyards by the same constructors as warships were. The emergence of iron and steel in the nineteenth and twentieth centuries led to a divergence again, as merchant naval architects and naval constructors approached structural design from two very different perspectives.[33]

Figures 4.7 and 4.8 show a very specific contrast between the perspectives of commercial and naval practice during the early twentieth century, in this case concerning the design of web frames. Web frames are structures transverse to the main longitudinal axis. In ships, they reinforce the hull to resist hydrostatic loads, wave loading, and contact with piers, and they also stiffen the side shell plating against buckling.[34] As noted many times in previous chapters, ship structural engineering evolved in parallel with civil engineering, because many of the same people—Isambard Kingdom Brunel and William Fairbairn, for example—worked on both simultaneously. The civil engineering heritage of ship structures is nowhere more visible than a side-by-side comparison in figure 4.7 of American Bureau of Shipping Steel Vessel Rules (1922) concerning web frames and the 1910 edition of General Specifications for Steel Railway Bridges concerning web stiffeners (which are comparable to web frames as they provide transverse reinforcement of the structure). A close inspection shows how remarkably similar they were. Both the American Bureau of Shipping Rules and the Railway Bridges Specifications present formulas for sizing the scantlings and determining the frame spacing. In both sets of standards, the formulas were derived empirically using rules of thumb based on gross dimensions (length, depth, and so on) without any direct foundation in the theoretical principles of mechanics or statics. They were straightforward, comparatively easy to use, and based upon experience derived from comparable structures.[35]

By contrast, in the standard text for U.S. naval constructors during the same period, Hovgaard's *Structural Design of Warships* (1915), the analysis of transverse web frames was carried out from first principles. As shown in figure 4.8, scantlings were determined from the combined external loads of weight, hydrostatic force, shearing force from adjacent frames, and vertical forces due to longitudinal bending, all of which are integrated to develop a combined bending moment. The method was complex, required time-consuming calculations, and was based upon the assumed loading conditions in service. Although such methods were typically made by constructors in naval dockyards, the more sophisticated commercial shipyards like Denny, Cramp, and Napier were fully capable of carrying out these complex calculations for their naval customers.[36]

The divergence of design practices also extended across the Atlantic Ocean, also among navies and in the commercial world. When during World War I the British naval constructor Stanley V. Goodall was seconded to the U.S. Navy's Bureau of Construction and Repair, he reported back that British battleships had closer framing and were more tightly subdivided against damage than their American counterparts. In the lead-up to World War II, American commercial shipbuilders bought the plans for a standard British cargo ship, *Empire Liberty*, upon which the U.S. Maritime Commission would base its enormous fleet of *Liberty* ships. When the plans arrived from Clyde shipyards to America, it was quickly discovered that British engineers left many structural details to shop practice, whereas American engineers specified connections, tolerances, fit, and so on, to a much greater degree for use on the shop floor. Whereas 80 working drawings for an engine would suffice in a British yard, the Americans ended up drawing 550 plans for the same thing.[37]

The continued pattern of divergence in design practices continued after World War II and into the Cold War. During the 1970s and 1980s, the U.S. Navy made a concerted effort to study the comparative naval architecture of Soviet Union and NATO ships to better understand how design standards affected the operational characteristics of warships and improve the understanding of trade-offs between technical attributes such as seakeeping and survivability. Toward the end of the Cold War, several NATO countries embarked on a project (later abandoned) to develop a common frigate, NFR 90, and as part of that effort the U.S. and British navies studied in detail the differing design standards between their warships. The study highlighted several areas where long-standing practice differed considerably between the two nations. For example, American warships have long been stiffened with a pair of longitudinal bulkheads aft, nicknamed Hovgaard bulkheads for their developer, while British practice did not include them. Another example, following on from the experience of each navy's welding investigation during World War II, was that, while both navies used high-strength steel in their hulls, U.S. warships were fitted with special HY-80 high-tensile alloy steel as crack arrestors at the sheer strake, while British naval architects considered the regular hull steel to be effective enough for the purpose.[38]

Even as the Cold War was winding down, the beginnings of another cycle of convergence in ship structural design standards was beginning to emerge. Classification societies began responding to the increasingly interconnected nature of the market, and in 1968 several joined together to form the nongovernmental International Association

## SECTION 9

## WEB FRAMES

(1) **Web Frames** fitted in association with side stringers and hold framing in accordance with Section 8, paragraph 4, are to be of the sizes given in Table **6**. The length ($l$) to be used with the Table is the distance in feet from top of double bottom, or from the intersection of face of web with top of bracket floors to the underside of deck beams plus ·002**L**+·2; forward of the midship one-third length in Vessels having sheer, the actual distance is to be taken. The value of **W** is obtained from the formula:

$$*W = s \times h \times l_1 \times \cdot 03.$$

Where **s** = spacing of webs, in feet.

**h** = the distance in feet from the load line or ·66**D** to the middle of $l$.

$l = l_1$, except in cases where the upper end of the length is above the load line; in such cases $l$ is the distance from the heel of web to the load line or ·66**D**, whichever is the greater.

When the decks are supported by strong beams fitted at the head of the web frames, the sizes of the latter are to be suitably increased.

* The value of **W** is not to be less than two-thirds of that which would be obtained with a draft corresponding to **D**.

(2) **Attachments** to beam and at heel are to be at least equal to those required by Table **8**.

(3) **Bracketed Web Frames** may be adopted in conjunction with a modified co-efficient of ·02 in the above formula, provided the brackets extend for a distance exceeding ·12$l$ below the beam and above the top of the connecting angle on level tank tops, or ·15$l$ above the junction of the moulded line of web with the bracket floor; the brackets are to be properly stiffened on the edge and to resist buckling; they are to have attachments not less efficient than required for brackets under Table **8**.

(4) **Frames** on webs are to be ·10″ thicker than the web plate, and are to be of the sizes required by Table **8** for the sizes of rivets in use; frames may be single riveted on webs without brackets, and on bracketed webs which are not more than 24 inches deep; bracketed webs having depths between 24 inches and 36 inches are to have zig-zag riveting; those of 36 inches depth and upwards are to have chain riveting.

(5) **Face Bars** are to be of the sizes given in Table **6** or of equally efficient sections; the area of the outstanding flange is not to be less than that of the Table section. The face bar is to overlap the deck beam and the floor bracket; in the case of bracketed webs it should be carried along the face of the brackets, which may be slightly hollow for the purpose; the gusset attachments to tank tops are to be based upon the requirements of Table **8**. The thickness of the face bar is to be increased ·08″ in the boiler space and the side bunkers in way of same.

**Figure 4.7**
Web frame design commercial practice, American Bureau of Shipping Rules 1922 (this page) and Railway Bridges Specifications 1910 (opposite page).
Credits: University of Michigan (this page); Ohio State University (opposite page).

**Web Stiffeners.** 79. There shall be web stiffeners, generally in pairs, over bearings, at points of concentrated loading, and at other points where the thickness of the web is less than 1/60 of the unsupported distance between flange angles. The distance between stiffeners shall not exceed that given by the following formula, with a maximum limit of six feet (and not greater than the clear depth of the web):

$$d = \frac{t}{40}(12{,}000 - s),$$

Where $d$ = clear distance, between stiffeners of flange angles.

$t$ = thickness of web.

$s$ = shear per sq. in.

The stiffeners at ends and at points of concentrated loads shall be proportioned by the formula of paragraph 16, the effective length being assumed as one-half the depth of girders. End stiffeners and those under concentrated loads shall be on fillers and have their outstanding legs as wide as the flange angles will allow and shall fit tightly against them. Intermediate stiffeners may be offset or on fillers, and their outstanding legs shall be not less than one-thirtieth of the depth of girder plus 2 in.

**Stays for Top Flanges.** 80. Through plate girders shall have their top flanges stayed at each end of every floor beam, or in case of solid floors, at distances not exceeding 12 ft., by knee braces or gusset plates.

TRUSSES.

**Camber.** 81. Truss spans shall be given a camber by so proportioning the length of the members that the stringers will be straight when the bridge is fully loaded.

**Rigid Members.** 82. Hip verticals and similar members, and the two end panels of the bottom chords of single track pin-connected trusses shall be rigid.

**Eye-Bars.** 83. The eye-bars composing a member shall be so arranged that adjacent bars shall not have their surfaces in contact; they shall be as nearly parallel to the axis of the truss as possible, the maximum inclination of any bar being limited to one inch in 16 ft.

**Pony Trusses.** 84. Pony trusses shall be riveted structures, with double webbed chords, and shall have all web members latticed or otherwise effectively stiffened.

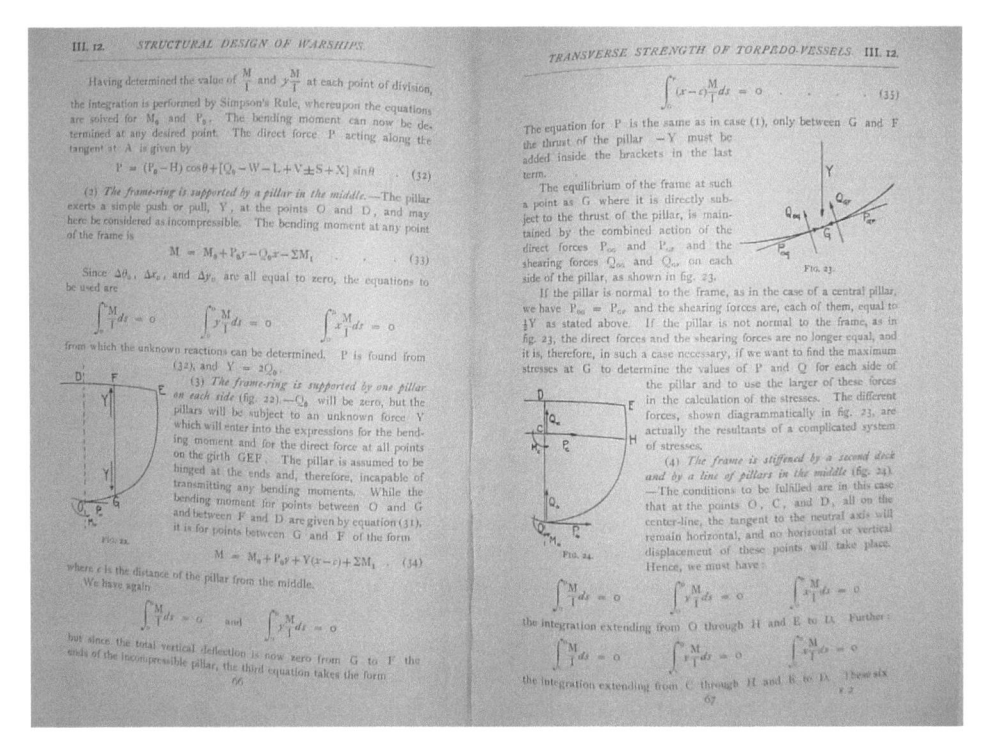

**Figure 4.8**
Web frame design naval practice, 1915.
Credit: Author's collection.

of Classification Societies to ensure a uniformly high level of safety across the industry. The association, which represents classification societies to the IMO, regularly publishes a common set of minimum standards ("Unified Requirements") in the IACS Green Book, IACS Blue Book, and Common Structural Rules. Also starting in the 1960s, the rise of computer-aided design techniques—which is explored in chapter 7—allowed classification societies to move away from empirically based rules and toward rules based on first principles.[39] By the turn of the twenty-first century, a convergence of commercial and military shipbuilders' design practices and many engineering standards had given rise to the increasing employment of classification societies to develop and enforce rules for naval vessels, which is further described in the epilogue.

As with structural design practices, the criteria and requirements for intact and damaged stability varied between commercial and naval practice and also among navies themselves. The evolution of these stability standards, however, has tended to be driven by responses to specific stability accidents and incidents. The capsize and sinking of HMS *Captain* in 1870, as noted, led to the widespread use of the $\overline{GZ}$ curve to measure the full range of stability over all angles of heel. In 1883 the steamship *Daphne* sank at launch on the River Clyde, with over 100 workers and family members lost, in part because its stability at very light load conditions had never been properly established. This led William Denny to develop and popularize "cross-curves of stability," a method of quickly establishing stability conditions at various drafts and angles of heel.[40]

During the 1880s, the loss of many grain-laden British vessels persuaded the Glasgow professor Philip Jenkins to investigate the effect of shifting cargos on stability. At the same time a new breed of ship had arrived on the scene, the oil tanker, which at first transported crude oil for the Nobel Brothers petroleum company from Azerbaijan to Europe and had suffered reduced stability due to sloshing of cargo. Jenkins characterized the physics of what is now called the free surface effect, in which the movement of liquids or free-flowing solids toward the low side of a rolling ship reduces the righting arm, thus decreasing stability. Meanwhile, in 1889 William H. White, while designing HMS *Royal Sovereign*, resolved the problems of a number of warships in the predreadnought era which had been completed far heavier than their design displacement, by ushering in a new policy of including in the weight estimate a contingency Board Margin (at the time, 500 tonnes) to account for growth in armaments, armor, and so on, during design and construction, a practice still in wide use today.[41]

The loss of *Titanic* in 1912 was of course the most notorious of these stability accidents and led to the 1914 SOLAS convention that elevated ship stability as an international concern. Yet even at this trailblazing conference, the great problem still facing the industry was clearly stated by John H. Biles, a member of the Safety of Construction committee and one of the preeminent naval architects at the time: "It is not practicable to determine the amount of stability which a vessel ought to have in order to be safe."[42] Nor was it always useful to rely on the observations of ship captains and masters, who might characterize their vessels as "crank" or "stiff" (respectively, subject to long slow rolls or short fast rolls) without quantifiable data to back it up. Despite the advances in the theory of static and dynamic stability and increasing use of these theoretical calculations in shipyard design offices, it was still not clear what criteria naval architects and shipbuilders should use to assess their designs.

In the years after the first SOLAS, a flurry of initiatives on both sides of the Atlantic aimed at establishing a minimum set of criteria for stability. In Britain, Charles Fordham Holt focused on metacentric height and proposed guidelines for both minimum and maximum values. The Italian naval architect Angelo Scribanti suggested that the criteria be based on minimum values for the dynamic work (energy) of righting the ship against wind, crowding of passengers to one side, and other conditions. Similar work was done in Germany by Ludwig Benjamin. In the United States, the American Marine Standards Committee, a short-lived government agency, commissioned the consulting naval architect William F. Gibbs to chair a four-year study of regulating the stability of new passenger vessels, which developed a wide-ranging series of criteria that covered metacentric height, righting arms for wind heel and passenger crowding to one side, flooding in case of two-compartment damage, and avoidance of synchronous roll (i.e., when the ship's roll period coincides with and is amplified by wave periods).[43]

These various criteria were placed on the table for the next SOLAS convention in 1929. As the committee members were preparing for the convention, the shipping world was stung by another major stability disaster. On 12 November 1928, the passenger liner *Vestris* foundered and sank in a moderate storm off Virginia, with the loss of 110 lives. The inquiry was even longer than the one for *Titanic* and eventually blamed overloading and downflooding through coaling ports as the reason. There was every reason to expect that the SOLAS conference now would endorse some kind of minimum stability requirements. But this did not happen; claiming that "it was not ready to establish standards of stability," the conference pointedly rejected the American Marine Standards Committee proposals, weakly lamenting that "it [was] not practicable to lay down stability regulations" despite ample evidence to the contrary.[44]

That did not stop the work by naval architects and constructors from continuing through the 1930s to pursue such criteria. In Italy, Ernesto Pierrottet continued Scribanti's work in developing criteria for righting energy, which he called reserve of stability, against combined wind, rudder action, and passenger movement. In the United States, the naval constructors John C. Niedermair and Edward L. Cochrane, who had participated in the 1929 SOLAS convention, developed multiple criteria for a range of stability issues, including wind, waves, and damage. They candidly noted the empirical nature of their work, stating, "These standards have been developed to a large extent from practical experience, they have not in some respects followed purely scientific lines." In fact, their criteria were often arbitrarily arrived at. For example, when they proposed that the criterion for adequate metacentric height $\overline{GM}$ be $0.06B$, it was to ensure that the roll of the ship was "comfortable while at sea" without defining what "comfortable" meant. Likewise, the maximum angle of heel under survivable damage was set to 30 degrees, with little explanation except that "excessive heeling is objectionable" because of boat handling and shifting of weight. Meanwhile, in Britain, a 22-year old student named Lennard Burrill was given a scholarship to study the stability of colliers (which had notoriously poor safety records) in the North Sea and other locations, using firsthand observational data. Burrill decided that $\overline{GM}$ alone was an inadequate measure of stability, proposing instead a "stiffness factor," or the percentage of displacement needed to heel the ship 15 degrees, also an arbitrary value that Burrill had placed as a "reasonable upper limit to the rolling angle" expected in ships.[45]

It took the work of another student to pull together all the threads of these stability studies and weave a coherent set of criteria for determining the minimum acceptable stability for ships in various conditions. Jaakko Juhani Rahola Jr. (figure 4.9) was the youngest of 11 children, born in Mänttä, Finland, on 1 June 1902, shortly after his father had died. Then a part of the Russian Empire, in 1917 Finland declared independence, and the young Rahola joined the new nation's navy as a constructor designing

**Figure 4.9**
Jaakko Rahola.
Credit: Armas Rahola and Jouni Arjava.

gunboats and small craft, by 1933 rising to head of the Construction Office at Navy Headquarters. In 1937 the shipbuilding professorship came open at the Helsinki University of Technology (today Aalto University), and to compete for the position against another candidate, Rahola needed to complete his doctorate. He was given 18 months to do so, during which time he investigated the spate of stability disasters that had lately befallen Finnish shipping, including the capsize in heavy winds of SS *Kuru* in Lake Näsijärvi in 1929, taking 138 lives, among which were members of his family. He published his doctoral thesis "The Judging of the Stability of Ships" in 1939, but by the time he could apply for the professorship, World War II had begun and Helsinki was being hit by Russian bombs. Rahola finally obtained his post in 1941 but continued designing ships for the Finnish navy even while teaching, becoming a full-time academic only after the war ended in 1945.[46]

Rahola's thesis, though subtitled "Especially Considering the Vessels Navigating Finnish Waters," was written in English and clearly aimed at a global audience. Rahola began with a comprehensive survey of stability theory since Bouguer's day, with commentary on the major efforts. He carefully defined the potential foundations for judging a vessel's stability, notably initial metacentric height, righting arm curve, and dynamic lever curve. After concluding that the dynamic lever curve (which describes the work, or energy, needed to right the ship) was "a particularly suitable foundation for judging the stability," he laid out three methods by which to judge stability: (1) comparison with vessels that have sailed successfully for many years, (2) comparison with vessels that have capsized because of poor stability, and (3) direct calculation of forces and moments. Rahola determined that method 2 was most favorable for establishing criteria, given

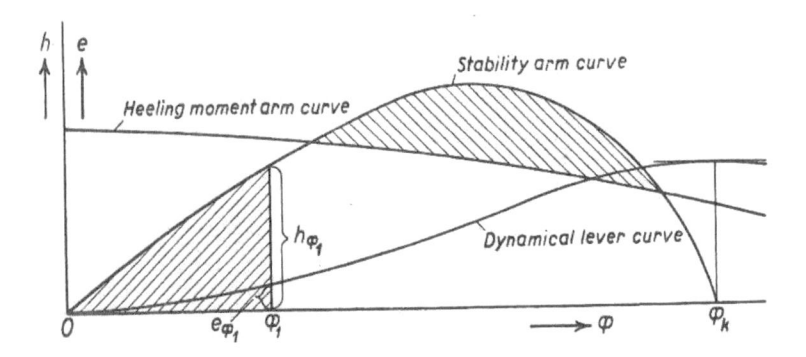

**Figure 4.10**
Rahola's dynamic lever curve, 1939 (this page), and IMO Resolution A.749(18) beam wind and rolling criteria, 1993 (opposite page).
Credits: Author's collection (this page); IMO (opposite page).

# The Demand for Standardization

that method 1 required extensive knowledge of many ship properties and method 3 could not take into account all factors such as wave motion. He then subdivided the problem of stability criteria into those for seagoing vessels and those on inner waters such as lakes. Finally, Rahola synthesized the various criteria proposed over the years, as well as stability data from over 30 capsizings and sinkings, to arrive at a simple but comprehensive "new minimum rule for the dynamical stability" of vessels, which he claimed was "more favorable" than any previous rule.[47]

The Rahola criteria, as they soon became known worldwide, were based on comparison of the righting arm curve (which he called the stability arm curve) and the heeling moment arm curve to develop the dynamic stability curve (figure 4.10, left figure). The criteria were the following:

- Righting arm $\overline{GZ}$ at least 0.14 meters at a heeling angle $\varphi$ of 20 degrees
- Righting arm $\overline{GZ}$ at least 0.20 meters at a heeling angle $\varphi$ of 30 degrees
- Critical heeling angle (at which ship capsizes) greater than 35 degrees

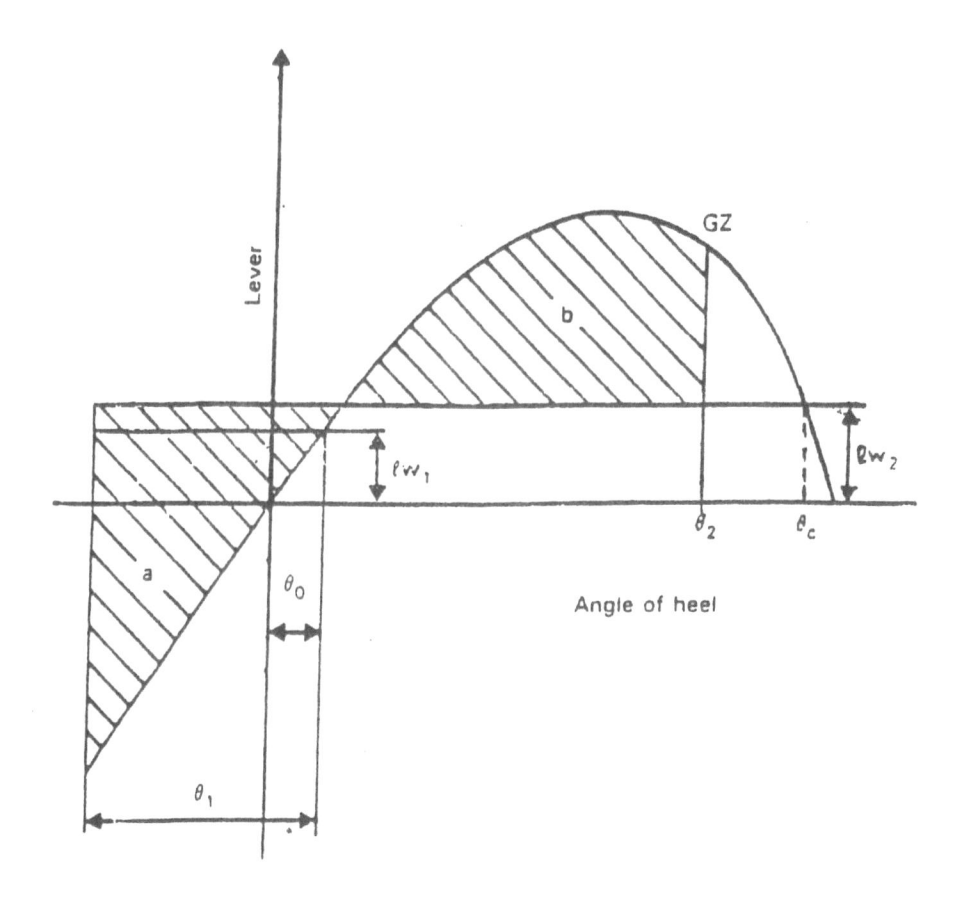

- The shaded area (energy) under the righting arm curve (left side) should be equal to the shaded area (energy) between the righting and heeling arm curves (right side)—that is, hull has enough energy to right itself
- Other criteria cover factors such as passenger crowding, rudder-induced heel, and flooding

Although Rahola's analysis was based as much on his personal judgment as on the data, the Rahola criteria have had a profound impact in the shipping world that is felt even in the present day. It became the basis for the stability criteria of many nations after World War II, starting with Russia (USSR) in 1948, then Poland, West Germany (German Democratic Republic), and Japan in the 1950s. In the early 1960s, the Rahola criteria were adopted by a slew of Western nations, including France, Britain, and the United States, adopted by IMCO in 1968, and finally made part of SOLAS in 1974. At the same time, further research resulted in modifications to the Rahola criteria, although they still remain substantially the same; figure 4.10, right, shows the current IMO standard adopted in 1993, IMO Resolution A.749(18), with a modified wind heeling arm and rollback (or rolling to the opposite side of the initial heel) taken into account for the area under the righting arm curve.[48]

The experiences of modern navies in the two world wars greatly affected their development of stability criteria, which was especially influenced by the standards of the German battle fleet. Before the outbreak of war in 1914, the naval buildup under Admiral Alfred von Tirpitz emphasized the importance of damaged stability and damage control. Tirpitz firmly stated that "the supreme quality of a ship is that she should remain afloat, and, by preserving a vertical position, continue to put up a fight," and he leaned heavily on his naval constructor Hans Bürkner to translate this philosophy into steam and hardened steel. Bürkner's designs, which included the *Bayern* class of dreadnoughts and the *Derfflinger* class of battlecruisers, were marked by tight subdivision to control flooding, heavy plating, and extensive armor belts and side protection systems.[49] Although Germany lost World War I, it lost only one modern capital ship to enemy action, which happened during the Battle of Jutland in 1916. By contrast, the British navy lost three capital ships and three armored cruisers in that one battle alone, leading Vice Admiral David Beatty to utter his damning statement, "There seems to be something wrong with our bloody ships today." At the end of the war, Germany ceded its capital ships to Britain, the United States, and France as part of its war reparations.

The Greek playwright Aristophanes famously noted that "men of sense often learn from their enemies ... it is from their foes, not their friends, that states learn the lesson of building high walls and ships of war."[50] The Allied constructors followed Aristophanes's dictum, intensively studying their war prizes and examining drawings and regulations to understand the philosophy behind Bürkner's designs. The British constructor Stanley V. Goodall made special note of the extensive subdivision of German battleships but also noted areas for improvement. The Americans were especially impressed with the damaged stability and damage control arrangements, so much so that the Bureau of Construction and Repair issued new instructions in 1931 and 1938 specifically stating that "the Germans were right and rational in treating the control of damage as they did.... The experience and results attained during the World War are a lesson we cannot afford to ignore." Translated copies of German damage control regulations were also issued to all capital ships in the U.S. fleet.[51]

Allied navies had learned some lessons during the interwar period, but World War II provided additional ones that had to be learned in a hurry. By this time, Stanley Goodall was the Director of Naval Construction in Britain, overseeing a harried group of constructors who worked seven days a week. In 1941 it was apparent that the destroyers, which had been fitted with additional equipment topside, were becoming dangerously top heavy, so Goodall approved a rough-and-ready stability criterion of a minimum 0.375-meter metacentric height in the deep condition as a way to ensure an acceptable level of safety while sparing the already-overworked constructors from complex calculations using slide rules and mechanical calculators. In 1942 Goodall's department produced a booklet, *Stability of Ships*, that established a complete set of rules for assessing ship stability in intact and damaged conditions. Even with these measures, warships often operated perilously close to the edge of safety; on one North Atlantic convoy run in 1943, HMS *Duncan*'s rolling so alarmed its commanding officer that he personally helped chip paint to get rid of topside weight. In the United States, the situation was equally precarious—Goodall's counterpart John Niedermair of the Bureau of Ships (BuShips), equally overworked, assigned a young naval architect, Theodore "Ted" Sarchin, to oversee the tasks of developing stability policy for new ships and ships in the fleet (it helped that BuShips had a copy of the British stability booklet) while also assessing war damage reports. The worst stability disaster occurred in December 1944, when American Task Force 38 off the Philippines ran into a massive typhoon that capsized three destroyers with the loss of 800 men and severely damaged another nine ships. Although too late in the war to have further affected stability policy for wartime operations, it signaled the danger of combined rolling and beam winds in heavy seas.[52]

The postwar period gave naval architects both the time and the data to formulate longer-term policies for ship stability. In BuShips, this fell under the purview of Ted Sarchin in the Ship Design Division, who established overall guidelines, and Lawrence Goldberg in the Hull Division, who implemented them on individual ships and classes. For many years, stability policy was tackled piecemeal. For example, the December 1944 typhoon resulted in a spate of studies on the stability of destroyers and other vessels in those conditions. In February 1946 BuShips issued a memorandum (figure 4.11) establishing the policy that "the combined effect of beam winds and rolling, rather than damaged conditions, will ordinarily govern the required stability of smaller vessels" and that determined minimum wind velocities for design purposes (e.g., 100 knots for new oceangoing vessels, 60 knots for harbor craft). In 1947, BuShips studied the torpedo damage to 24 ships, combat ships designed for battle as well as auxiliary ships that were not, which was supplemented by testing of scale-model ships simulating various degrees of damage. The study determined that warships would typically sink after a breach greater than 15 percent of the hull length, while auxiliaries could survive up to a 12.5 percent length of hull opening. Thus, the policy was established for bulkhead separation so that naval combatants could withstand a 15 percent hull breach, while auxiliaries could survive a 12.5 percent breach. In 1957 Captain Charles J. Palmer, head of the Hull Division, assigned Sarchin and Goldberg to consolidate these criteria in a typewritten memo titled "Ship Design Division Stability Criteria," but this was made available to only a few individuals and proved to be of only limited utility.[53]

In 1961, this somewhat haphazard state of affairs impelled the director of Ship Design, Captain John H. McQuilkin, to ask Sarchin and Goldberg to collect all the disparate guidelines—including the 1946 wind heel memorandum and the 1947 damaged length study—and create a single coherent policy document that would be given wide distribution in a public forum. The resulting paper, "Stability and Buoyancy Criteria for U. S. Naval Surface Ships," was read before the Society of Naval Architects and Marine Engineers (SNAME) in 1962. It gave precise details of the criteria for intact stability (beam winds and rolling, lifting heavy weights over the side, turns, and so on) and damaged stability, including length of underwater damage, as well as some indications of how the calculations could be accomplished by the still-novel electronic computer. The paper drew the immediate attention of the wider naval community as the first comprehensive expression of the governing stability standards for warship design. The phrase "Sarchin and Goldberg" became shorthand for naval stability criteria, in much the same way as "Rahola criteria" was instantly recognized by the commercial shipbuilding community. The Sarchin and Goldberg criteria became the basis for the U.S. Navy's official stability policy (Design Data Sheet 079) in 1975 and was adopted by many other navies as well—such as the German Bureau Veritas 1030 in 1965, UK Naval Engineering Standard 109 in 1980, and French Instruction Technique 6014. Even so, the wartime and peacetime experience of each navy shaped the way it developed these standards. For example, the U.S. and French navies established a margin line eight centimeters below the main deck as the limit to flooding, while the British navy had no

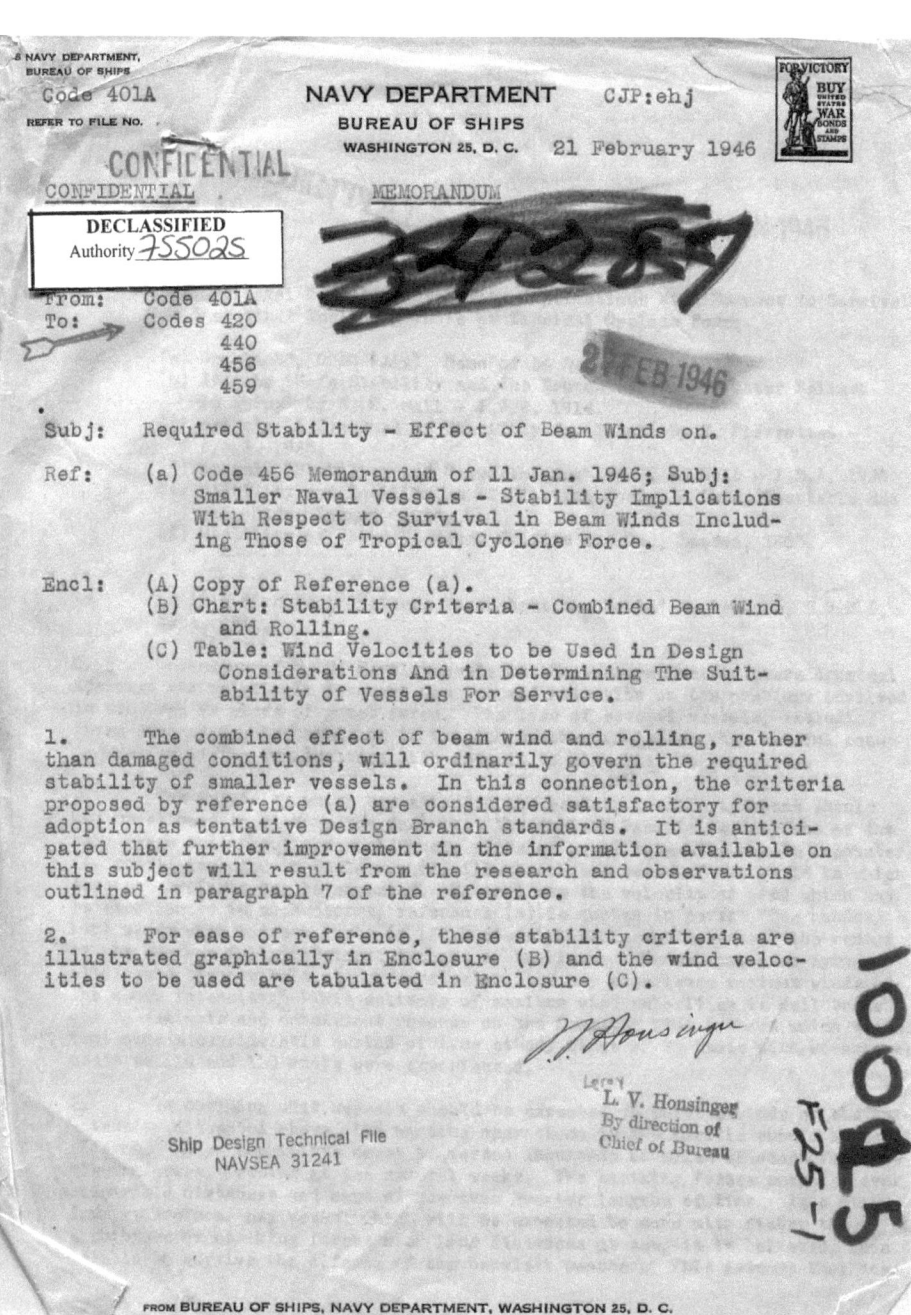

```
NAVY DEPARTMENT,
BUREAU OF SHIPS
Code 401A                NAVY DEPARTMENT         CJP:ehj
REFER TO FILE NO.         BUREAU OF SHIPS
                         WASHINGTON 25, D. C.    21 February 1946

CONFIDENTIAL
CONFIDENTIAL             MEMORANDUM

DECLASSIFIED
Authority 755025

From:   Code 401A
To:     Codes 420
              440
              456
              459

Subj:   Required Stability - Effect of Beam Winds on.

Ref:    (a) Code 456 Memorandum of 11 Jan. 1946; Subj:
            Smaller Naval Vessels - Stability Implications
            With Respect to Survival in Beam Winds Includ-
            ing Those of Tropical Cyclone Force.

Encl:   (A) Copy of Reference (a).
        (B) Chart: Stability Criteria - Combined Beam Wind
            and Rolling.
        (C) Table: Wind Velocities to be Used in Design
            Considerations And in Determining The Suit-
            ability of Vessels For Service.

1.      The combined effect of beam wind and rolling, rather
than damaged conditions, will ordinarily govern the required
stability of smaller vessels.  In this connection, the criteria
proposed by reference (a) are considered satisfactory for
adoption as tentative Design Branch standards.  It is antici-
pated that further improvement in the information available on
this subject will result from the research and observations
outlined in paragraph 7 of the reference.

2.      For ease of reference, these stability criteria are
illustrated graphically in Enclosure (B) and the wind veloc-
ities to be used are tabulated in Enclosure (C).

                                        L. V. Honsinger
                                        By direction of
                                        Chief of Bureau
Ship Design Technical File
    NAVSEA 31241

FROM BUREAU OF SHIPS, NAVY DEPARTMENT, WASHINGTON 25, D. C.
```

Figure 4.11
Bureau of Construction and Repair 1946 memorandum on the effect of beam winds on required stability.
Credit: Author's collection.

such limit, and each navy had differing criteria for minimum intact $\overline{GM}$. In practice, these stability criteria have tangible effects on ship design, affecting size and placement of engines, hull equipment, and more, which also explains some of the differences in shipbuilding standards and practices among navies.[54]

**Standardization in Context**

The demand for standardization in shipbuilding and ship design largely followed the pattern seen in other industries. For example, the development of structural standards in shipbuilding mirrored those in the civil and mechanical engineering disciplines, where standards-setting bodies had been established throughout the twentieth century aimed at achieving either minimum or acceptable levels of quality and safety. In nations like Britain and France, those bodies were generally governmental or government chartered, such as the British Standards Institution and the German Institute for Standardization. In the United States, most standards-setting bodies were private not-for-profit institutions, such as the American Society of Mechanical Engineers (which famously produced the Boiler and Pressure Vessel Code in 1914 after a series of terrible boiler explosions) and the American National Standards Institute. After World War II, international standards-setting bodies arose, such as the International Organization for Standardization (ISO), to serve the increasingly interconnected global markets for goods and services.

The demand for acceptable stability standards also had its parallel in the development of objective flying quality specifications for aircraft, from soon after the Wright brothers made their first flights through World War II. Aeronautical engineers (many of whom were former naval architects) attempted to work closely with early pilots to establish the engineering criteria for good flying and handling characteristics but could not always articulate what that meant. Just as ships' captains freely used the terms "crank" and "stiff," pilots might call the controls "heavy" or "light" without quantifying what that meant. An extended period of research was carried out by government and commercial entities, subdividing flying qualities into longitudinal and lateral stability and control, stall, maneuverability, and so on, in much the same way as naval architects individually addressed stability characteristics of wind-induced heel, rolling, passenger crowding, and more. The slow accumulation of flight data, like the process employed by Rahola, Sarchin, Goldberg, and others, allowed aeronautical engineers by the end of World War II to develop standardized handling requirements for military and commercial aircraft. But as with ship stability standards, even today there remains considerable variance in aircraft handling standards among nations and companies, such as with the Airbus "hard envelope" fly-by-wire system versus the Boeing "soft envelope" approach.[55]

# The Demand for Standardization

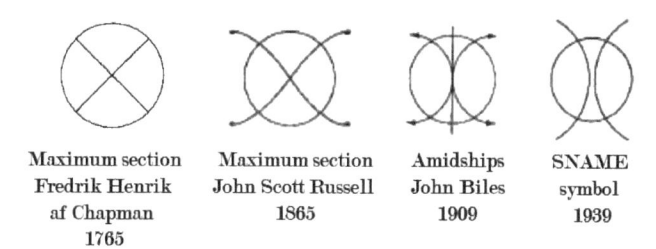

| Maximum section Fredrik Henrik af Chapman 1765 | Maximum section John Scott Russell 1865 | Amidships John Biles 1909 | SNAME symbol 1939 |

**Figure 4.12**
Evolution of the amidships symbol.
Credit: Author, after Matthew Forrest (SNAME).

On a personal level, I have often found variations in ship design standards and practices among navies and from commercial to military ships that, while minor, can still be exasperating. As a naval architecture student, I learned that the defining characteristic of a ship, length between perpendiculars, varies in meaning between commercial and military vessels. The forward perpendicular for both is the intersection of the bow with the waterline; but the after perpendicular on most commercial ships is drawn through the rudder post, whereas on naval vessels it is the intersection of the waterline with the stern or transom. Twenty years later, when I worked on exchange from the U.S. Navy with the French navy's design bureau, Direction des Constructions Navales, I was surprised to see on the ship-weight estimates the notation "air de la carène," or air in the hull. My coworkers explained that French practice, which dates from the early twentieth century, was to use the absolute density of seawater compared with a vacuum (1,026 kilograms per cubic meter), while Americans and British used the relative density compared with the atmosphere (1,025 kilograms per cubic meter). The difference, which was not found in U.S. or British weight estimates, is the "air in the hull."[56]

Even something as basic as the symbology on ship design drawings and plans can vary from nation to nation. Americans and British today recognize the two right-hand symbols in figure 4.12 as marking amidships, or center between the perpendiculars. This was not always the case; it began with the Swedish constructor Frederick Henrik af Chapman in the eighteenth century as the section of maximum area, but only in the twentieth century did it attain its present meaning (and become the symbol of SNAME). Yet there is no privileged symbol for amidships in many other nations; the French navy simply labels it PPM, or *perpendiculaire milieu*.[57] Culture more than technology, it would appear, is the most resistant to change.

# 5   The Need for Professionalization

> I must study politics and war, that my sons may have the liberty to study mathematics and philosophy, geography, natural history, and naval architecture, navigation, commerce, and agriculture, in order to give their children a right to study painting, poetry, music, architecture, statuary, tapestry and porcelain.
>
> —John Adams to Abigail Adams, Paris, 12 May 1780

John Adams was in France, America's formal ally during the War of American Independence, as the Congress's minister plenipotentiary appointed to negotiate a peace with Great Britain. But with the war going so badly for the Americans at that point, there was no peace to be had, so Adams instead filled his time wandering the sumptuous gardens of Paris and reflecting on the future of his uncertain country. He favored a nation with a strong navy, backed by an economy rooted in manufactures, industry, and agriculture and underpinned with policies that supported science and technology to further these disciplines. Adams clearly saw naval architecture not as a future luxury like poetry and civil architecture but rather as one of the critical disciplines needed by his new nation for its economic prosperity, which in turn would foster a distinctly American culture.[1]

Neither Adams nor his contemporaries used the word "professional" to describe the men who used naval architecture. At that time, only lawyers, doctors, clergy, and military officers fit the social model of professionals—that is, a body of men having both specialized knowledge and the autonomy to exercise that knowledge for the benefit of others. The Industrial Revolution changed all that. By the middle of the nineteenth century, several innovations had come together to give rise to new types of hierarchical, science-driven industries. Reliable power in the form of steam, reliable transportation in the form of railroads, and reliable communications by telegraph all contributed to the expansion of large, capital-intensive businesses and trade networks. As chapter 4 demonstrates, this period saw the Industrial Revolution merging with a managerial revolution that resulted in standards-setting bodies and vertically integrated enterprises and with a scientific revolution that demanded accurate prediction of end products and processes. These in turn required business-savvy managers, and in particular scientifically trained engineers, to administer the scale and complexity of these new enterprises and to meet increasingly stringent demands of the customer.

Although the term "naval architecture," as John Adams used it, had existed as far back as 1610, the title of naval architect as a profession was the product of the industrial age. During the age of sail, the most commonly used terms were "shipwright" in commercial shipbuilding and "constructor" or "surveyor" for naval designers. The advent of steam and iron in the nineteenth century changed the image of the ship designer from a craft tradesman to a professional. This evolution can be seen in how a New York ship designer, John Willis Griffiths, titled himself in the business directories and census forms of the time. In 1845, when he first conceived his clipper ship form, he listed himself as a "ship carpenter." In the censuses of 1850 and 1860, he was an "architect." Only in the 1870 census did he assume the title "naval architect." Even in Britain, where the Institution of Naval Architects had been formed in 1860, it still took time for the title to catch on. The heads of design in the British Admiralty (Isaac Watts, Edward Reed) had been called the chief constructor up until 1870; in that year Reed's successor Nathaniel Barnaby was referred to for the first time as the chief naval architect. Even at Britain's Clyde shipyards, the term "naval architect" did not achieve widespread use until the 1880s.[2]

As noted in chapter 4, the vertical integration of manufacturing industries during the nineteenth century gave rise to the professionalization of railroad, civil, and chemical engineering disciplines, a pattern that was followed by that of naval architecture. The first step in this professionalization was the organization of the workforce in specialized occupations, such as happened when industrial firms began dividing front-office functions into accounting, engineering, and so on. The next step was the establishment of formal training and education to provide engineers the opportunity to acquire the scientific and practical knowledge needed to carry out their duties and the establishment of criteria for qualifying the individuals as engineers in the profession. The third and final step was the establishment of formal professional associations to codify the requirements of the profession, develop codes of ethics and conduct, and provide venues to share information. During these stages, professions often engaged in long, drawn-out struggles to obtain support and recognition at both the legal and the public levels and sometimes fought these battles multiple times over the course of years and decades.[3]

In fact, this very same process of professionalization had been carried out a century earlier by the naval constructors of France, Spain, Britain, and elsewhere, which coincided with the demand for standardization of their naval ships and equipment. Since they had a 100-year head start in professionalization over their commercial brethren, this chapter begins by examining how naval constructors evolved as professional bodies from the age of sail through the industrial age.

## Naval Constructors Corps

It was not by coincidence that John Adams was in Paris in 1780 and was thinking of naval architecture amid the turmoil of the War of American Independence. The United States had gone to war against Britain, which had the mightiest fleet in the world, without a navy of its own to protect its coast and interdict supplies. The fledgling nation knew it had to depend upon France and Spain, which had pledged to fight alongside the Americans for their own political agendas—France allied with the Americans and Spain allied with France—to win the war at sea and gain independence. But the only hope for France and Spain to prevail was for their combined navies to defeat the British at sea. They had been preparing for this eventuality for 15 years by combining the shipbuilding skills of the two nations. By 1780, the French and Spanish navies had developed highly professional corps of naval constructors who were well trained in naval architecture and could apply the latest science to build fast, sturdy, maneuverable ships.[4]

France and Spain had been among the first to create a national corps of naval constructors, which provided the template for a professional body of engineers—usually part of the naval officer's corps or carrying equivalent military ranks—who were trained in science and mathematics to design and build warships. As I describe in my previous work *Ships and Science*, the impetus for this was the poor showing by their navies in the Seven Years' War, which ended in 1763 with humiliating losses by both nations to Britain. The Corps du Génie Maritime (Naval Engineering Corps) was established in 1765 with a centralized control structure from Paris to the shipyards, a firm military hierarchy, and a specialized school housed in the Louvre palace, next to the French Academy of Sciences, which supported and published research into naval architecture. France subsequently sent one of its own constructors, Jean-François Gautier, to Spain, where he established the Cuerpo de Ingenieros de Marina in 1770 along the same lines (and with the same name) as the French corps and with a specialized engineering school in El Ferrol.[5]

These efforts paid off handsomely, for during many battles and naval engagements the French and Spanish fleets came out victorious, thanks in part to the designs of their highly effective and interconnected naval construction corps. These victories forced the British to the peace table and the acknowledgment of the independence of the United States in 1783. In 1786, the French navy began a program of standardized designs under

Jacques-Noël Sané and Jean-Charles de Borda that would last through the French Revolution and the Napoleonic Wars. Even after the decimation of the French naval officer corps during the Reign of Terror and the overwhelming British victory at the Battle of Trafalgar in 1805, the French navy remained a considerable fighting force through the end of the war. By contrast, Trafalgar took the wind out of the Spanish navy's sails, and the subsequent French occupation of Spain from 1808 until 1813, followed by the loss of Spain's overseas empire—and thus the raison d'être of its navy—resulted in the abolition of the cuerpo in 1828.

In France, the years following the Napoleonic Wars saw a rebirth of the Corps du Génie Maritime not as an insular bastion within the navy but rather as part of France's emerging technocratic elite. During the old regime, state engineers had been partitioned into separate corps for navy, artillery, roads, and so on. The Revolution had led in 1794 to the creation of a common educational stream for all state engineers, starting with the prestigious École Polytechnique (Polytechnic School) with its emphasis on mathematics and pure sciences and then passage to an *école d'application* (application school) for specialized training. Graduates from the École d'application du Génie Maritime went on to serve in technical positions of increasing responsibility, from dockyards to factories to the central design bureau, with some moving on to become professors and instructors. Others were seconded to up-and-coming navies to assist in their development, most notably Louis-Émile Bertin (figure 5.1), who from 1886 to 1890 was sent to Tokyo. There he designed the entire Japanese fleet (which won the critical Battle of Yalu against China in 1894), before returning to France and becoming chief constructor in 1895, a position he held for 10 years as France seriously challenged British naval supremacy.[6]

These highly educated engineer savants, who took the lead over the British navy in introducing new technologies such as ironclads and steel hulls, were seen as more than mere technicians in their respective fields; they were true national leaders. Charles Dupin and Henri Dupuy de Lôme, for example, who both considerably advanced the state of naval architecture, each became *hommes politiques* in the French government. The iron triangle of the Corps du Génie Maritime, École Polytechnique, and its application school remained in roughly the same form, and with the same level of professional prestige, until the 1960s and 1970s, when the Corps du Génie Maritime was subsumed under the Corps de l'armement (France's centralized armaments acquisition body). Engineers and constructors were classed either as ingénieurs d'armament or ingénieurs des études et techniques d'armament, each with a corresponding military rank, though only the ingénieurs d'armament could occupy senior-level positions.[7]

By contrast with France's steady policy, Spanish naval constructors went through a series of vicissitudes throughout the nineteenth and twentieth centuries, as the nation

**Figure 5.1**
Louis-Émile Bertin.
Credit: Bibliothèque Nationale de France/Gallica.

itself experienced waves of political upheavals. The extinction of the Cuerpo de Ingenieros de Marina in 1828 gave way to a combined corps of constructors and civil engineers. In the wake of reforms aimed at modernizing Spain, in 1848 the civil engineer's corps was disestablished and in 1851 the constructors followed, which were then replaced in 1852 by a new combined Cuerpo de Ingenieros de la Armada (Corps of Navy Engineers) that incorporated machinery and naval engineers. Its academy was in El Ferrol from 1860 until 1932, with several closures between. The cuerpo itself was reorganized several times, but its numbers were never significant in the nineteenth century; from 1834 to 1885, only 101 naval engineers entered Spanish service, and from 1885 until 1909, no new naval engineers entered at all.

Even so, Spain was entering the industrial age in fits and starts. At first it had little confidence in its own engineers, buying many of its steam ironclads from abroad, like *Numancia* from France in 1863 and *Vitoria* from Britain in 1865. It also largely rejected homegrown attempts to develop submarines, like Narcís Monturiol's *Ictineo I* and *II* submersibles in the 1860s and Isaac Peral's 1887 battery-powered boat *Peral*.[8] The devastating losses during the Spanish-American War of 1898 demonstrated how ineffective these approaches were. In 1900 a political group, the Liga Marítima Española (Spanish Maritime League), began lobbying the government for fundamental reforms, and in 1908 a British-led shipbuilding consortium was created to modernize the Spanish fleet. The following year (1909) the Spanish navy reinvigorated its various technical cuerpos (naval, electrical, machinery) and in 1914 it reestablished the engineering school at El Ferrol. This lasted until the Second Republic came to power in 1931 and disestablished the cuerpos, with civilian engineers instead taking the role of technical leadership. A new series of cuerpos were established in 1948 along military lines and a new school established in Madrid. In 1967 all the technical corps were combined into a single Cuerpo de Ingenieros de la Armada with its own distinctive military rank and insignia, thus returning to its nineteenth-century roots.[9]

The French Génie Maritime also greatly influenced other navies around Europe. During the Napoleonic Wars when the Netherlands was first a vassal state of France and then part of the French empire, Dutch constructors from the two main shipbuilding centers—Rotterdam and Amsterdam—were trained in the French engineering schools and learned practical shipbuilding directly from Jacques-Noël Sané. Even with the establishment of the Kingdom of the Netherlands in 1815, naval ship design and construction remained divided between the two ports. It was not until 1843 that a unified Korps Ingenieurs der Marine (Naval Engineers Corps) was created to manage the transition of the Dutch navy to iron and steam, with most of the design and shipbuilding consolidated at the Amsterdam *rijkswerf*. However, the naval engineers did not have the same level of prestige as the state hydraulic engineers (*ingenieurs van den waterstaat*) in charge of keeping the Netherlands dry and who received their training at the prestigious Polytechnic School of Delft (today's Delft University of Technology). It was only in 1868 that Delft established a curriculum for naval architects, with Bruno Tideman at its helm. He resigned his professorship in 1873 to become the chief naval constructor. Several years later he embarked on the ship model experimentation and collaboration with Robert Froude (discussed in chapter 3) that led to the development of nondimensional notation for scaling results from model to full size. The korps, with some changes in name, remained small but effective through the twentieth century.[10]

Jacques Balthazar Le Brun de Sainte Catherine was a member of France's Génie Maritime, and a disciple of Sané, when he was dispatched to assist the Ottoman navy in 1792. In 1799 he was enticed to go to Saint Petersburg, where Russia was building a modern navy with the help of many foreign officers. There, he rose quickly through the ranks to become the director of shipbuilding and was given a seat on the Admiralty Board. In 1826 he was made superintendent of the newly formed Korpus Korabel'nyh Inzhenerov (Corps of Ships Engineers), which had ranks equivalent to other naval officers. The Russian defeat in the Crimean War (1853–1856) led to a wholesale reorganization of the navy, with the role of the korpus largely replaced by technical committees principally staffed by naval officers who spent time at sea as well as at the drawing board. Some of these officers produced truly innovative designs, such the circular ironclads developed in 1870 by Andrei Alexandrovich Popov and models of which were tested by Froude and Tideman. Other officers were highly regarded as mathematicians and theoreticians, most notably Aleksey Nikolaevich Krylov. But though these officers were well trained and innovative, the lack of an effective bureau structure meant that warship designs were often delayed or badly executed. In 1886 the korpus was reorganized to have five special ranks for naval engineers, but after the Russian Revolution of 1905, the government returned the korpus to traditional naval ranks. The Russian Revolution of 1917 sent many naval engineers fleeing the country, most famously Vladimir Yourkevitch, who emigrated to France and designed the high-speed hull of the passenger liner *Normandie*. A generation later, Joseph Stalin's Great Purge of 1936–1938 further decimated the ranks of qualified leaders. It would not be until after World War II that the Soviet navy (and later, the Russian navy) would rebuild the ranks of its naval constructors.[11]

The creation of the Italian Corpo del Genio Navale (Corps of Naval Engineers) in April 1861 was the product of the unification of Italy, which had occurred just weeks earlier. Prior to unification, only the Kingdom of Sardinia (the most powerful of the Italian states) had a professional naval constructor corps; one of the first tasks of the corpo was to consolidate and stabilize power between the shipbuilding centers in Genoa, Naples, and Ancona and later in Venice, each of which had its own traditions. In 1871, Benedetto Brin became the director of naval construction. Educated at the French École d'application du Génie Maritime, Brin presided over the designs that would bring the Italian navy close to the French navy in size and strength. The *Caio Duilio* class of ships, in particular, was highly innovative and represented the apogee of the popular Jeune École of strategic thought: a big-gun battleship that also carried its own high-speed torpedo boat in a stern compartment. But Brin's tenure as director did not last long;

in 1876 he was appointed as minister of the navy, a post he held intermittently until 1891.

Meanwhile, the Corpo del Genio Navale underwent several changes in status, consolidating with and then separating from other naval engineering corps until the early twentieth century, when it was stabilized under the reign of Mussolini as an independent body of naval architects and marine engineers. During that time, other innovative constructors gained notoriety on the world stage. The most famous of these was Vittorio Cuniberti, the developer of a revolutionary battleship concept that would forever change naval warfare. Until Cuniberti, most capital ships were armed with several calibers of gun, to be effective at different ranges. Cuniberti recognized that this mix of gun calibers not only presented logistical problems (e.g., different ammunition) but also made it impossible to accurately target the enemy, since range was determined by spotting the splashes of shells, and the rangefinders could not distinguish between splashes of different calibers.

In 1900 Cuniberti proposed a high-speed, all-big-gun battleship for the Italian navy's 1902 program, but it was considered too expensive to build. He nevertheless published his concepts in several naval journals, including *Rivista Marittima*, but it was his article in the 1903 edition of the widely read *All the World's Fighting Ships*, provocatively titled "An Ideal Battleship for the British Navy," that caught the attention of John Arbuthnot "Jacky" Fisher, First Sea Lord and in overall military command of the British navy. Fisher embraced Cuniberti's concept, and in just two years' time—a breakneck speed in that era—he pushed the design and construction of HMS *Dreadnought* to completion. *Dreadnought* immediately gave its name to a new breed of steam-turbine-powered, all-big-gun ship that instantly made every other warship in the world obsolete.[12]

Great Britain was not the only nation to put Cuniberti's vision of the all-big-gun battleship into practice; Japan and the United States also had such designs on their drawing boards. But *Dreadnought*—which cost the equivalent of $2.5 billion today—was first off the ways, because Jackie Fisher had at his disposal the Royal Corps of Naval Constructors (RCNC), a tight-knit body of naval architects and engineers whose members individually packed a generation's worth of experience into a single decade. Fisher's Committee on Designs (later dubbed the Dreadnought Committee), which consisted of 16 officers and civilians, including Director of Construction Phillip Watts and the engineer Robert E. Froude, met for only seven weeks in early 1905, but in that time they were able to converge on a design for the new battleship and begin detailed drawings for its construction. The RCNC was crucial to this process, producing eight concept designs for consideration just days apart from one another. The vast experience of the constructors also allowed them to make critical decisions on the fly; when the committee was faced with the choice for propulsion machinery between reliable but clunky steam reciprocating engines or compact but still unproven steam turbines, Watts unhesitatingly stated, "If you fit reciprocating engines, these ships will be out of date within five years." Once the final configuration was agreed to, the RCNC worked closely with Froude to test seven hull models before settling on a final set of lines that allowed *Dreadnought* to make 21 knots with far less horsepower than originally predicted, trading engine weight for armor, a decision that made the ship not only fast and heavily gunned but superbly protected as well.[13]

Yet the RCNC, formed in 1883, was a relative latecomer compared with other European corps of constructors and also distinct in that it was created as a civilian and not a military body of naval architects. The reason that Britain lagged in forming a professional corps is that, since the time of Henry VIII, it had successfully employed a series of civilian surveyors such as Thomas Slade (builder of Nelson's flagship HMS *Victory*) to oversee the design and construction of the world's most powerful navy. Another factor that held back any British interest in developing a system of professionalization was that up to half its warships were built not in naval dockyards but in commercial shipyards, which had little financial interest in the long education and training periods needed for professionalization.

The advent of iron and steam placed surveyors under increasing pressure to develop accurate estimates of both ship characteristics (weight, speed, horsepower) and costs. Understrength staffs under Isaac Watts and Edward Reed struggled to keep up with the mounting workload, while the public, the Parliament, and even naval officers attacked them as backward thinking and resistant to change. This came to a head with the HMS *Captain* affair, wherein the low-freeboard turret ship, designed by Cowper Coles, capsized with almost all hands lost in a strong gale in 1870. While the blunders and missteps in the ship's stability calculations (described in chapter 3) were evident to any trained professional, the political calculations that lay behind its authorization carried greater weight. Coles, a naval officer and hero of the Crimean War, "had many friends" (according to William Froude) in Parliament, newspapers, and the Admiralty who trusted his technical authority over that of less socially connected engineers like Reed. When Coles began stumping for his idea of a turret ship in 1865, his supporters mounted a campaign in the popular press (London *Times*, the *Standard*) and in military forums like the Royal United Services Institute and on the floor of the House of Commons, all denouncing Reed and other constructors as "bunglers," even though these "bunglers" were rightly concerned about the vessel's safety.

Despite the misgivings of the naval constructors, Cole's camp had the louder voice, and so—in the words of the report of the court-martial conducted after the ship's loss—"Her Majesty's ship *Captain* was built in deference to public opinion expressed in Parliament and other channels." Yet even after the court-martial established the folly of Coles's idea and validated the design authority of the Admiralty's technical staff, some naval officers and public figures continued to accuse Reed—who had resigned well before the ship had been launched—of secretly conspiring to bring about *Captain*'s loss. The ensuing national uproar depressed and demoralized the vaunted British navy, even as it faced resurgent naval powers in Europe and an emerging American threat across the Atlantic.[14]

The British naval historian Oscar Parkes memorably described the 1870s, the 10-year period after the HMS *Captain* disaster, as "the dark ages of the Victorian navy." Propelled by the general depression that gripped the British economy, it was marked by reduced naval budgets and a naval administration adrift and poorly organized. Other navies, and even commercial shipbuilders, appeared to have stolen a march on the Admiralty's technological superiority: France was the first to build steel hulls, Italian and German armor was seen to be more effective, and the compound steam engine appeared to be in wide use everywhere but in British warships. With the loss of both HMS *Captain* and the Admiralty's credibility, the system that had ensured for centuries that Britannia ruled the waves, that had brought victory at Trafalgar and created HMS *Warrior*, was now perceived to be breaking down. The only part of the system that had been demonstrably trustworthy was the body of naval constructors. That their voice had carried so little influence needed to be rectified, and that could happen only if they were elevated to a position of authority on par with that of senior naval officers.[15]

And yet the creation of the RCNC was proposed not as a means of attaining professional recognition but rather as a cost-saving measure under the notoriously tight-fisted Gladstone government (which, remember, also funded Froude's towing tank for reasons of economy). In 1878, William Henry White (figure 5.2), a graduate of the School of Naval Architecture in Kensington and by then assistant to Director of Naval Construction Nathaniel Barnaby, submitted memoranda to the controller of the navy asking for pay rises for the draftsmen (then making £500 per year) and more suitable titles, on the grounds that "the salaries are very low for the work which is now required to be done ... [and] which requires us to be naval architects or marine engineers in the fullest sense, demanding special qualifications both professional and mathematical.... Mere mechanical drawing is not our work. The duties performed by us would be more generally appreciated ... if we were known as assistant constructors." The controller demurred, suggesting a committee of inquiry be established.

Meanwhile, White began developing a "scheme leading to the establishment of a Constructive Corps for the Royal Navy, closely resembling the French *Génie Maritime*."

**Figure 5.2**
William Henry White.
Credit: National Maritime Museum, Greenwich, London.

In 1881 he drew up a pamphlet, signed by Thomas Brassey, Gladstone's civil head of the navy, that outlined the reorganization of the construction and engineering staff to create a corps of constructors, with a corresponding education and training system that began in the naval dockyards, led through the Royal Naval College at Greenwich, and resulted in a position as assistant naval constructor. Although the constructors themselves would be civilian employees, they would be considered of equivalent rank and stature to their naval counterparts. In 1882, the controller finally established his long-promised committee to examine the proposal, but much of the correspondence from that period focused on costs rather than benefits; one letter from Brassey insisted that the salary increases for the constructors "are largely met by a reduction in numbers" of draftsmen. Then, in a not-so-subtle allusion to HMS *Captain*, he went on to argue for greater professionalization of the constructor corps, since "the failure of design in a single ironclad may involve a loss far in excess of their entire salaries."[16]

In the end, professionalization won out over cost savings. Although the Treasury was unenthusiastic about the utility or cost effectiveness of the scheme, in August 1883, Queen Victoria signed an order-in-council that established the RCNC. Nathaniel Barnaby was the first head of the corps, followed in 1885 by its mastermind William H. White, who went on to oversee the construction of almost 250 warships of every size. The RCNC became a crucial factor in bringing the British navy back to technological predominance; not only dreadnought battleships but also battlecruisers, aircraft carriers, and amphibious landing ships were first created and later refined by British naval constructors—who among other things invented the angled flight deck and aircraft ski jump. Despite its late start, the RCNC also became one of the most respected professional engineering bodies in world; when Winston Churchill established the Landship Committee in 1915 to develop what is now known as the armored tank, he called upon the director of naval construction (and head of the RCNC) Eustace Tennyson d'Eyncourt to head up the design, which is why tanks have nautical-sounding hulls, turrets, and ports. The reputation of the RCNC, which continues even today, albeit in a much-reduced form, is perhaps attributable to the corps' mind-set, expressed by one of the other great naval constructors, Stanley Goodall, in response to the gunpowder magazine explosions that resulted in the loss of three battlecruisers during the Battle of Jutland:

> In the Navy List after the name of each ship is a column headed—By whom constructed. In that column is the name of the Director of Naval Construction. That fixes the individual responsibility for the ship, mark you, not for part of the ship.[17]

Across the Atlantic, the U.S. Construction Corps was the shortest-lived of all the naval construction corps—its heyday would last just 50 years—yet during that brief period it conceived and designed the most powerful navy that had ever existed. The first American constructors, such as Josiah Humphreys and his son Samuel Humphreys (see chapter 1), were civilian employees who had come up through the dockyards and sometimes maintained a private shipbuilding practice on the side. That did not mean they were any less committed to their government position; a telling story from 1824 recounts Samuel Humphreys's response to a well-remunerated offer, complete with several homes and servants, to build ships for Russia's Czar Alexander I:

## The Need for Professionalization

> The salary is greater than I could earn; more than I need; more than I want; more than I could use. As to the town house and country house, I need but one. As to the coaches and servants, I always walk and wait upon myself.... I do not know if I have the talents ascribed to me, but I do know that whether my merit be great or small, I owe it all to the flag of my country and that is a debt I must pay.[18]

In 1842, the U.S. Navy established the bureau system, with five bureaus covering everything from yards and docks to clothing and steam engines. The Bureau of Construction and Repair (which changed names several times) oversaw the building of ships, and though it was to have been headed by a "skillful naval constructor," this did not happen until 1853 when John Lenthall, an experienced constructor with both practical and theoretical training in engineering, assumed the position of chief constructor. Assisted by the great marine engineer Benjamin Isherwood, head of the rival Bureau of Steam Engineering, they designed the first mass-produced gunboats and sloops, as well as many of the Union's other warships, during the Civil War. Even so, the war's most famous ship, USS *Monitor*, was designed not by the navy but by the émigré Swedish engineer John Ericsson, with only scant oversight by Lenthall and Isherwood.

In 1866, after the nation had passed through the Civil War and once again become a fully United States of America, the navy instated a Construction Corps with naval architects coming through the Naval Academy in Annapolis and commissioned as officers (Lenthall was given the rank of commodore). The purpose was to develop a coherent system of education and professionalism in the same manner as the French (and later, British) system. In fact, there would be no effective training or professionalization system for over a decade, which contributed to the overall difficulties of the corps. Although the various bureaus occupied the same buildings, they might have been in different countries for their lack of communication. This was largely the result of the congressional patronage system, in which each bureau was controlled by a different faction of the Senate, fighting the other bureaus for money and authority, with no overall coordination by the secretary of the navy. It was not uncommon for the Bureau of Construction and Repair to design a hull for a ship that could not accommodate the engines or coal supply required by the Bureau of Engineering. Navy leadership was no better; when one secretary of the navy, Richard Wigginton Thompson from landlocked Indiana, was led down the hatch of a battleship, he exclaimed in surprise, "The damned thing's hollow!"[19]

This postbellum "dark age" was met with a sustained push to professionalize the entire U.S. Navy. Education and training for the maritime establishment was put on a more elevated footing, starting with the first engineering courses at the Naval Academy in 1874 and culminating in the establishment of the Naval War College in 1884. Meanwhile, in 1879 the Construction Corps began sending its junior officers to receive naval architecture training at the Royal Naval College, the École d'application du Génie Maritime, or Glasgow University, where they also saw the most advanced warships then under construction. At the same time, a series of laws and instructions came into effect that began winding down the congressional patronage system and brought the naval bureaus under greater centralized authority.

These initiatives first bore fruit when the Congress and the navy took the decision to reverse the post–Civil War decline, which had diminished the fleet to a position behind that of Belgium. Starting in 1882, the navy began rebuilding its fleet with the all-steel ABCD ships (*Atlanta*, *Boston*, and *Chicago* cruisers and *Dolphin* dispatch ship). It depended heavily on technology transfer from overseas to jump-start its rebirth; not only were the constructors who oversaw the program trained in Britain and France but the Congress authorized a payment of $2,500 ($3 million today) for the plans of British warships (e.g., HMS *Leander*) upon which the ABCD ships were based. As the U.S. Navy extended its reach overseas, first during the Spanish-American War of 1898 and then with the Great White Fleet circumnavigation of 1906–1907, the Construction Corps assumed greater authority over the design process, while new academic programs at the Naval Academy and Massachusetts Institute of Technology ensured that the corps was continually replenished with a stream of well-educated officers.[20]

By the outbreak of World War I, the U.S. Construction Corps had attained global stature, due in no small part to the renown of its chief constructor, David W. Taylor (figure 5.3). A top-flight graduate of the Royal Naval College in Greenwich, in 1898 he established the first American experimental model basin, carried out a series of experiments that literally rewrote the book on naval hydrodynamics, and was called by the British government as an expert witness in the case of the *Hawke-Olympic* collision in 1911. Under Taylor's tenure from 1914 until 1922, the Bureau of Construction and Repair carried out the designs for almost 1,000 new ships, which brought the U.S. Navy to near parity with Great Britain.

During the interwar period, rivalry between the various bureaus increased, while actual control of the ship design process was shifting from the naval officers of the Construction Corps to civilian draftsmen (later designated as naval architects), led first by James L. Bates and then by John C. Niedermair. The same lack of coordination that plagued the post–Civil War navy now made itself felt in the halls of the newly constructed Main Navy Building in Washington, D.C. The situation came to a head in 1939, when the destroyer USS *Sims* (DD-409) was built overweight and without sufficient reserve stability. The U.S. Navy used this problem as the pretext, in 1940, to consolidate the Bureau of Construction and Repair and the Bureau of Engineering into a

**Figure 5.3**
David Watson Taylor, with his Experimental Model Basin in the background, 1910.
Credit: David Walden.

single Bureau of Ships (BuShips), at the same time eliminating the Construction Corps and bringing all the BuShips naval officers into the Engineering Duty Officer (EDO) corps. From that point until the present day, EDOs managed the ship design projects but did no actual design work, which instead was left to the civilian naval architects and engineers. Ironically, despite their greater responsibility for the technical performance of naval vessels, the professionalization of these civilian naval architects, in terms of education, training, and career path, would continue to run a distant second to that of EDOs.[21]

**Education and Formation**

In the past as well as in modern times, the professionalization of civilian naval architects had always been well behind that of naval constructors, and this was nowhere more evident than when comparing the education and formation of the two. Through much of the age of sail, shipyards and their guilds fostered a system of apprenticeship to pass knowledge along, but because private yards and even guilds could disappear from one generation to the next meant that these were never robust fixtures in the shipbuilding community. By the 1600s, some shipbuilders were writing and publishing treatises that explained the art and practice of their trade, but this also remained a haphazard affair. Only with the advent of permanent navies did systematic organizations for educating constructors begin to arise. As this development coincided with the Scientific Revolution, naval administrators often turned to science and mathematics as the way to communicate with and control their constructors, and they accordingly sponsored instruction in those subjects. The first school of naval architecture, teaching mathematics, physics, and practical aspects of shipbuilding, began under the auspices of the French navy in 1741, and within a generation, similar schools were operating in the navies of Spain, Denmark, Sweden, and Venice. In short, navies had the permanency and the bureaucratic structures needed to allow such education to flourish.[22]

With the end of the Napoleonic Wars and the coming of steam and iron ships, navies began overhauling their existing systems of education and formation (respectively, academics and training) or creating new ones. Naval architecture education began with European navies as a means of professionalizing their constructor corps and soon spread overseas to the navies of the United States, Japan, and other nations. Civilian schools for naval architects did not begin appearing until much later in the nineteenth century, when the expansion and vertical integration of shipyards demanded a new breed of professional naval architects. This corresponded with the more general wave of professionalization that was sweeping many nations in the throes of becoming industrialized economies. As small businesses grew to become complex, vertically integrated corporations, the need for higher technical education of middle and upper management increased dramatically. From 1850 until 1900, the number of European engineers with university diplomas and degrees increased tenfold, even while population numbers barely budged. While the overall trend was exponentially upward, each nation took a different path to professionalizing its naval and civilian engineers.[23]

France led the way in engineering professionalization with its state-sponsored École Polytechnique and the specialized application schools for mines, roads, etc. It is hard to overstate the importance to France's educational system of the École Polytechnique, whose name is often rendered as "X" for its symbol of crossed cannons. It was opened in 1794 during the French Revolution under the direction of Gaspard Monge, who developed the field of descriptive geometry that became the basis for engineering technical drawing. Located in the Latin Quarter of Paris for most of its existence, it moved to the southern suburb of Palaiseau in 1976. The École Polytechnique has formed a central part of the French intellectual, professional, and military communities since its inception. For over 200 years, its entrance examinations and curriculum have been heavily oriented to mathematics, on the grounds that this encourages analytic thinking. It has graduated world-renowned scientists, economists, engineers, heads of state and government, military leaders, and founders of major corporations. Many prominent French constructors and naval architects, such as Henri Dupuy de Lôme, Louis-Émile Bertin, Émile-Georges Barrillon, Roger Brard, Jacques Stosskopf and Jean Dieudonné, put the prestigious "X" after their names.[24]

By contrast with the stability of École Polytechnique, France's naval application school suffered numerous vicissitudes throughout its life. The old École des Élèves Ingénieurs-Constructeurs de la Marine (School of Student Engineer-Constructors of the Navy), which had begun in 1741 and been at the Louvre since 1765, was closed in 1793 during the French Revolution. Its last director, the naval constructor and mathematician Jean-Charles de Borda, opened the new naval application school in 1795. Originally called the École Spéciale du Génie Maritime while it moved from Paris to Brest and then to various ports around the French empire, it returned to Paris in 1882 under the name École d'application du Génie Maritime. The school, as the name implied, applied practical and scientific principles of ship design and construction to the mathematical foundations built at the École Polytechnique. Professors like Frédéric Reech, Alphonse Hauser, and Charles Doyère regularly published textbooks for the institution, although the spectacular (for a naval architect) four-volume *Architecture navale; Théorie du Navire* (Naval architecture: Theory of the ship) by Jules Pollard and Auguste Dudebout remained the standard work for generations. Before 1882, only military officers, including those from foreign nations, had been able to pursue an education leading to naval constructor; after that date, civilians were admitted to the program and went on to careers in commercial shipyards and establishments. The school changed names again, first in 1948 to the École nationale supérieure du Génie Maritime and then in 1970 to ENSTA (École nationale supérieure de techniques avancées, Higher National School of Advanced Technologies). For many years, the école d'application of ENSTA had been the only university in France that offered an education in naval architecture, although shipyards like Penhoët in Saint-Nazaire offered apprentice training. In 1934, a technical university was created in Brest, in 1975 renamed ENSIETA (École nationale supérieure des ingénieurs des études et techniques avancées, Higher National Engineering School of Advanced Studies and Technologies), which offered a second route to becoming a naval architect. A third route opened in 1991 through the École Centrale de Nantes (and the only one not directly under the Ministry of Defense). In 2000, ENSIETA was absorbed and renamed ENSTA Bretagne, and in 2012 ENSTA moved from its Paris location to the Palaiseau campus of the École Polytechnique.[25]

The French system of naval architecture education heavily influenced Spain, which like France tied it directly to the naval engineer corps. The Academia de Ingenieros de la Armada (Naval Engineers Academy), which had existed in El Ferrol since 1772, closed its doors in 1819, well before the final suppression of the Cuerpo de Ingenieros de Marina. Starting in 1852, the first naval engineers of the reconstituted Cuerpo de Ingenieros de la Armada completed their courses of study at the French École Spéciale du Génie Maritime. In 1860, El Ferrol again was the site of the navy's Escuela de Ingenieros Navales (School of Naval Engineers), which used a mixture of books from France, Britain, Sweden, and the United States, as well as the Spanish text *Curso métodica de arquitectura naval* (Methodical course of naval architecture) by Joan Monjo i Pons. The school was closed in 1885 after graduating just 76 naval constructors and 49 civilian naval architects during three decades. Even though new cuerpos were established in 1909, aspiring naval constructors were educated in France and Italy until a new Academia de Ingenieros y Maquinistas de la Armada (Academy of Naval Engineers and Machinists) opened in El Ferrol in 1914. As noted earlier, the Second Republic extinguished the cuerpos but in 1932 established a new Escuela Especial de Ingenieros Navales (Special School of Naval Engineers) in Madrid. In 1948, the Escuela Técnica Superior de Ingenieros Navales (Technical University of Naval Engineers) was established in a new multiuniversity campus on the outskirts of Madrid. In 1971 it was incorporated under the Universidad Politécnica de Madrid (Polytechnic University of Madrid) and remains one of the largest schools of naval architecture in Europe, with a student population that has approached 1,000 in some years.

As Spain began its late but successful entry into industrialization starting in the 1960s—often termed *el milagro español*, the Spanish Miracle—the need for more schools of engineering became apparent, especially naval architecture schools to handle the burgeoning civilian shipbuilding sector. In 1962, a naval engineering technical school was established in Cádiz, which became the Escuela de Ingeniería Naval y Oceánica (University of Naval and Ocean Engineering) when the Universidad de Cádiz was created in 1979. Many more schools teach naval architecture and ocean engineering: Escuela Politécnica Superior (Higher Polytechnic University) at the Universidade da Coruña, which opened in El Ferrol in 1991; Escuela Técnica Superior de Náutica (Higher Nautical

Technical University) at Universidad de Cantabria, which was established in 2004 in Santander; and more recent degree programs at Universidad Politécnica de Catalunya in Barcelona and Universidad Politécnica de Cartagena.[26]

The Netherlands also closely followed the French model for naval architecture education and formation. Several Dutch constructors attended the École Spéciale du Génie Maritime during its brief stay in French-occupied Antwerp from 1810 to 1814. One of them, Cornelis Jan Glavimans, established the School voor scheepsbouw (School for Shipbuilding) in Rotterdam in 1821. The first graduate of the school was August Elize Tromp, who went on to become the director general of the Dutch navy from 1857 to 1866. The school moved to the Amsterdam *rijkswerf* in 1850, where Bruno Tideman took over the training in 1857. Meanwhile, the Polytechnische School te Delft (Polytechnic School of Delft) was opened in 1864, and instruction of naval architecture was transferred there in 1868. The Delft school differed from the French École Polytechnique in that it emphasized practical engineering as well as mathematical sciences. Tideman continued to teach at Delft alongside civil architect Adrianus Cornelis van Loo, who taught drafting and navigation, though both left the school in the mid-1870s. In 1905 its charter was changed to a degree-granting academic university, and it was renamed the Technische Hogeschool te Delft (Technical University of Delft). For many years its naval architecture department was led by the German émigré Ernst Vossnack. Although the student population was small, the department graduated many prominent naval architects such as Laurens Troost and Wilhelmus van Lammeren. In 1986 it became known as the Delft University of Technology, with programs in naval architecture up through the doctorate level.[27]

From its inception in 1827, the members of the Russian Korpus Korabel'nyh Inzhenerov received their naval architecture education at the Naval Academy in Saint Petersburg (today called the Kuznetsov Naval Academy), which educated junior and midlevel officers, not cadets. In 1904 the newly formed Sankt-Peterburgskiy politekhnicheskiy institut (Saint Petersburg Polytechnic Institute) established a department of shipbuilding under Ivan Grigoryevich Bubnov, himself a Naval Academy graduate. Following the French education model, the polytechnic offered a highly mathematical curriculum for cadets and civilians. World War I and the Russian Revolution radically altered the curriculum, admissions, and professorships in favor of the procommunist ideals of the Soviet Union, which led to the emigration of many faculty, the most famous of which was Stepan (Stephen) Timoshenko, the father of elastic beam theory. Starting in the 1930s under Stalin, the Russian economy was organized into five-year plans, which extended to the number and types of engineers and naval architects entering Saint Petersburg Polytechnic and the Naval Academy. During the Cold War, academic and research developments in Russian naval architecture were relatively unknown in the West, since professors, students, and journal papers only rarely crossed the Iron Curtain. But since the fall of the Berlin Wall in 1989, advances in such technologies as wing-in-ground-effect craft (*Ekranoplan*) have been shared widely.[28]

Neither the unification of Italian states nor the creation of the Corpo del Genio Navale, both of which occurred in 1861, immediately led to a school of naval architecture. For many years Italian constructors like Benedetto Brin had been educated in the French École d'application du Génie Maritime. In 1871, under the impulsion of Brin, then director of naval construction, the Italian government created the Regia Scuola Superiore Navale di Genova (Royal Naval College of Genoa), based in part on Brin's alma mater in Paris. It offered instruction for aspirants to the Corpo del Genio Navale as well as commercial naval architects. In 1936 it was absorbed as part of the engineering faculty of the Università degli Studi di Genova (University of Genoa). Meanwhile, in 1913, the Accademia Navale (Naval Academy) in Livorno also opened a curriculum for naval constructors. In 1942, the Università degli Studi di Trieste (University of Trieste) established a naval engineering department. Both these universities continue today to offer degree programs in naval architecture, along with Università degli Studi di Napoli Federico II (University of Naples) and Università degli Studi di Messina (University of Messina).[29]

The École Polytechnique model also extended into northern European nations. The 1871 unification of Germany spurred an accelerated program of industrialization, which led the government in 1879 to create a new breed of institution, the *Technische Hochschulen*, or polytechnic institutes. These were degree-granting institutions that specialized in engineering and sciences; before that, such education was relegated to trade schools (shipbuilding trade schools went as far back as 1834), while only classical universities like Heidelberg were permitted to grant degrees. The "German genius," however, was to go beyond the French model and create what we know today as the modern research university, in which classical education is combined with an institutionalized system of scientific and technical research that exposes students and faculty to hands-on laboratory work. Further, this research university model was tied directly to industrial concerns; new research would lead to novel industries, and industrial companies themselves benefited by hiring engineers with solid theoretical education and practical formation.[30]

The Technische Hochschulen in Berlin-Charlottenburg and Hannover were the first to offer higher education in naval architecture. Hannover became the center for hydrodynamic research at the turn of the twentieth century; one of its most celebrated professors, Ludwig Prandtl, developed the boundary layer theory in 1905 as part of his laboratory work. A third Technische Hochschule that taught naval architecture opened in 1905 at Danzig (now the Politechnika Gdańska, University of Technology at Gdańsk), Poland, and counted the naval-architect-turned-airship-designer Johann Schütte among its faculty. Intermediate-tier engineering schools (*ingenierschulen*) also opened, offering more practical and less scientifically rigorous education. For naval architects, these included schools in Hamburg, Bremen, and Kiel, which opened from 1895 to 1905. After World War II, new campuses opened at Aachen, Duisburg, and Rostock. Since 1999, the distinctions between higher-level and intermediate-level schools have been vanishing as the result of European Union agreements on postsecondary education; today most are known as *Technischen Universität* (TU, or technical universities). Whatever their designations, German engineering schools have always punched above their weight class, in large part owing to the close connections between academics, research, and industry. Naval architecture students around the world today instantly recognize the names of Günther Kempf, Hermann Lerbs (both professors at TU Hamburg), and Horst Nowacki (TU Berlin) as pioneers in the fields of, respectively, hydrodynamics, propeller cavitation, and computer-aided design.[31]

The Scandinavian nations—Denmark, Sweden, and Norway—also followed the École Polytechnique model and at the same time were heavily influenced by the research university model of the German Technische Hochschule. In Sweden, for example, technical schools for shipbuilders had flourished since 1843, providing practical skills and a modicum of theoretical education for civilian naval architects as well as members of Sweden's Mariningenjörskåren (Naval Engineers Corps), which from 1867 to 1955 replaced a series of prior naval constructor's corps dating back to Fredrik Henrik af Chapman during the age of sail. The rapid pace of industrialization led Sweden in 1876 to create the Kungliga Tekniska högskolan (Royal Institute of Technology) in Stockholm. Naval engineers were required to graduate from Kungliga Tekniska högskolan, although the school did not open a department of naval architecture (for both civilian and military students) until 1898. As in Germany, intermediate-tier engineering schools offered more practical education; for civilian naval architects, this was the Chalmers Tekniska Högskola (Chalmers Institute of Technology) in Göteborg, founded 1829, which became a university in 1937. The Swedish system produced many naval architects who went abroad to achieve renown, among them John Ericsson and John W. Nystrom, the former being the designer of USS *Monitor* and the latter serving during the Civil War as one of the Union Navy's chief engineers. At this time, it was quite common for Swedish naval architects to complete their education and formation abroad—especially in the United States and Britain—where, in the words of one engineer, they could "see how things were done in a big way."[32]

Shipbuilding was certainly "done in a big way" in Britain, which as noted earlier, accounted for 75 percent of the world's ship construction by 1890. Yet compared with its European neighbors, Britain took a far more laissez-faire approach to the higher technical education of the men who designed and built those ships. The whole concept of engineering as a profession had a completely different meaning in Britain than it did in Germany and France. For much of the nineteenth century, engineers (including naval architects) followed the time-honored apprentice system to learn the ropes, rather than focusing on the scientific and theoretical aspects as was done in the École Polytechnique model. And even when higher-level engineering schools were established, it was done in a piecemeal fashion at privately or locally funded universities or by industry-sponsored mechanics' institutes for workingmen, instead of at the national systems of universities found across the European continent. This was in keeping with the general reluctance of the British government to involve itself with what it considered to be market-driven sectors of the economy like education. Apprentice systems, run by private companies or by government industries like dockyards, were considered to have both social as well as educational benefits, leading to "better, i.e., more intelligent, men" within the British working class. For most of these workers, industrial magnates argued, "a very high degree of technical education is not essential."[33]

The Admiralty's extensive network of naval dockyards—such as Chatham and Portsmouth—constituted the largest shipbuilders and were among the largest manufacturers in Britain during much of the industrial age, but their apprentice systems, which had worked tolerably well in the age of sail, came under increasing criticism for corruption and technological backwardness. In 1806, a parliamentary commission on naval revision under Charles Middleton, Lord Barham, recommended a universal two-tiered education system—dockyard schools for apprentices and a separate school (with competitive entrance exams) for constructors.

The navy, acting on Barham's recommendations, opened a School of Naval Architecture in 1811 at the newly established Royal Naval College in Portsmouth, with the mathematician James Inman at its head and shipwright John Fincham providing practical instruction. This was the first British school, apart from military and naval academies, to offer both theoretical and practical instruction to its students, and it served as a model for future British institutions. However, because constructor graduates were immediately put in charge of more experienced shipwrights at naval dockyards, they were uniformly ignored (though many of them, like Thomas Lloyd and Isaac Watts, eventually went on to illustrious careers), and the school itself was increasingly despised. The school stopped accepting applicants in 1822 and closed in 1832.

The political backlash against the first School of Naval Architecture was part of the reason that William Symonds, a naval officer without technical training, was appointed surveyor in 1832 (see chapter 1). One of the school graduates, Henry Chatfield, engaged

## The Need for Professionalization

in a decadelong war of pamphlets against Symonds; he argued that constructors should have technical authority over ship designs, while Symonds acolytes argued that only naval officers possessed the moral credibility for that authority. In response to this challenge, Chatfield and several other graduates successfully petitioned the Admiralty to accept their designs for new ship classes, among them the fast and maneuverable HMS *Thetis* in 1846.[34]

By 1848, the Admiralty was willing to give professionalization another chance, in the form of a second school of naval architecture. It was established at the Portsmouth dockyard under the more formal title Central School of Mathematics and Naval Construction and under the direction of Joseph Woolley, formerly a church deacon, with close oversight by another clergyman, Henry Moseley. The mathematician Robert Rawson and shipwright John Fincham provided theoretical and practical instruction. It was while at the Central School that Moseley and his colleagues developed the first comprehensive theory of the dynamics of rolling (see chapter 3). But as with the first School of Naval Architecture, the resentment of shipyard workers against its graduates led the Admiralty to shutter the school after just five years, in 1853 (though it is worth noting that two chief naval constructors, Edward Reed and Nathaniel Barnaby, were among those graduates). By contrast, a series of dockyard apprentice schools, which began opening in 1843, were so successful in creating a cadre of skilled workers that the naval constructor William White, in 1911, declared them "the best investment of money that has ever been made in this country." These dockyard apprentice schools endured until 1971, when they were integrated into local colleges and universities.[35]

But as skilled as these dockyard workers were, they were not trained to design and calculate the performance of highly complex warships, and the Admiralty still needed a cadre of such naval architects. John Scott Russell, vice president of the newly formed Institution of Naval Architects (of which more later), lamented this in an 1863 speech to the institution, pointedly contrasting Britain's lack of a naval architecture school with the well-regarded École d'application du Génie Maritime that he had recently visited. This was at the height of the naval arms race between France and Britain—epitomized by the ironclads *Gloire* and *Warrior*—and Parliament quickly took up the cause for fear of Britain lagging its competitor. The third school of naval architecture would not be in the dockyards as the two previous schools had been; rather, London was preferred, as it was close to the centers of power in the sciences and engineering. The location in the nation's capital was also an acknowledgment of warship design being now a political and not waterfront activity; besides, Parliament wanted the school to educate both naval constructors and civilian naval architects, so they sought to distance it from the overt influence that the Admiralty would exert if it were in a naval dockyard.

In a nod to Russell, the school would be located in South Kensington, the site of the 1851 Great Exhibition that he helped create. Henry Cole, a civil servant who worked with Russell in the exhibition planning, was placed in charge of constructing a new science building along Exhibition Road, which would house the School of Naval Architecture and Science along with other schools of chemistry, physics, and art. The school had its first intake of students in 1864, even though construction of the new science building did not start until 1867. Joseph Woolley was again called to direct the school, now with a full-time faculty of seven instructors and numerous lecturers from dockyards, industry, and even the Royal Observatory. For several years classes were held in wooden sheds on the construction site, neither heated nor properly ventilated. By 1872 the science building was roofed over and the terracotta façade largely completed, and though the interior was barely furnished, the School of Naval Architecture and Science finally moved into proper classrooms and laboratories on the ground floor (figure 5.4). This arrangement would last less than a year.[36]

While the 40-odd students—a third from industry or abroad—were just settling into their new rooms on Exhibition Road, the Admiralty was already making other plans for them. It had decided to move the Royal Naval College, where naval officers received

**Figure 5.4**
Author's son at the Henry Cole Wing of the Victoria and Albert Museum, which was the site of the South Kensington School of Naval Architecture, 1872–1873. The classrooms and laboratories were on the ground floor behind the colonnades.
Credit: Author's photo.

# The Need for Professionalization

advanced training, from its decrepit facilities in Portsmouth to the palatial Christopher Wren–designed buildings at Greenwich. It now saw the opportunity to create a great naval university by combining officer training with other academics, including the School of Naval Architecture. In 1873, the Royal School of Naval Architecture and Marine Engineering, as it was now known, moved to Greenwich, where it fell under the auspices of the Royal Naval College. When the RCNC was established in 1883, its members all were required to be graduates of Greenwich. Even though it was an Admiralty school, it continued to admit private and foreign students; the future chief constructor of the U.S. Navy, David W. Taylor, graduated from the Royal Naval College in 1888 with the highest honors received to that date. For generations its standard textbook was William White's highly influential *A Manual of Naval Architecture*, which went through several editions, although Edward Attwood's *Text-Book of Theoretical Naval Architecture* filled in some of the gaps. For almost a century the school continued at Greenwich, but in 1967, following the Robbins Commission reforms for higher education, the British government opted to move it to University College London. Louis Rydill, a naval constructor who had been a professor at Greenwich, revised the curriculum to create a master of science course for naval constructors, where it is taught to this day.[37]

Even though private students could study at Greenwich, the commercial shipbuilding industry needed more naval architects to fill their burgeoning middle-management needs. This was especially true among the vertically integrated Clyde shipyards. The first two departments of naval architecture were opened in Glasgow—at Anderson's College (later the University of Strathclyde) in 1882, followed in 1883 by the University of Glasgow, with the John Elder Chair of Naval Architecture endowed by his widow, the Clyde shipbuilder Isabella Elder (the two universities merged their naval architecture departments in 2001 under the auspices of Strathclyde). In 1890, the Durham College of Physical Science (later University of Newcastle upon Tyne) established a combined chair of naval architecture and marine engineering, followed in 1908 by the University of Liverpool. Even with this growth in technical education, Britain—the world's largest shipbuilder—still lagged other nations in terms of graduating engineers. In 1907, one observer noted that only 90 naval architecture students total were at all British universities, while TU Berlin alone had over 200 students. Instead, most naval architects in Britain and its Commonwealth were trained in apprentice schemes at commercial shipyards; Thomas Andrews, chief designer of RMS *Titanic*, came up through the ranks that way at Harland and Wolff. This situation changed only with the Robbins Commission reforms of the 1960s that gradually led to the elimination of many apprentice schemes in favor of university programs—for example, at the University of Southampton, which in 1968 developed a ship science curriculum.[38]

Although Britain dominated industrial shipbuilding during much of the nineteenth century, the United States began to catch up by the 1880s, followed by Japan in the early twentieth century. As these nations rose in industrial stature, they looked to British technical education as a model for their own. The Americans began the process in 1879, as noted earlier, when the navy began sending its constructors to receive an education in naval architecture at the Royal Naval College, the École d'application du Génie Maritime, and Glasgow University. In 1881, just as the first groups were returning from abroad, the navy ordered a young officer, Mortimer Cooley, to teach steam engineering and iron shipbuilding at the University of Michigan in Ann Arbor. After just four years Cooley resigned his commission and became one of four engineering professors there, and for 20 years he was the only person to teach naval subjects to military and civilian students. In 1899 the university agreed to establish a formal department of naval architecture and marine engineering. Cooley agreed to run marine engineering, while he recruited Herbert C. Sadler, an assistant professor at the University of Glasgow, to chair the department and reorganize the naval architecture curriculum along Glasgow's lines. Under Sadler's leadership, in 1904 the department opened the very first university model basin for academic instruction, and from 1914 it provided courses in aeronautics until a new department was established several years later.[39]

Soon after the University of Michigan program began, William H. Webb, one of New York's most important shipbuilders, addressed the issue of professionalization in commercial shipyards, where apprenticeships were still the only route to advancement. In 1889 he endowed Webb's Academy and Home for Shipbuilders in the Bronx borough of New York City as both a retirement home and a place for higher education for shipbuilders. Instruction was not formalized into a degree-granting program until 1933, and the campus moved in 1947 to its present location in a *Great Gatsby*-style mansion in Glen Cove, Long Island. In 1890, a naval architecture program was also established at Cornell University in upstate New York in the hopes of providing education for the U.S. Navy's Construction Corps, but it closed just 13 years later. The reason for this closure was that, in 1901, the navy bypassed Cornell and chose the nascent program in naval architecture at the Massachusetts Institute of Technology as the home for its graduate naval constructor's course, which was relabeled Course XIII. That choice had been left to William Hovgaard, a captain in the Danish Navy and former classmate at Greenwich of David W. Taylor, who was now head of the Experimental Model Basin and instrumental getting Hovgaard selected to be head of the new program. After a few more changes, the three-year master's degree Course XIII-A for naval officers remained consistent for over a century (as of 2005 offered under the title 2N), and is considerably more thorough than civilian naval architecture programs. Those civilian programs have included graduate studies at the University of California at Berkeley, undergraduate programs at maritime universities (like the State University of New York Maritime College), and more recently, undergraduate and graduate programs at the Stevens Institute of Technology in New Jersey, Virginia Tech, and University of New Orleans.[40]

Japan began its drive toward naval architecture professionalization at about the same time as the United States but by far different means. After the Meiji government consolidated power in 1868, it pushed the nation to rapidly integrate with the rest of the world, bringing politics, education, and industry up to Western standards in a breathtakingly short period known today as the Meiji Restoration. For shipbuilding, the government sent students to the Royal Naval College and the École d'application du Génie Maritime to learn naval architecture and marine engineering while bringing expert talent to Japan to build up its military, educational, and industrial capabilities. While the French constructor Louis-Émile Bertin oversaw the creation of a new imperial navy, British engineers and professors helped establish naval architecture training and degree programs in industry and academia. In 1884, Tōkyō teikoku daigaku (Tokyo Imperial University, now University of Tokyo) established a naval architecture course and by 1898 had a full program in place. From 1902 to 1918 its chair was Frank Purvis, who had been one of William Froude's assistants at his tank in Torquay. Purvis oversaw the rapid development of the university's research capabilities, an example of what the historian Miwao Matsumoto calls Japan's "military-industrial-university complex," which gave the Japanese navy a significant advantage in the early adoption of the steam turbine built under license by the Mitsubishi Company. Similar programs were soon established at Kyushu Imperial University and the Polytechnic Universities in Osaka and Yokohama. Meanwhile, Mitsubishi Nagasaki established apprenticeship programs along British lines that trained shipyard workers and hull draftsmen (one of them, Tsutomu Yamaguchi, famously survived both the Hiroshima and the Nagasaki atomic blasts). The rapid expansion of this education and training meant that by 1941 Japan had developed naval and maritime capabilities that rivalled those of Britain and the United States and seriously threatened them at the outset of World War II.[41]

The landscape of naval architecture education and training changed considerably after World War II. As noted, apprentice schemes largely gave way to formal university education. Meanwhile, the numbers of programs and students in Europe and the Americas have dwindled in proportion to their loss of shipbuilding capacity, while the rise of shipbuilding nations in Asia has fostered large networks of university programs there. In China, the Shanghai Jiao Tong University established a program in 1943, followed by universities in Dalian, Tianjin, and Harbin, and this list continues to grow almost every year. In South Korea, naval architecture programs are offered in Seoul,

Busan, Chosun, and Ulsan, while India has a half dozen programs in Chennai, Cochin, and Kharagpur. New programs continue to emerge around the world.

**Professional Societies and Publications**

The third and final step of professionalization was the establishment of formal professional associations, which serve as venues to share information. Naval constructors during the age of sail relied upon the scholarly structures of the Scientific Revolution, notably the French, Russian, Swedish, and Danish academies of sciences, to support and publish research in naval architecture. The French revolutionary and Napoleonic Wars closed or disrupted almost all the academies' activities from 1792 through 1815, and there was little to take their place. The Society for the Improvement of Naval Architecture lasted only a decade, from 1791 to 1801, and its most important legacy, the resistance experiments of Mark Beaufoy, would not be published until 1834. The first professional societies devoted to naval architecture would not arise until 1860. So for much of the nineteenth century, naval architects relied on a hodgepodge of sources—scientific and engineering associations, encyclopedias, trade journals, and world's fairs and expositions—to learn about advancements in their field.[42]

The British scientific establishment was not affected to the extent of its Continental counterparts and continued to function reasonably well. The Royal Society continued to publish papers on fluid resistance and ship stability (notably Atwood and Vial du Clairbois on righting moments) and supported the Admiralty in its work on the Seppings structural system and on hull coppering. From 1837 onward, the British Association for the Advancement of Science began funding and publishing John Scott Russell's wave-line research, Admiralty investigations into stability and rolling, and commissions investigating the powering of ships. But engineers still wanted their own professional associations, separate from those of scientists. In 1818 a small group formed the Institution of Civil Engineers in London, with the purpose of "facilitating the acquirement of knowledge" and "promoting mechanical philosophy." Its members included men like Brunel, Russell, Fairbairn, and Froude, who made the bridge (so to speak) between civil and naval engineering. The institution's *Transactions* (later *Minutes of Proceedings*), begun in 1836, published articles on naval architecture; the very first volume contained a paper by the engineer John Farey Jr., "An Approximative Rule for Calculating the Velocity with Which a Steam Vessel Will Be Impelled through Still Water." Even so, relatively few papers were published in comparison with the great advancements taking place in the field. Another professional body, the Institution of Mechanical Engineers, followed in 1847, but this body favored marine engineering more than naval architecture.[43]

Augustin Francis Bullock Creuze, a graduate of the first Portsmouth School of Naval Architecture, almost single-handedly filled in the gap in naval architecture knowledge during this period. He had been a Royal Navy surveyor until 1844, when Lloyd's Register hired him to be their chief surveyor in London. Creuze filled his nonworking hours amassing information and writing articles on naval architecture, some of which appeared in the *Encyclopedia Britannica* and others in the collection *Papers on Naval Architecture*, which he published serially between 1829 and 1842. He also wrote many articles on the subject in trade publications and was instrumental in establishing the naval architecture displays for the Great Exhibition of 1851. Shortly before his death in 1852, he donated his collection to Lloyd's Register, which became the start of its now extensive library.[44]

Creuze's articles, along with many others on naval topics, appeared in the engineering trade journals that were established to meet the demand for information by the burgeoning population of workingmen attending classes and lectures at industry-sponsored mechanics' institutes. Among the most widely read of those industrial journals was *Mechanics' Magazine*, which was published from 1823 to 1872. One of its editors, Edward Reed, went on to become the navy's chief constructor. After resigning in 1871, he established a consultancy and for several years published the short-lived quarterly journal *Naval Science* (1872–1875). Although it mixed technical articles by the likes of Froude and Russell with topical news, polemics, and editorials, it was never commercially viable and abruptly stopped after four years.

Continental navies took a more dirigiste approach to technical journals, which were often official publications that promulgated laws, ordinances, and regulations yet also carried cutting-edge research papers. In France, the *Annales Maritimes et Coloniales* (Maritime and colonial yearbook, 1816–1964) was the first to print Taurines's theory of propeller action in 1842, while *Revue Maritime et Coloniale* (Maritime and colonial review, 1861–1896) carried many articles on maneuvering and rudder theory. The French constructors corps also published annually the *Mémorial du Génie Maritime* (Memoirs of the naval engineering service) from 1847 to 1914. This was almost exclusively devoted to technical articles on propulsion, hydrodynamics, and maneuvering (it was the first to report Joëssel's method for predicting rudder torque in 1873) and articles on naval developments, and it came with a more famous atlas of ships' plans. The Italian *Rivista Marittima* (Maritime journal, 1868–present) and the Russian publication *Morskoĭ sbornik* (Marine anthology, 1848–present) have been the longest-running naval journals and carry a mix of scientific, technical, and operational articles.

In the United States, the professionalization of engineers in general, and naval architects in particular, happened at a slower pace. As in Britain, the establishment of industry-sponsored mechanics' institutes was one of the first steps in this process.

The most influential and longest-lived of these was the Franklin Institute in Philadelphia (1824–present), whose *Journal of the Franklin Institute* was, for a long time, the primary source of detailed technical information on developments in naval and marine engineering available to American practitioners. The journal tended to focus on steam power—it carried a slew of articles on steamboat boiler explosions from 1836 to 1874—and on the battles between screw and paddle wheel propulsion. Notably absent were articles on naval architecture theory. The clipper ship designer John W. Griffiths was the first to fill this information gap with his 1850 book *A Treatise on Marine and Naval Architecture*, followed in 1853 by *The Ship-Builders Manual*. He then published two short-lived technical journals, *Nautical Magazine* (1854–1858) and *American Ship* (1878–1880). Few other publications filled this gap during the nineteenth century; *Scientific American* (1842–present) mostly dealt with public spectacles like the wave-line theory employed in the yacht *America*, while naval journals like the *Proceedings of Unites States Naval Institute* (1874–present) were almost entirely operational in nature, with few articles on technical subjects.[45]

On both sides of the Atlantic, the Industrial Revolution had spurred increasing public interest in technical matters, which was accompanied by national interest in using technology as a tool of diplomacy. These needs converged in a series of international exhibitions and congresses in which countries from Europe and the Americas—and increasingly, Asia—would exhibit their wares and hold lectures and discussions on technical matters. The first, most famous of these was the Great Exhibition of 1851 in London, featuring the Crystal Palace in Hyde Park. In addition to an entire section of the display halls devoted to advances in naval architecture, the Hundred Sovereigns Cup regatta was also held in conjunction with the exhibition, in which the yacht *America* showed off the United States' ability to adapt and improve British ship design. From that point on there were international world's fairs, engineering conferences, and exhibitions somewhere in Europe or the Americas every few years, in which exhibits from around the globe on the latest developments in naval architecture and marine engineering figured prominently. A few were the American Crystal Palace Exhibit in New York City, 1853; the 1876 Centennial International Exhibition in Philadelphia; the 1888 International Exhibition of Science, Art and Industry in Glasgow, followed by another in 1901; the 1889 Exposition Universelle in Paris, on the anniversary of the French Revolution and featuring the Eiffel Tower; the International Congress of Naval Architects and Marine Engineers in London in 1897; Exposition maritime internationale in Bordeaux, 1907; and the International Engineering Congress in San Francisco, 1915.

One of the largest, and for American naval architects most important, was the Chicago World's Fair Columbian Exposition, held in 1893 on the 400th anniversary of Columbus's voyage. Whereas the London Crystal Palace Exhibition had 6 million visitors, Chicago hosted 27 million, literally dazzling them with the first large-scale use of alternating-current electricity to light the grounds and sporting a full-scale section of a transatlantic passenger liner. In conjunction with the Columbian Exposition was an International Engineering Congress, with a special section on naval architecture hosted by the U.S. Navy and having its chief engineer George W. Melville as the chair. This truly marked the coming of age of the American navy and its industries, on the heels of the new steel warships now building across the country. Other navies took note of America's presence on the world stage; the French constructor Louis-Émile Bertin wrote a detailed description of the rising naval power across the Atlantic.[46]

Yet these scientific associations, societies, journals, congresses, and expositions still left a technical void in the naval architecture community; they were often ad hoc and did not provide a systematic means of collecting and disseminating professional knowledge. The model for a professional society had already been demonstrated by the Institution of Civil Engineers, and it was in this vein that John Scott Russell proposed the establishment of the Institution of Naval Architects in the fall of 1859. Its first meeting was held in London in January 1860, and from the start its *Transactions* has published until this day some of the most important and influential papers on naval architecture, many of which are sources in this book's notes. At its centennial in 1960 it was designated by Queen Elizabeth II the *Royal* Institution of Naval Architects (RINA) and today is the largest professional body of naval architects in the world, with 11,000 members and a global reach into almost every maritime nation.[47]

Britain became host to other professional naval and marine societies. A few months after the Institution of Naval Architects was founded, a rival association was established in Glasgow, called the Scottish Shipbuilders Association, which merged in 1865 with the Institution of Engineers to become the Institution of Engineers and Shipbuilders in Scotland. The focus of its *Transactions*, then as now, has been on shipbuilding and construction. In a similar vein, the shipbuilders on the Tyne and Wear Rivers sought to get out from under the influence of London by establishing the North East Coast Institution of Engineers and Shipbuilders in 1884, which lasted more than a century and whose *Transactions* published important papers on cavitation research before folding in 1992. The Association of Engineering and Shipbuilding Draughtsmen (1913–1961) was more of a trade union than a professional society, although it did publish technical papers at annual sessions. Marine engineers established a separate society in 1888, the Institute of Marine Engineers, which in 1999 became IMarEST (Institute of Marine Engineering, Science, and Technology). When Britain established a chartered engineer scheme in the 1960s, only the RINA and Institute of Marine Engineers were given status as accrediting bodies for professional engineers.[48]

Despite the rapid success of the Institution of Naval Architects and its British cousins, the idea of a professional society of naval architects was slower to arrive elsewhere. In France, the Association Technique Maritime (Maritime Technical Association) was established during the Exposition Universelle in 1889 and published its first annual bulletin the following year. The association added the fledgling science of aeronautics in 1924, becoming Association Technique Maritime et Aéronautique, as it is known today, devoting many pages of research to analyses of aircraft aerodynamics and ship hydrodynamics. It has always been a small organization, with fewer than 500 members. In Germany, the STG (Schiffbautechnische Gesellschaft, Society of Shipbuilders) was formed in 1899 and today boasts over 1,500 members. Other European societies were not established until well into the twentieth century: for example, Asociacion de Ingenieros Navales de España (Association of Naval Engineers of Spain) in 1929 and in postwar Italy the Associazione Italiana di Tecnica Navale (Italian Association of Marine Technology) in 1947. Collaborations between European societies have become more common under various frameworks, notably the Confederation of European Maritime Technology Societies, originally founded in 1971 and today comprising over 35,000 participants. Japan formed what would become known as the Nihon Zōsen Gakkai (Society of Naval Architects of Japan) in 1897, which consolidated with several other Japanese societies in 2005 into a single Japan Society of Naval Architects and Ocean Engineers.[49]

The creation of naval architecture societies in the United States corresponded to its rise as a naval power and the increased professionalization of its engineering workforce. In 1888, marine engineers and naval constructors in, respectively, the fractious Bureaus of Steam Engineering and of Construction and Repair decided to create the American Society of Naval Engineers to serve their common interests. Although not an official arm of the navy, its offices were in navy headquarters buildings until World War II, after which it moved to private digs in Washington, D.C., and Alexandria, Virginia. Its journal has generally favored articles about naval ship design, construction, and operations, although its almost 10,000 members come from many engineering backgrounds.[50]

The aforementioned 1893 Chicago World's Fair prompted the formation of two civilian naval architecture societies, only one of which survived. The Penn Institute of Engineers and Naval Architects was created as a section of the Franklin Institute in 1893, though it met for only two sessions, in 1893 and 1894, before disappearing without a trace. The far-better-known Society of Naval Architects and Marine Engineers was formed in late 1892 by the many of the same members who were at the time planning the International Engineering Congress that accompanied the Columbian Exposition. Its *Transactions* have, over the years, contained many ground-breaking articles of original research, and the Society of Naval Architects and Marine Engineers is famous for

*Principles of Naval Architecture*, which has been the standard reference bible for naval architects around the world since 1939 (now in its fourth edition). Its headquarters has moved from New York City to New Jersey to Alexandria, Virginia, over the years, and today counts 6,000 members, smaller than either the Royal Institution of Naval Architects or the American Society of Naval Engineers. Yet it is worth noting that all the naval architecture societies put together are still dwarfed by those of other professions; the American Society of Mechanical Engineers and the American Society of Civil Engineers each have 150,000 members, and the Institute of Electrical and Electronics Engineers is half a million strong.[51]

## Naval Architecture Consultancies and Design Firms

The word "autonomy" is an important but often overlooked part of the definition of a profession. During the nineteenth century, it was rare for naval architects who designed large oceangoing vessels to work separately from the shipyards that built them. That was not the case for the designers of yachts (and later, pleasure craft), many of whom worked as independent consultants, in much the same way as did home architects, to develop unique designs for their clients and then worked hand in hand with the shipyards to oversee construction. For example, George Steers, in 1850, designed the yacht *America* for his client John Cox Stevens, but it was built at the William H. Brown shipyard. As the professional stature of naval architects grew in the twentieth century, it became increasingly common for naval architects to strike out on their own, establishing consultancies and design firms that could work autonomously from the shipbuilders to serve larger clientele like governments or industrial consortia. Like many business trends of the twentieth century, this one began in the United States.

William Francis Gibbs was a sickly child in Philadelphia when in 1894 he first saw a steamship being launched, which crystallized in him the dream to design the world's fastest passenger liner. Even though he studied political science at Harvard College (he never graduated and later claimed he studied "science") and began his working career in 1913 as a lawyer, he was driven to learn everything he could about naval architecture. His interest even attracted the attention of David W. Taylor, who in 1919 authorized tests of Gibbs's passenger liner design at the navy's Experimental Model Basin. It was never clear how Gibbs was able to develop such a professional-level understanding of ship design, but he got his chance to demonstrate it in 1922 when he and his brother Frederic were given the contract to refurbish a German World War I prize, SS *Vaterland*, to become the new American passenger liner SS *Leviathan*. Gibbs got the contract because he had been the assistant to the chair of the U.S. Shipping Board (a government agency that oversaw shipping and shipbuilding contracts during World War I) and had written the specifications for the refurbishment contract he had then bid on and won. The newly formed Gibbs Brothers oversaw the refit of *Leviathan* and several other ships at Newport News Shipbuilding. By 1929, just before the Great Depression, they had hired David Taylor, designed the passenger ship *Malolo*, and brought the yacht designer Daniel Cox into the business. The new firm Gibbs & Cox survived the financial downturn by becoming the U.S. Navy's premier design agent, a position that lasted throughout World War II and into the Cold War. Gibbs finally designed his dream passenger liner SS *United States* in 1952, but its success was short-lived as the jet passenger plane quickly replaced the steamship for transatlantic voyages.[52]

Gibbs & Cox soon had competition from other design agents, some of whom also got their start with U.S. government contracts. During World War I, Theodore E. Ferris had, like Gibbs, made a name for himself at the Shipping Board, designing wooden cargo ships that became the basis for a lucrative private career. George G. Sharp was designing river steamboats when, during the buildup to World War II, his firm began designing the merchant ships, notably the famous *Liberty* ships, that turned the tide of the war. Mandell Rosenblatt had been a draftsman and designer for the construction and conversion of many navy ships before and during World War II, but as his grandson Bruce Rosenblatt told me, he faced considerable anti-Semitism in the maritime industry. So in 1947, he struck out with his son Lester to form M. Rosenblatt and Son, and by the 1960s the firm was handling a large percentage of the design and conversion work for the U.S. Navy, as well as for many foreign navies. James J. Henry followed a similar path, resigning from the U.S. Maritime Commission (a follow-on to the Shipping Board), on which he had served during World War II, and forming JJ Henry Company in 1951. His firm focused more on cargo and container ships than on warships. In 1959, John J. McMullen, who also served on the Maritime Commission, formed John J. McMullen Associates, also dividing the work between military and commercial vessels (McMullen himself was better known as the owner of several U.S. sports teams like the New Jersey Devils and the Houston Astros). By the end of the Cold War, all these companies, except George G. Sharp and Gibbs & Cox, had been bought by larger technology firms or gone out of business.[53]

In Europe, the formation of naval architecture consultancies and design firms took a decidedly different and less entrepreneurial path than in the United States. In 1934, the severe economic downturn of the Great Depression led several Dutch maritime firms like the Rotterdamsche Droogdok Maatschappij (Rotterdam Drydock Company), Koninklijke Maatschappij De Schelde (Royal Schelde Company), and Werkspoor to pool their resources in a single naval design bureau called Nevesbu (Nederlandse Verenigde Scheepsbouw Bureaus, United Dutch Shipbuilding Bureaus). The new firm provided naval design and engineering services not just to Dutch shipyards but also for export; its first design contract in 1935 was for a pair of Polish submarines, and after World War II it became one of the leading submarine designers in Europe. During the Cold War it was the design agent for many of the major Dutch warship projects, and starting in the 1990s it branched out into the offshore energy sector. Similar to Nevesbu, the German company MTG Marinetechnik (Naval Engineering Company) was formed in 1966 by the German Ministry of Defense as a means to consolidate and separate naval ship design and engineering from shipyard production facilities. Since then, MTG has been at the center of design and integration studies for the German navy as well as other navies. In Britain, the consulting firm BMT (British Maritime Technologies) was formed in 1985 from the consolidation of two research bodies, the governmental National Maritime Institute and the private British Ship Research Association (of which more in chapter 6), to provide research and engineering services for the British navy and maritime industries. After the end of the Cold War it expanded worldwide into naval, energy, and commercial maritime sectors and is today an important design agent for the U.S. Navy.[54]

Several independent naval architecture firms have their roots in specific patents and technologies, of which the best known are Joseph Isherwood Company and Maierform. Isherwood was a surveyor for Lloyd's Register when he developed a novel way of arranging the iron hull structure, with lighter, closely spaced longitudinal stiffeners and heavy, widely spaced transverse frames. Even though this was only slightly different from other longitudinal framing systems, Isherwood patented it in Britain and in 1906 formed his own company to market the product. By the 1920s the Isherwood system had become the standard for many oil tankers, making Isherwood quite wealthy. By contrast, Fritz Franz Maier, an Austrian-Hungarian naval architect, patented in 1906 his idea for a V-shaped hull form with a sharp cutaway bow but was unsuccessful at selling the concept to shipbuilders. After his accidental death in 1927, his son Erich Maier formed the company Maier-Schiffsform in Bremen, Germany, and was successful with the fishing industry. His biggest contracts came from the German navy during World War II, which built over 600 fisheries and coast guard vessels having the Maierform bow. After the war, both companies expanded their naval architecture services well beyond their original patents and remain in operation today.[55]

### Women in Naval Architecture

By now, readers will have noticed that almost all the pronouns in this book are masculine. As with almost every engineering discipline, women held regrettably few positions of note in naval architecture until the late twentieth century, but they have always been there; sometimes behind the scenes, more often deliberately hidden by others in plain sight. It has been pointed out that while *Titanic*'s designer Thomas Andrews is well known, the woman who invented its life rafts—Maria Beasley—is almost unknown.[56] In this short section, I highlight the contributions of a few of these women, who stand in for a larger, as yet unrecognized populace.

One of the first women we know of to follow this career was Mary Lacy, who had disguised herself as a man during the Seven Years' War and signed on to a British warship as a ship's carpenter. Still in disguise, she then worked at the Portsmouth dockyard as a shipwright, a physically demanding job that involved not just laying down hull lines in a lofting shed but also sawing, drilling, and joining enormous timbers. After she retired in 1772, she caused quite a stir when she published her diaries as *The History of the Female Shipwright*. More common was the case of women who owned and ran shipyards, usually because they had inherited them from their husbands. The historian Helen Doe has documented over a dozen who ran British yards throughout the nineteenth century; in most cases, she argues, they did not give up control even to their sons, for "once these women were in charge they liked to stay that way." And they were quite successful at it—the most famous of these women, Isabella Elder, ran the highly profitable Govan Shipyard after her husband John's death in 1869 and later endowed the University of Glasgow with a naval architecture chair named in his honor.[57]

Women frequently worked behind the scenes as tracers and computers, supporting the naval architects who designed the ships and performed the engineering analyses. In the 1880s, William Denny hired young women "between the ages of fourteen and twenty-two" to work in the drawing office of his shipyard and engine works, tracing and inking the design drawings that would be used on the shop floor (see figure 5.5). Denny claimed to be proud of these women—"The tracing department as a branch of female labour was a new development, and its success was a source of special gratification to him," said his biographer—yet he also harbored a deep prejudice, stating that women "are deficient in the judicial faculty … and unfit to judge the affairs of practical life."

Computers were a later development in ship design, the word originally referring to the people—often women—who carried out the numerical calculations for engineers as part of their analyses. One of these women, Frances (Betty) Holberton, was originally hired by the U.S. Army in 1943 to program the analog ENIAC for computing ballistic equations (figure 5.6). Ten years later, Holberton started work at the David Taylor Experimental Model Basin, where she soon rose to become chief of the Programming Research Branch in the Applied Mathematics Laboratory. As one of the original developers of the UNIVAC, she oversaw its use in a variety of tasks around the laboratory. Other women worked as computers in calculating hydrodynamic forces and structural loads.[58]

**Figure 5.5**
Mockup of Denny drawing office with women tracers.
Credit: Author's photo.

Women also held positions as naval architects and ship designers. One of the first on record was Lydia G. Weld, who graduated from the MIT Course XIII in 1903 and worked at Newport News Shipbuilding for many years, in charge of producing machinery installation plans. Eily Keary worked as junior assistant to George S. Baker at the UK National Physical Laboratory experimental tank (Teddington) during and after World War I, researching the hydrodynamics of destroyers and seaplanes. Elsie Mackay, a British stage actress and pilot who never trained as an engineer, designed the interiors of many passenger ships in the 1920s, though her career was cut short when her plane was lost over the Atlantic in 1928. Jean Monro, an interior designer, followed in her footsteps in the 1950s and 1960s.[59]

World War II saw women enter traditionally men's roles in engineering, designing, and building ships, which went far beyond the iconic images of Rosie the Riveter and Wendy Welder. Probably the best known of these women engineers was the British

**Figure 5.6**
Betty Holberton (right foreground) with ENIAC.
Credit: U.S. Army photo.

engineer Constance Tipper, who as noted in chapter 4, helped solve the cracking problem in welded ships during World War II. During that same war but across the Atlantic, Elaine Scholley had just graduated with a mathematics degree from Hunter College in New York City when she was hired at Gibbs & Cox to work on the dozens of naval ship designs that came through the office at breakneck speed. She married fellow engineer Howard Kaplan in 1945, and after the war the company placed Elaine Kaplan in charge of the propeller design team for its superliner SS *United States*, winning an award for her work that helped the ship capture the Blue Riband in 1952, a record still unbroken.

Meanwhile, near Washington, D.C., the David Taylor Experimental Model Basin routinely employed women in traditionally male roles. Figure 5.7 shows the women who ran the towing carriage during the unrelenting, three-shifts-per-day routine at the model basin described in chapter 6. Another woman, Rachel S. Welch, was working for the Army Transport Corps as a draftsman while taking naval architecture courses at

# The Need for Professionalization

night at George Washington University. Her drive and talents brought her to work in 1945 as a naval architect in the stability branch of the BuShips, where she completed crucial studies for wartime operations. After the war she moved to the model basin, where she worked with Morton Gertler on resistance experiments for high-speed tankers. Although she took early retirement in 1953, it is worth noting that Welch—like Kaplan and Tipper—was highly regarded by her male peers and felt that her work was both important and respected.[60]

Yet this apparently enlightened attitude toward female engineers did not last much past the end of World War II, as women returned or were forced back to their more traditional roles as housewives and mothers. The attitudes of the day were amusingly expressed in the University of Michigan's 1961 pamphlet "What Every Young Girl

**Figure 5.7**
Women run the carriage, David Taylor Experimental Model Basin, 1943. Left to right: Grace Ann Snipes, Mildred Smith, Mary Cavanaugh, Helen Burnison, and Arvella Brak.
Credit: Naval History and Heritage Command.

Should Know Before She Marries a Naval Architect," written by Harry Benford under the nom de plume Dame Pamela Ritter (figure 5.8). It is the advice of a mother to her daughter, who is intent on marrying a naval architect. Nowhere in the text is there any assumption that the young woman could possibly *become* one. The closest Benford/Ritter comes is when the mother warns that "it won't be long until he's teaching you how to use a slide rule ... doing endless computations for him." (In fact, Mark Beaufoy credited his wife, Margaretta, "a good mathematician," with performing many of the computations for his seminal experiments in naval architecture during the 1790s, discussed in chapter 1.)

Since that time, women have been entering the profession in greater numbers and even going beyond it. Milka Duno is today a well-known Indy and NASCAR race-car driver, but she got her start as a naval architect in Venezuela in the 1990s and then graduated with her master's degree in the subject from Escuela Técnica Superior de Ingenieros Navales in Madrid. Her naval architecture experience, she told me, gave her the knowledge and confidence not only to get behind the wheel but to lead her engineering team to get the maximum performance from her race cars.[61]

**The Professional Naval Architect in Popular Culture**

Milka Duno may be famous today, but it is for being a race-car driver and not for being a naval architect. Few naval architects ever became household names, and the profession of naval architecture itself is often invisible to the general public, in factual representation and in fiction. The reasons for this are by no means clear, but comparatively few people around the globe having regular contact with ships and ocean transportation must play a considerable role.

The man at the center of this book, Isambard Kingdom Brunel, was voted number two in the 100 Greatest Britons polling contest run by the BBC in 2002. But he was primarily acknowledged—in 2002 as well as back in 1852—for his bridges, railways, and tunnels that revolutionized public transportation, receiving only a head nod for his SS *Great Britain*. In roughly that same era, Donald McKay was briefly celebrated for his legendary clipper ships *Stag Hound*, *Flying Cloud*, and *Sovereign of the Seas*, while John Ericsson was lauded during the Civil War as the man who built USS *Monitor*, which stopped the Confederate ironclad navy in its tracks, then promptly forgotten. Even the most noteworthy naval architect of the era, Henri Dupuy de Lôme, was recognized during his lifetime for building a dirigible airship to help relieve the Siege of Paris during the Franco-Prussian War in the 1870s. At the turn of the twentieth century, Nathaniel Herreshoff did become a household name every few years when his yachts brought

**Figure 5.8**
What every young girl should know before she marries a naval architect, 1961.
Credit: Author's collection.

back the America's Cup five consecutive times. And in 1942 during World War II, William Francis Gibbs briefly achieved fame when he was featured on the cover of *Time* magazine as the naval architect responsible for the design of three-quarters of the U.S. Navy's ships.[62]

By way of comparison, civil architects have routinely become long-lived household names for the buildings they created—Frank Lloyd Wright with his Prairie style designs, and Frank Gehry and his Guggenheim Museum in Bilbao, to name but two. Even in fiction, civil architects are well-recognized icons of stoic individualism; Howard Roark in Ayn Rand's *The Fountainhead* and the lead romantic figure of Daniel, played by Liam Neeson, in the 2003 film *Love Actually*. I have found no significant representations of naval architects in works of fiction. In film, the best known was based on the real-life naval architect of *Titanic* in the 1997 movie of the same name. Toward the end of the film, Thomas Andrews apologizes to the lead character Rose, "I'm sorry that I didn't build you a stronger ship," before going down with the vessel he designed.

Hardly a figure to encourage future generations to enter the profession of naval architecture. For me, the inspirational image of the naval architect must be C. K. Dexter Haven, the suave yacht designer played by Cary Grant in the 1940 movie *The Philadelphia Story*. His yachts, as Katherine Hepburn's character notes, were "yar—easy to handle, quick to the helm, fast, bright, everything a boat should be." At the end of the movie, after a series of revelations that redefine their beliefs, both characters learn to adapt to each other and to the changing times—everything a professional naval architect should be.

# 6   Laboratory Life

When William Froude privately built a model tank in Torquay, detailed in chapter 3, this was the norm rather than the exception for nineteenth-century British scientists. As the mantle of scientific research passed from France to Britain after the French Revolution and the Napoleonic Wars, the British government took a decidedly laissez-faire approach to scientific endeavors. The Royal Society and the British Association for the Advancement of Science were privately funded organizations, and much of the research carried out under their auspices or reported in their journals occurred in the laboratories and workshops of individual scientists and inventors. Charles Parsons, who as noted earlier developed the steam turbine, was the first to study the problems of propeller cavitation—the phenomenon in which water vaporizes into bubbles and then collapses onto the propeller blade, causing erosion, noise, and loss of thrust—using a small circulating water channel he built in his kitchen at Holeyn Hall near Newcastle in 1895. The problems of cavitation were mathematically modeled in 1917 by John William Strutt, Baron Rayleigh, in the laboratory he constructed in a stable of his country estate in Essex, where he also investigated electromagnetism, color, and acoustics.[1]

The same was true in the United States, where before World War II, the majority of government-sponsored scientific research went to agriculture. During the Civil War, the many innovations that went into John Ericsson's USS *Monitor* came from his own workshops. The only research carried out directly by the U.S. Navy was focused on gunnery experiments by John Dahlgren at the Washington Navy Yard; even the experiments on the magnetic compass in iron ships, noted in chapter 2, were carried out under the private National Academy of Sciences. Forty years later, Elmer A. Sperry developed his navigational gyrocompass and gyroscope autopilot (for ships and aircraft) in his home laboratories in New York and Washington, D.C. In the early twentieth century, the Wall Street financier Alfred Lee Loomis established his noted laboratories at Tuxedo Park in upstate New York, where scientists devised new technologies in radar, atomic physics, and navigation. Only during World War II, with the advent of what became known as the military-industrial complex (a phrase coined by President Dwight Eisenhower), did large-scale government and industry sponsorship of scientific research replace private laboratories as the primary source of discovery and invention.[2]

The shift from privately funded research into naval architecture to research funded by government and industry happened first in Britain at the end of the nineteenth century, as the nation was rapidly losing its standing as the uncontested naval and maritime superpower in the face of competition from Continental powers like Germany and France, from rising Asian powers like Japan, and from the Americans across the Atlantic. Almost as soon as the British government and maritime industries began building model basins, the practice spread among competing nations as they all wrestled for primacy on the high seas. Despite the military and economic rivalry between these countries, laboratory technologies and scientific discoveries flowed quite freely across their borders, an illustration of the statement made by Joseph Banks (who oversaw exchanges between the British Royal Society and French Academy of Sciences even during the Napoleonic Wars) that "the sciences were never at war."[3]

**Model Basins around the World**

When Robert Edmund Froude moved his staff from the old Torquay tank to the new Haslar facility near Portsmouth in February 1886, it was only the third government establishment dedicated to research created in the United Kingdom in two centuries, and all had their origins with the Royal Navy. The first laboratory was the Royal Observatory in Greenwich, establish in 1675 to observe the skies in order to improve ocean navigation. The second was the Royal Botanic Gardens at Kew, created in 1759 to research and supply seeds and crops for Britain's expanding overseas empire.[4] The third, the Admiralty Experimental Works (AEW) Haslar model basin—at 120 meters, considerably longer than the one at Torquay—was officially christened in April 1887 when Robert Edmund ceremoniously poured in a flask of water taken from the Torquay tank before it had been demolished. This began the practice of christening new model basins with a flask of water from the older basin or, in certain instances, from the original filling of water in Haslar Number 1 Ship Tank (figure 6.1).

Robert Froude's work at Haslar established many other procedures that are now standard practice in model basins around the world. He designed as large a basin as feasible, to allow for running bigger models at higher speeds, which gave results more relevant to full scale. This required solid foundations so that the tank and its sensitive equipment would not shift or vibrate, and Portsmouth had a firm clay soil ideal for the job. Once the Haslar tank was in operation, Froude noted that tests there of the model of

**Figure 6.1**
Haslar Number 1 Ship Tank.
Credit: RINA.

the cruiser HMS *Iris* showed considerably higher frictional resistance than previous tests at Torquay. He began using the *Iris* model as the standard to test for consistency in experimental results every few years, a practice now followed in every model basin and often using the same model of HMS *Iris*. Every so often tanks experience what are called "*Iris* storms," in which frictional resistance dramatically decreases, apparently because of the presence of long-chain organic polymers in the water from algal blooms. At Haslar, the tank was kept clean of these polymers by hosting live eels until chlorination eventually was introduced. Robert Edmund also built the first wavemaker in 1904, employing a motor-driven flapping board to generate waves for testing resistance and propulsion characteristics in a seaway. Another now-common practice that the younger Froude introduced was the use of bicycles to get from one end of the tank to the other.

Haslar Number 1 long outlived Robert Froude, who died in 1924. It was the workhorse for the British navy and maritime industry for almost half a century, testing more than a thousand models in hundreds of thousands of carriage runs, even using some of the original equipment from Torquay. In the 1930s, the AEW constructed the larger 270-meter Haslar Number 2 tank to accommodate models needed for bigger warships and a two-story-tall cavitation channel to manage the increasingly common problems of cavitation on more powerful, higher-speed propellers. Later, new model basins were added to the facility to study maneuvering and seakeeping. In 2001, AEW Haslar, along with much of the British military research enterprise, was privatized under a new company, QinetiQ.[5]

AEW Haslar was too busy with naval work to accommodate the needs of commercial shipbuilders, who also wanted to specify and predict the performance of their ships before construction. At first, these commercial shipyards built their own model basins. As previously noted, William Denny opened his Ship Model Experiment Tank in Dumbarton, Scotland, in 1883. In 1904, another Clyde shipbuilder company, John Brown, built a model basin in the shipyard and was followed by Vickers in 1912 with a tank in Saint Albans north of its London office. At the same time, the Southampton shipbuilder John I. Thornycroft built a boat testing tank at his estate (a disused gun battery) in Bembridge on the Isle of Wight. In 1908 the Clyde shipbuilder Alfred Yarrow offered to rationalize this plethora of individual tanks by pledging £20,000 ($20 million today) to build a national model test basin under the auspices of the Royal Society. In 1911 the new facility was unveiled at the society's National Physical Laboratory (NPL) in Teddington, outside London, established a decade earlier to act as the nation's standardization and testing laboratory. Shortly thereafter, NPL came under government control. George S. Baker, the first superintendent of the 150-meter tank, oversaw basic research such as the testing of a methodical series of hull forms to establish the effects of, for example, variations in length and beam on resistance. At the same time, the laboratory was servicing the industry with predictive model tests for ships like RMS *Mauretania*. By 1935 the workload had grown so great that a new, larger model basin and a cavitation channel were built, and the whole facility was renamed the William Froude Laboratory.[6]

World War II brought the pace of work to a fever pitch, and after the war a wholesale rebuilding program for the British merchant fleet was envisioned that would further tax the NPL's capabilities. British shipbuilders agreed to pool their resources to form a cooperative research association, which was becoming popular in Britain as a way of leveraging scant governmental funds with private monies. The British Shipbuilding Research Association (BSRA, later the British Ship Research Association) was formed in 1944 and served as the coordinating body to fund laboratories like the NPL on a wide range of basic and applied research, including methodical hull series testing (the BSRA Series), propeller testing, and jointly with the British Welding Research Association, finding improvements in welding techniques. One of the most famous BSRA projects was the *Lucy Ashton* trials. These were carried out in 1950–1951 to establish an accurate correlation between hull model tests and full-scale trials, which had not been done systematically since Froude's *Greyhound* experiments. *Lucy Ashton*, an old (1888) river paddle wheeler, was stripped to its bare hull, fitted with instruments, and propelled with four deck-mounted jet engines above the waterline. The data were compared with a series of models tested at the NPL. Among the many results were assessments of various frictional coefficients and corrections between model and full scale, and a better understanding of blockage effects to correct for the flow of water being choked off around large models in towing tanks. Although new research facilities were added to the NPL at the nearby Feltham site in the 1960s, the downturn in shipbuilding led to a contraction in research. The NPL ship facilities were spun off into the National Maritime Institute in the late 1970s, but that was short-lived. Under the sweeping privatization initiatives of the UK government under Margaret Thatcher, the National Maritime Institute was combined with BSRA in 1985 and privatized to form British Maritime Technologies at Teddington. The Feltham facilities were subsequently razed, and a supermarket now covers the site.[7]

The first model basin built outside the United Kingdom was constructed in the Italian naval base of La Spezia from 1887 to 1889 by direction of Minister of the Navy Benedetto Brin. Designed by Robert E. Froude, the tank was 146 meters long and was fitted with the very latest instrumentation from the same companies that furnished the tanks of Haslar (Munro of London) and Denny (Kelso of Glasgow). From the beginning, the basin was used for both commercial and naval testing. Its first director, Giuseppe Rota, was not even 30 years old when he began the testing program. Control of the basin passed in 1910 to Angelo Scribanti, even as he carried out his duties as director of the Royal Naval College of Genoa. In 1929 Rota oversaw the construction of a second model basin in San Paolo, Rome, near the Tiber River, the Istituto Nazionale per Studi ed Esperienze di Architettura Navale (National Institute for the Research and Study of Naval Architecture). The bombings of World War II destroyed the La Spezia tank but left this second one intact. During the postwar reconstruction, new facilities were built south of Rome, while nearby Lake Nemi—famous as the site of Roman emperor Nero's pleasure barges—was used to carry out maneuvering experiments.[8]

Germany was the next nation to build model basins, and it did so in a systematic fashion. It was not the performance of oceangoing ships that led Germany to place its faith in small-scale testing but rather that of its extensive system of riverine and canal boats. The first German basin, just 63 meters long, was constructed in 1892 by Ewald Bellingrath in his Übigau shipyard in Dresden along the Elbe River. Although his basin was modeled on the Denny tank, Bellingrath had his instruments and equipment built in France. In 1903 he constructed a second, longer (88 meters) basin, which was used to test the hull forms of river and oceangoing ships, such as the renowned passenger liners *Imperator* and *Vaterland*. Friedrich Gebers, who got his start at the Dresden-Übigau tank, went on to found the Vienna Model Basin in Austria in 1916, in response to the increasing demands from Danube River shipping firms and the Austro-Hungarian navy. Around the same time, the shipping company Norddeutscher Lloyd ( North German Lloyd), in answer to problems with its steamships not reaching contractually required speed, also established an ultimately short-lived research institute in Bremerhaven with a 164-meter tank, under the direction of Johann Schütte,

The first government-funded model basin opened in 1903 in Berlin, on a lock island in the Landwehr Canal in the Great Tiergarten park, a site personally selected by Kaiser Wilhelm II. The Königliche Versuchsanstalt für Wasserbau und Schiffbau (Royal Research Institute for Hydraulic Engineering and Shipbuilding) grew over the years and was used for ship model testing, notably by the German Imperial Navy, as well as for hydraulic testing of locks, canals, and waterways. To accommodate its massive increase in shipbuilding, in 1906 the German navy opened its own towing tank in the Lichtenrade district of Berlin. The director, Hermann Wellenkamp, devised a novel falling-weight system (similar to that used by Beaufoy a hundred years earlier) that allowed models to accelerate faster, thus permitting the model basin to be much smaller and less expensive (45 meters versus 150 meters for the kaiser's first basin). In 1915, a legacy of 200,000 marks ($20 million today) from Ernst Otto Schlick, best known for inventing shipboard gyroscopic stabilization systems, allowed the City of Hamburg to construct the Hamburgische Schiffbau-Versuchsanstalt (HSVA, Hamburg Ship Model Basin) with a massive 350-meter tank. From 1922 until 1955 the HSVA was led by Günther Kempf (who, like Gebers, got his start at Dresden-Übigau). After 1930 it expanded its work in propeller cavitation under the direction of Kempf's deputy, Hermann Lerbs, with the first fully enclosed cavitation tunnel. It also built the world's first high-speed towing tank, originally for studying flying boats but later for fast attack craft (figure 6.2). HSVA, like many other German facilities, was heavily damaged by bombings during World War II and slowly rebuilt itself into a leading research facility through the latter half of the twentieth century and into the twenty-first century.[9]

The Russian Empire was next after Germany to build model basins. Andrei Alexandrovich Popov, the Russian admiral who worked with Froude on his circular ironclads, led one of the navy's technical committees that in 1882 recommended a towing tank similar to Froude's. It was not until 1894 that the 120-meter basin, designed with input from Edmund Froude and fitted with Munro and Kelso equipment, was commissioned at New Holland Island in Saint Petersburg. In January 1900, Aleksey Nikolaevich Krylov was appointed as head of the model basin, but his first months were marred when a fire

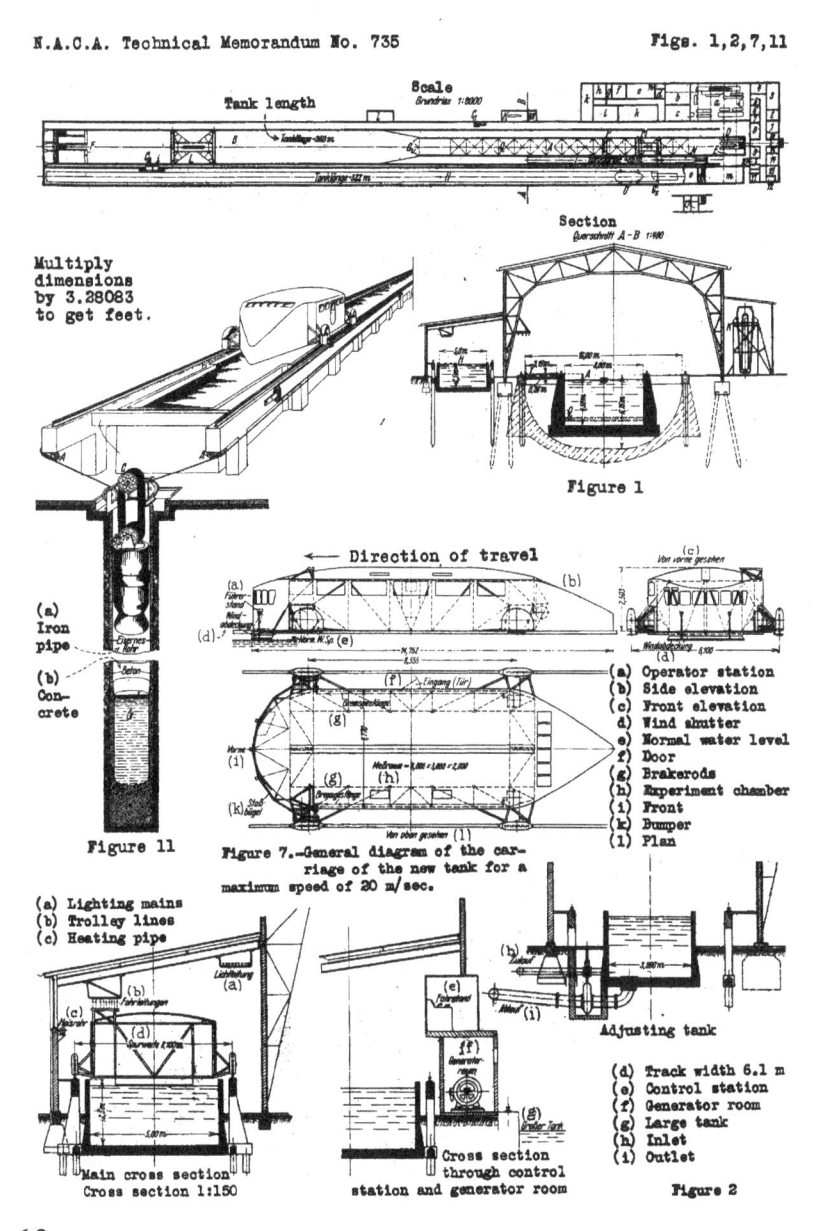

**Figure 6.2**
HSVA high-speed towing tank with aerodynamic carriage.
Credit: Author's collection.

nearly destroyed the facility and valuable equipment had to be reordered from Britain. After Krylov's departure to become chief constructor in 1908, control passed to Ivan Grigoryevich Bubnov. From that point, through the Russian Revolution and many civil disruptions, the model basin somehow survived in often appalling conditions, even if one of its directors—Nikolai Vladimirovich Alyakrinsky—was baselessly executed "for sabotage" in 1937, in part because he made possible the basin's reconstruction using German and Italian know-how. That same year, a new, much larger research facility, the Nauchnogo instituta voyennogo korablestroyeniya (Scientific Institute of Military Shipbuilding), was constructed on the southern outskirts of Saint Petersburg, then called Leningrad. It survived the three-year Siege of Leningrad and in 1944 was renamed for Krylov, the year before his death. Now known as the Krylovskiy Gosudarstvennyy Nauchnyy Tsentr (Krylov State Research Center), it was significantly enlarged in the 1950s and 1960s. It is the world's largest naval architecture research facility at almost one square kilometer and today houses a nuclear reactor, the world's longest deepwater towing tank (1,300 meters), an array of maneuvering basins and circulating channels, and extensive structural testing facilities.[10]

The French navy's Bassin d'essais des Carènes (Hull Model Basin) was conceived almost as soon as Louis-Émile Bertin became chief naval constructor and head of the new Service Technique des Constructions Navales (Naval Construction Technical Service) in 1896. He knew the particulars of the model basins already in operation in France's potential adversaries Britain, Italy, and Germany and was well aware that the United States was also planning a new towing tank. Bertin fought to obtain funds for the construction of a French basin, which were approved in 1902, though the facility would not be opened until 1906, a year after he had stepped down from his post. It was located on military property on Boulevard Victor, in one of the southern bastions along the Thiers defensive wall surrounding Paris (since torn down), next to the Seine River (figure 6.3). The size of the bastion limited the length of the model basin to 160 meters, the minimum Bertin believed necessary compared with other tanks in Europe and America. From the beginning, the Paris basin was used exclusively for naval testing and basic research; French commercial shipyards had to use other European basins, such as when Penhoët shipyard in Saint-Nazaire went to HSVA in 1931 to test its new *Normandie* hull, as described later in this chapter.

Soon after the initial construction, new facilities were added to the laboratory while design bureaus were installed around it. In 1935 a small wave tank, 30 meters long, was built to test seakeeping. In 1939, the world's first rotating-arm basin for maneuverability testing began construction (more on this later), though the outbreak of World War II, the Nazi occupation, and subsequent bombings delayed its opening until 1942. After

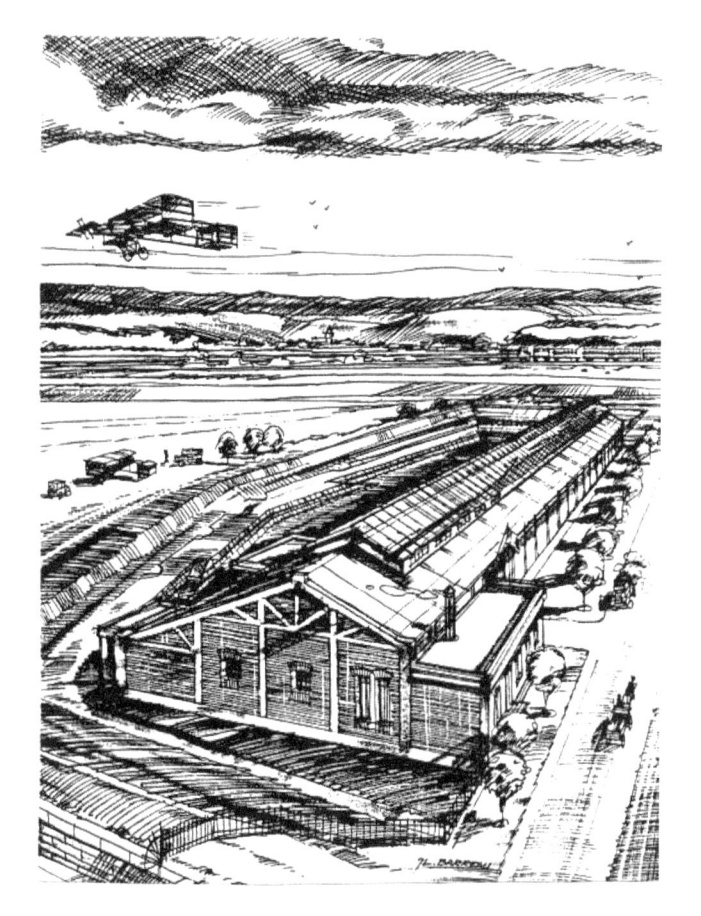

**Figure 6.3**
Paris Model Basin circa 1908, located in bastion 69 of the Thiers fortifications. In the background is the Issy-les-Moulineaux airfield with Voisin Brothers biplanes, the first manned aircraft in Europe. Credit: Author's collection.

the war, larger towing tanks were installed and used by France's maritime industries. For almost half a century (1920–1969), the model basin was under the direction of just two men, Émile-Georges Barrillon and Roger Brard, both of whom carried out groundbreaking fundamental research on seakeeping and hydrodynamics. By the 1980s it was obvious that the site was too small for further expansion, and in 1996 a new, more modern complex was opened in Val-de-Reuil, Normandy, with three towing tanks (the largest 220 meters long), plus cavitation tunnels, maneuvering, and seakeeping basins. In 2010 the old Paris test facilities were demolished to make way for the consolidated

Ministry of Defense headquarters, puckishly nicknamed "L'Hexagone" as the French counterpart to the American Pentagon.[11]

The advent of World War I and the postwar reconstruction of shattered nations considerably slowed the development of naval research facilities. Back in 1900, the Netherlands had shuttered its open-air testing facility in Amsterdam created by Bruno Tideman, and from 1916 its shipbuilders had primarily depended on the Vienna Model Basin. It was only in 1934 that the nation opened a new, state-of-the-art research center in the town of Wageningen, profiting from the hard clay soil at the center of the country, which gave far better structural support than the marshy ground in the port cities. Even from the start it was often referred to by its English-language name, the Netherlands Ship Model Basin, later the Maritime Research Institute Netherlands (MARIN). Its superintendent, Laurens Troost, aided by Wilhelmus Petrus Antonius van Lammeren, who took over Troost's job in 1952, immediately made the facility famous with a systematic series of screw propellers having airfoil sections, later dubbed the Wageningen B-series propellers, that became an industry standard. But World War II hit the facility hard, and not until 1947 was it able to regain its bearings and even add to its capabilities. Propeller research continued to be the basin's strong suit, and over the years it added several cavitation channels to its portfolio, culminating in 1972 with the construction of a 240-meter vacuum tank (now called the depressurized towing tank or wave basin), which controls atmospheric pressure to more accurately model propeller cavitation at scale.[12]

The construction of the Spanish model basin began with postwar initiatives to wean its navy from dependence on foreign (mostly British) technical assistance and achieve a measure of self-sufficiency. A Canal de Experiencias hidrodinámicas (Hydrodynamic Experimental Basin) was proposed as early as 1921, but it was not begun until 1928 and not completed until 1934. Instead of installing it at the principal naval dockyard at El Ferrol, the government decided that the model basin should serve the entire nation and chose the ancient royal hunting grounds of El Pardo, about 15 kilometers north of Madrid, as the site for the 320-meter canal. Damage during the Civil War (1936–1939) kept the basin from operating until 1940. Since that time the facility has been expanded to include maneuvering and cavitation testing and been used to test the hull form of almost all naval and commercial vessels built in Spain. In 2013 it was administratively moved from the naval defense ministries to INTA, the Spanish aerospace agency equivalent to NASA.[13]

In Scandinavia, model basin construction was cut short by World War II. The Norwegian Towing Tank (now the Norwegian Marine Technology Research Institute, or Marintek) was inauspiciously opened in Trondheim on 1 September 1939, the day

Germany invaded Poland and began the war in Europe, and its operations were severely constrained by the German occupation. Sweden remained neutral during the war, so when it opened its model basin in Göteborg in 1940 (today SSPA, the Swedish Maritime Research Center), it could not obtain precision instruments from abroad and had to make do with ones cobbled together from scratch. After the war, the two nations—along with Denmark's Skibsteknisk Laboratorium (now the Danish Maritime Institute), which opened in 1959—developed into highly regarded hydrodynamics research facilities. Many other European nations—Turkey, Poland, Finland, Greece, and Belgium, to name a few—have opened model basins in the years since World War II,

By contrast to the rapid spread of model basins in Europe, the notion of a scale model facility for testing ships was accepted only gradually overseas. Leading the way was the nascent U.S. Navy, which by the 1890s, as previously noted, was undergoing a renaissance of professionalization and industrialization. In 1898, after several years of budget fights and problems of shifting foundations (both geological and political) and in the middle of the Spanish-American War, David W. Taylor, who had been trained as a British naval constructor and studied British experimental techniques at firsthand, finally oversaw the completion of the 140-meter Experimental Model Basin (EMB) at the Washington, D.C., Navy Yard. Although naval work took priority, Taylor made time for commercial work, such as in 1919 when he tested a model of William Gibbs's concept for a passenger liner. Taylor also saw the promise of the airplane, and adjacent to the tank he built one of the first wind tunnels in the United States. The commandant of the Navy Yard, however, neither understood nor appreciated the importance of the scientific work conducted at the facility; on 19 June 1919, he issued an order "permitting the use of the model basin as a swimming pool by the officers of the yard and their families."[14] Despite this ignominy, the EMB became best known for the Taylor Standard Series, a methodical series of hull forms (like the later BSRA Series) based on models of the British cruiser HMS *Leviathan*. When the series was published by Taylor in his 1910 magnum opus *The Speed and Power of Ships*, it instantly became an industry standard for quickly deriving powering characteristics of hull forms.

The increasing size and speed of naval ships meant that a new tank was needed. In 1939, under the direction of Harold E. Saunders, the David Taylor Model Basin was opened in Carderock, Maryland, just 10 miles outside the nation's capital, on a solid granite plateau overlooking the Potomac River (figure 6.4). At 360 meters, it was at the time the largest basin in the world, and so long that the tracks had to be adjusted to the curvature of the Earth. Taylor himself was then 75, just months away from death, and for the last eight years had been confined to a wheelchair by left-side hemiplegia (paralysis). Nevertheless, he attended the opening of basin and admired the handsome

**Figure 6.4**
Harold Edward Saunders (this page) and the David Taylor Model Basin (opposite page).
Credit: Naval History and Heritage Command.

art deco architecture of the surrounding buildings. The opening was just in time; the basin was soon placed on a war footing, testing hundreds of ships, submarines, and weapons that would be used in World War II. During and after the Cold War, laboratories for research ranging from maneuvering and seakeeping to structures to acoustics were built and continue to carry out groundbreaking research. Today, after several rebaptisms, the facility bears the unwieldy title Naval Surface Warfare Center Carderock Division, but old-timers, including me, still refer to it simply as David Taylor.[15]

The first towing tank built for academic purposes was opened in 1904 at the University of Michigan in Ann Arbor. Mortimer Cooley worked with Herbert Sadler to design a 90-meter tank that would be integrated into the foundations of the new West Engineering Building at little extra cost. The tank found immediate use as an instructional and basic research facility, as well as a place for testing by commercial shipbuilders in the notably underserved Great Lakes region. Since then, most university schools of naval architecture around the world have incorporated towing tanks in their facilities (such as the Davidson Laboratory at the Stevens Institute of Technology and the University of Tokyo Experimental Tank, of which more later) to instruct students in the practice of ship model testing, to carry out fundamental research in ship theory, and to perform testing for clients. The United States was also home to a one-of-a-kind, independent model basin (not affiliated with government, university, or shipbuilder) in Laurel, Maryland. Hydronautics, Inc. was a naval architecture consultancy established in 1959 by two former navy scientists, Marshall Tulin and Phillip Eisenberg, who used their personal savings to build a 126-meter tank. For over 40 years, Hydronautics carried out fundamental research on cavitation, viscous drag, and propeller design for government and commercial clients. The facility closed in 2001, and the site is now covered by suburban homes.[16]

Japan was the next non-European nation to develop an indigenous model-testing capability and followed an accelerated path in transferring technical know-how from Britain. Early in the Meiji Restoration, Japan asked the French constructor Louis-Émile Bertin to help lead the renaissance of the Nippon Kaigun (Imperial Japanese Navy) for Japan to become a strong regional power. By the turn of the twentieth century, Japan had turned its sights to Britain as its role model, since like Japan it was a small island nation with few resources yet still managed to dominate an entire hemisphere. To design and build its fleet, Japan consciously and carefully set up technology transfer programs to wean it off imports and develop its own critical industries. By 1907 it had set up joint venture operations with the British companies Armstrong and Vickers to fabricate steel, armor plate, and guns and was building steam turbines under license from Charles Parsons's company. The Mitsubishi Nagasaki Shipyard, which had started as a small ironworks in 1857 to become Japan's naval industrial powerhouse, was the first to build an experimental tank. After learning about the advantages of such a facility from Frank Purvis, then chair of naval architecture at Tokyo Imperial University and who had worked at both Froude's original tank and Denny's, the Mitsubishi shipyard carried out a series of technical visits to the United Kingdom in 1905. By 1907 had brought back instruments from Kelso in Glasgow and several Denny engineers to help outfit the 130-meter tank, which made its first experimental runs in 1908. By 1912 it had tested over 100 models of both commercial and naval vessels.

At exactly the same time that Mitsubishi was building its Nagasaki tank, Japan's navy was constructing its own 137-meter model basin in the Tsukiji district of Tokyo. The navy first carried out a systematic analysis of its capabilities, deciding which parts of the tank it could build (concrete tank, motors, buildings) and which it needed to import (main carriage, model shaping machines, instrumentation from Munro of London). By 1909 the facility was carrying out hull and propeller tests for naval ships. In 1930, the Japanese government established a National Experimental Tank (139 meters) in Meguro, Tokyo, along the lines of the British NPL at Teddington, for carrying out basic research as well as for use by industry. In 1937 Tokyo Imperial University built a tank for teaching and basic research. Most of these facilities were heavily damaged or destroyed during the intense conventional and atomic bombings of World War II, but rebuilding was carried out in earnest during the 1950s, which was key to the astonishing rise of Japan as the world's leading shipbuilder from the mid-1960s until the present day. Japan today has over 60 model basins at almost two dozen facilities around the nation.[17]

The pace of building model basins accelerated after World War II and spread across the globe. South Korea and China established testing tanks in 1953, India opened its first facility in 1956, Brazil in 1956, Finland in 1969, Chile in 1973, Canada in 1977, Australia in 1982, Indonesia in 1986, and Turkey in 1988. Today, according to the International Towing Tank Conference (of which more later), over 100 model basin facilities carrying out research and development in naval architecture are in operation in 30 countries.[18]

## Two Faces of the Same Problem: Ship Hydrodynamics and Aircraft Aerodynamics

The widespread development of model basins at the turn of the twentieth century coincided with the sudden appearance of aircraft. Right from the start, naval architects noted striking similarities between the dynamics of seagoing vessels and aerial vehicles. For example, Dupuy de Lôme's dirigible airship of the 1870s was a direct inspiration for the submersible *Gymnote* that he and his acolyte Gustave Zédé developed in the late 1880s. Two decades later, a French journalist named Henri Noalhat described the similarities in the two craft in a small book, *Aerial Navigation and Submarine Navigation: Two Faces of the Same Problem*.[19] Aircraft engineers dubbed the body of the airplane its hull and brought into use the nautical terms "port," "starboard," "roll," "pitch," and "yaw" and even used "wetted surface" for calculating drag and a form of the metacentric equation to determine aircraft stability. Even the problems of aircraft model scaling and interference effects with wind tunnels were initially solved using naval architectural methods. By the 1930s, however, the transfer of knowledge was irrevocably reversed as empiricism gave way to more fundamental, physics-based research.[20]

Some of the earliest research into aircraft aerodynamics took place at ship model basins. David W. Taylor, as already noted, built one of the first wind tunnels in America at the EMB in Washington, D.C., in 1911, and he populated the corresponding Aircraft Division with crossover naval architects like William McEntee and Jerome C. Hunsaker, who worked on flying boats. Similarly, in Britain, submarine designers like Charles I. R. Campbell were assigned to develop rigid airships like the ill-fated *R38*. In Germany, Johann Schütte of the Norddeutscher Lloyd Model Basin cofounded the famous Schütte-Lanz Airship Company, Germany's only real competitor to the Zeppelin Company. In fact, the streamlined shape of these German airships was derived directly from Rankine's 1863 study of sources, sinks, and streamlines, and these airships in turn led to the development of the teardrop shape for submarines adopted after World War II (figure 6.5).[21]

Aerodynamic research was given a major push in the United States in 1915 with the formation of the National Advisory Committee for Aeronautics (NACA, the precursor to NASA), which from the start was closely linked with the naval architecture community. For most of its first 25 years it was headquartered in the Main Navy Building in Washington, D.C. NACA immediately started funding hundreds of proposals for critical research into aerodynamics and aircraft structures, among which was a systematic series of tests of airplane propellers by William F. Durand, who in 1897 had conducted similar testing on ships' propellers to gain comparative results across a range of design parameters (figure 6.6).

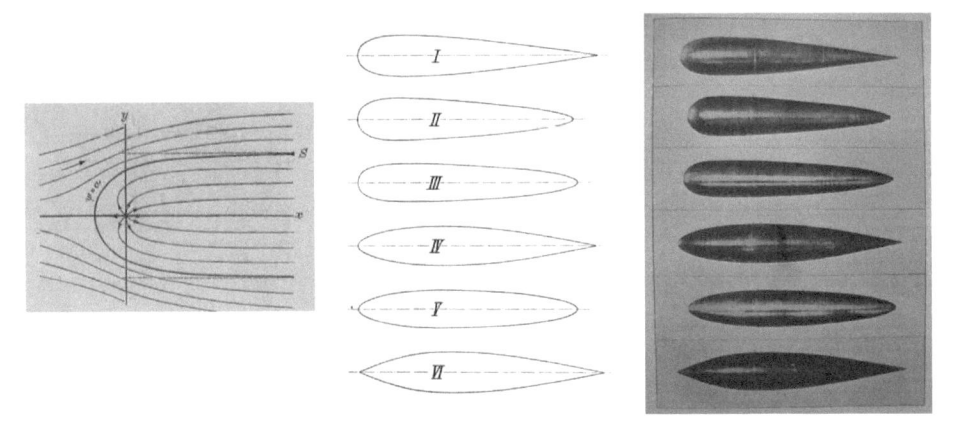

**Figure 6.5**
Left to right: Rankine-type streamlines, mathematical forms using sources and sinks, and wind tunnel models of streamlined dirigibles. The form of least resistance used for dirigibles is figure IV, which led to the teardrop shape for modern submarines.
Credit: Archive Deutsches Museum, Munich.

The early success of knowledge transfer from ships to aircraft led several classification societies, like the American Bureau of Shipping, Lloyd's Register, and Bureau Veritas, to develop classification rules for construction and survey of aircraft and for the issuance of airworthiness certificates in the same way they classed ships. In 1929, the Aircraft International Register was established to coordinate the development of an internationally accepted set of classification rules. But the use of a ship-based classification scheme for aircraft was never a good fit—ships relied on rule-of-thumb formulas and small-scale models, while aircraft designers were already employing first-principle calculations of physics and mechanics and prototyping them at full scale. Within a few years of their establishment, most classification societies shuttered their aeronautical branches, and by 1939 the Aircraft International Register was disestablished as governments took on the role of issuing airworthiness certificates.[22]

By the 1930s, much of the normal science of ship hydrodynamics had reached its limits, as empirical rules proved inadequate to explain what researchers were finding in their test results and full-scale predictions; naval architects complained that there was a "crisis in ship-powering prediction" and argued that "there should be less doctrine and more Physics" in ship design.[23] At the same time, aircraft designers were abandoning their naval architecture roots and using increasingly sophisticated tools and models. This was shown by governments pouring far more money and resources into aircraft aerodynamics than into naval hydrodynamics; in 1930, the United States spent just

**Ship propellers
Cornell (SNAME), 1897**

**Aircraft propellers
Stanford (NACA), 1916**

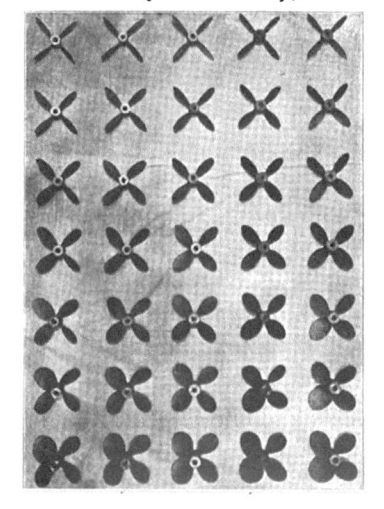

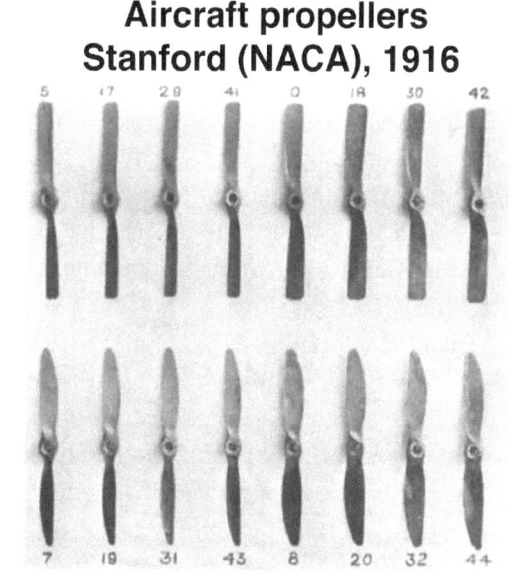

**Figure 6.6**
William Durand's systematic propeller series (left) and NACA propellers (right).
Credit: Author's collection.

$100,000 on its single EMB, while NACA was funding a dozen wind tunnels to the tune of $1.3 million.

It was apparent that the "Physics" necessary to build a solid theoretical foundation for naval architecture was to be found not simply by extending current work in ship hydrodynamics but by borrowing insights from the adolescent discipline of aircraft aerodynamics. This change in fundamental scientific assumptions is often characterized by the historian Thomas Kuhn's phrase "paradigm shift," but in this instance it is more accurate to call it a "flow reversal" between the two disciplines.[24] This shift or reversal occurred primarily in three domains: hull frictional resistance, propeller theory, and maneuvering theory. Many of these breakthroughs occurred in German laboratories, which as noted earlier had always punched above their weight in terms of worldwide impact on scientific developments, due to the close ties between academia, science, and industry.

The methods of predicting hull frictional resistance, based on the works of Froude, Reynolds, and Tideman, had changed little since the 1870s. Small models were still run in testing tanks, with correlation factors applied to bring the results to full scale.

It was not a completely satisfactory approach, as it did not accurately portray full-scale surface roughness or hull form effects. The reigning problem was that there was no significant understanding of what happened at the surface of the body that could be used to develop a more accurate picture of what contributed to frictional drag. Observant sailors and shipbuilders had known for ages that the water directly on the skin of a ship moved slower than the surrounding water, forming a sort of boundary layer; as one shipbuilder explained in 1836, "That portion of the water accompanying a ship which is immediately next to her body is not passed by so quickly as that a little distance further off, etc. Consequently, the velocity of a ship through the water is actually less than the velocity with which she moves forward."[25] Froude and Rankine also noted the existence of a boundary layer, but it would not be until the twentieth century that a new breed of mathematically trained physicists would develop a working theory of how fluids behaved right at the surface of a solid body.

The search for a new paradigm in frictional resistance took wing in Germany in 1904, the year after the Wright brothers' first flight, when Ludwig Prandtl developed his concept of the boundary layer in fluids. His great insight was that there existed not one but two regimes of frictional resistance, which could be analyzed separately: (1) a thin viscous region, the boundary layer itself, where frictional effects were felt, and (2) the region outside the boundary layer, which was essentially inviscid flow. The boundary layer and nature of resistance differed depending upon whether the flow was laminar (generally at lower Reynolds numbers, i.e., smaller scale, slower speeds, and lower specific resistance) or turbulent (larger scale, higher speeds, and higher specific resistance), as seen in figure 6.7.

Prandtl's revolutionary insight was expanded by his students at the University of Göttingen's aerodynamics research institute, notably Heinrich Blasius (who worked

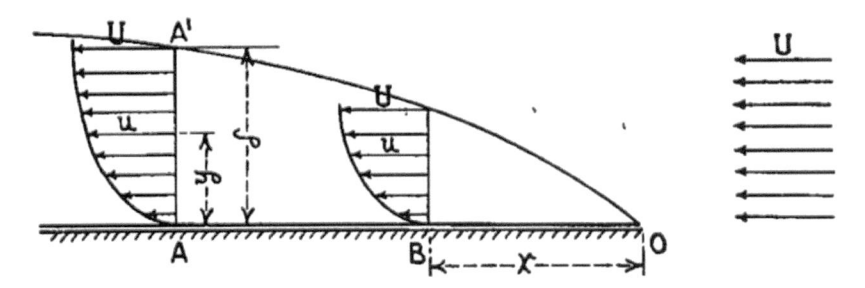

**Figure 6.7**
Boundary layer.
Credit: SNAME.

primarily in the laminar flow regime), Theodore von Kármán, and Hermann Schlichting (both of the latter explored turbulent flow). Turbulent flow, it was soon realized, was the critical domain to explore; although small-scale models often experience laminar flow conditions (and therefore lower resistance), at the scale of full-sized ships and aircraft the flow of air and water at the surface quickly becomes turbulent. Understanding the formation and characteristics of boundary layers in turbulent flow was crucial to accurate prediction of fluid-skin friction. All these researchers were in fact searching for a single nondimensional equation known as the friction line or skin friction coefficient ($C_f$) that relates drag force to surface area and Reynolds number ($R_n$), to predict fluid friction in turbulent flow.[26]

The German aerodynamics community had thus made enormous progress during the previous 25 years in the understanding of the boundary layer and its application to fluid friction, despite the upheavals of World War I and the subsequent deprivations under the Treaty of Versailles. Their work quickly influenced the conduct of aerodynamic research around the globe. The naval architecture community, however, demonstrated little interest in or knowledge of the subject, apart from work being done by Günther Kempf at HSVA and Friedrich Gebers at the Vienna Model Basin. Outside of these two centers, there was almost no mention of boundary layer theory in the papers of the Institution of Naval Architects, Society of Naval Architects and Marine Engineers, or Association Technique Maritime et Aéronautique.

That was about to change. In May 1932 Kempf and his colleague Ernst Foerster organized in Hamburg an international Conference on Hydromechanical Problems of Ship Propulsion (Konferenz über hydromechanische Probleme des Schiffsantriebs). It attracted enormous interest around the world, and the meeting room was filled with representatives from the major ship model testing basins in Germany, France, Sweden, Austria, the Netherlands, and the United States. This was the first time that the global community of naval architects received such wide exposure to the great theoretical leaps that had been made in the previous two decades by the German aerodynamic community. The primary subjects covered were frictional resistance and propeller hydrodynamics. This conference, more than any other event, served as the signal that a change in the direction of knowledge flow between the naval and aviation communities had begun, to where ship hydrodynamics began to be heavily influenced by aircraft aerodynamics.[27]

The first part of the two-day conference was devoted to current research on frictional resistance. Papers by Kempf and von Kármán dominated the proceedings, though there was also a short discussion of the effect of temperature on friction by a relatively unknown American naval architect of German origin, Karl E. Schoenherr. Schoenherr was at the time working at the EMB in Washington, D.C., while pursuing his doctorate on the subject of frictional resistance. Soon after the Hamburg conference, Schoenherr presented an equation that he stated best fit the hundreds of data points that he had laboriously plotted from model basins around the world, which both von Kármán and Prandtl agreed fit well with the data (see figure 6.11):[28]

$$\frac{0.242}{\sqrt{C_f}} = \log_{10}(R_n \cdot C_f).$$

Naval architects now took notice of the work on turbulent flow and boundary layer theory. Whereas before almost no naval hydrodynamic research had referred to studies at the Universities of Aachen and Göttingen, now those findings on frictional resistance took on great importance in response to the "crisis in ship-powering predictions." One direct result was that model basins recognized the need to stimulate turbulent flow in small-scale models (which, operating at lower Reynolds numbers, often had laminar flow over significant portions of the hull) in order to accurately predict full-scale, fully turbulent frictional resistance. On the basis of early work done by Kempf in Hamburg, model basins around the world began employing trip wires, pins, sandpaper, and other mechanisms near the bow of the ship model to generate turbulent flow over the entire body, a practice that significantly improved the correlation of model test data with full-scale data and is now taught to every first-year naval architect.

The second part of the 1932 Hamburg conference introduced naval architects to a new paradigm in propeller design, circulation. Prior to this conference, much of the research into shipboard screw propellers had been informed by the momentum theory and blade element theory developed by Rankine and Froude (discussed in chapter 3). Neither had proved satisfactory in calculating the actions of propellers in a way that allowed naval architects to overcome the "crisis in ship-powering prediction" that plagued them in the 1920s and early 1930s.

The concept of circulation had been developed in the early twentieth century almost simultaneously (as noted in chapter 3) by the aeronautical scientists Frederick Lanchester in Britain; Nikolai Joukowski in Russia; and Wilhelm Kutta, Ludwig Prandtl, and Max Munk in Germany. The key insight to their theories is that a two-dimensional airfoil generates lift from two sources: (1) asymptotically rectilinear uniform flow and (2) nonvanishing circulation around the foil. The circulation around a three-dimensional airfoil also induces drag as the airfoil sheds vortices from the tips (figure 6.8). These theories were given a boost after World War I by NACA, which paid expatriate German scientists handsomely as part of a concerted effort to transfer technical and scientific knowledge out of Europe and to the American aeronautical community. This continuing research into circulation theory became the basis for the now-famous NACA profiles,

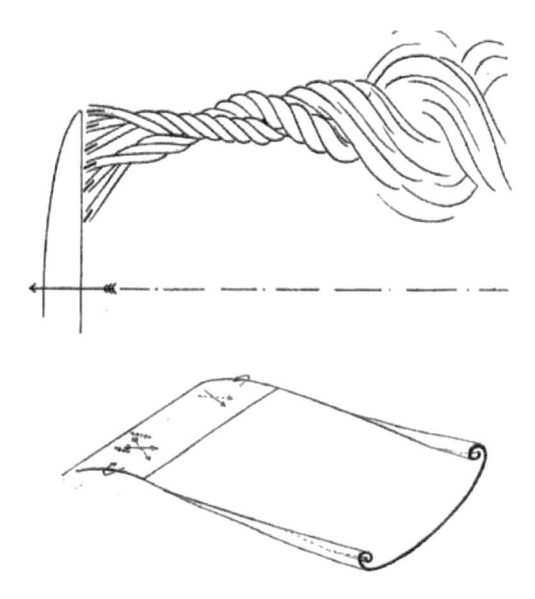

**Figure 6.8**
Fluid circulation and tip vortices around an airfoil.
Credit: Archive Deutsches Museum, Munich.

a systematic series of airfoil shapes that are widely used to develop aircraft wings and propellers, as well as ship propellers.

The 1932 Hamburg conference introduced circulation theory to the wider naval architecture community, whose main problem was understanding the action of propellers in nonuniform flow behind ships. Unlike most aircraft propellers, which are mounted at the front of the wing or nose (thus in a free stream), a ship's propeller is mounted behind the hull, so that the wake (flow into a ship's propeller) varies greatly across the disk. The Hamburg papers gave some early indications of how to use circulation theory to modify existing blade element theories to account for the nonuniformity of flow. This led to further developments in the 1940s and 1950s by Leonard Burrill in Britain and Hermann Lerbs (by then at the David Taylor Model Basin) to combine traditional propeller modeling techniques with lifting line theory—a three-dimensional subset of circulation theory—to more accurately predict propeller behavior in a ship's wake.[29]

The third flow reversal, maneuvering theory, was not a subject at the 1932 Hamburg conference but did occur roughly simultaneously with the shifts in resistance and propeller theories. As noted in chapter 3, early studies were concerned with simple measures such as tactical diameter, which addressed the kinematics but not the dynamics of the problem. By contrast, from the earliest days aeronautical engineers were modeling the aerodynamic forces on airplanes and airships both mathematically and empirically. In 1911 a British mathematics professor, George H. Bryan, worked out the six-degrees-of-freedom equations of motion for an aircraft based upon forces in straight-line motion and rotary motions of roll, pitch, and yaw. The rotary forces were particularly difficult to resolve, but Bryan suggested a method using a whirling arm to measure control derivatives in a nondimensional form. Most European aeronautical laboratories—for example, at Saint-Cyr in France and the NPL in Britain—were discontinuing the use of their whirling-arm apparatuses to measure airflow, but the Bryan equations gave them a new lease on life. In Britain, for example, Leonard Bairstow used one to develop equations for rolling during a banked turn.

In France, the whirling arm housed in the *manège* (roundabout) of Saint-Cyr became the focus of attention for aircraft stability. Maurice Roy built upon the work of Bryan and Bairstow to develop equations of motion that explicitly accounted for wind gusts. Both the *manège* and Roy's work came to the attention of French naval architects, who were searching for a methodology to more rigorously calculate a ship's course stability, or its ability to maintain course after a perturbation. In 1939, the planned expansion of the Paris model basin included a *manège* and *bassin de giration* (rotating-arm basin), based directly on the Saint-Cyr model (figure 6.9). The timing was terrible, of course, for the German occupation and bombings of World War II delayed its opening until 1942. After the war, course-stability and maneuvering research (employing Roy's original concepts of aircraft motion) went into full swing. These experiments were most famously performed under the naval architect Jean Dieudonné, who developed the eponymous spiral maneuver that is used even today to measure a ship's course stability.[30]

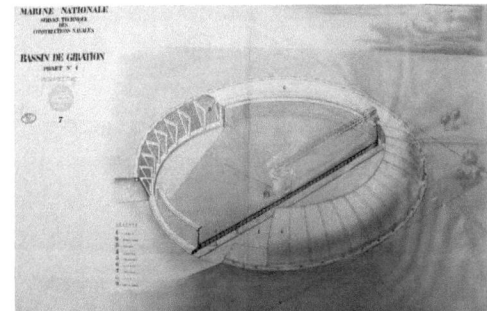

Manège à Saint-Cyr (1911)     Bassin de giration à Paris (1939-1942)

**Figure 6.9**
Whirling-arm facility (left) and rotating-arm basin (right).
Credit: Direction des Constructions Navales.

At roughly the same time in the United States, aircraft maneuvering theory was being adapted to naval use by Ken Davidson, a former military pilot and yacht racer, now at the Stevens Institute of Technology. Davidson's early work on the subject during the 1930s was limited to kinematic experiments he made in the swimming pools of Stevens Institute and Columbia University, though only on nights and weekends when they were not in use. During World War II, the need to more accurately predict the performance of destroyers led Davidson to the work of Leonard Schiff, a nuclear physicist who was examining antisubmarine operations. Schiff alerted Davidson to the large whirling arm at the Guggenheim Airship Institute in Akron, Ohio, which had been built a decade earlier to examine the rotary forces on rigid airships. Davidson realized that he needed a similar capability for his own work and applied to the U.S. Navy for funding. By 1945 a rotating-arm facility was up and running at Stevens Institute, which Davidson and Schiff used to further develop the theory of ship maneuvering and control.

The Davidson and Schiff paper that gave the results of these investigations, "Turning and Course-Keeping Qualities" (1946), explicitly showed the influence of the aerodynamics community. For the first time, forces and dynamics were directly addressed in ship maneuvering theory, by following aerodynamic practice using essentially the same equations first employed by Bryan. The hydrodynamic forces developed experimentally using the rotating arm provided systematic, quantitative results for control derivatives. This marked a turning point in the assessment of ship maneuvering and control; within a few years of the facilities at Paris and Stevens Institute began to use these equations, rotating-arm basins became regular fixtures at major naval architecture research establishments, and the aerodynamics-based representation of ship motions invoked by Davidson and Schiff became the standard for model basins around the world.[31]

## War Work

The work by Davidson and Schiff on destroyer maneuvering during World War II was but one of countless examples of naval architecture research contributing to the national defense in wartime. Most model basins and naval laboratories were of course built with navy (or other government) funds and therefore had defense work as part of their remit. But in wartime, this remit went well beyond the usual routine of hull resistance and propulsion testing, often involving near-impossible deadlines, tests of seemingly outlandish ideas, or attempts to extend the state of knowledge far beyond its known boundaries.

World War I was the first major conflict in which scientific research was placed on a war footing at the national and international scale. In Britain, for example, considerable effort was expended to defeat the emerging U-boat and sea mine threat from Germany, most famously in the development of ASDIC (a precursor to sonar), which navies in Europe and the Americas also adopted. NPL Teddington carried out important research into antisubmarine nets and tank tests of towed minesweeping gear. In other countries, however, lack of manpower due to mobilization of the workforce had different impacts. In France, work at the Paris basin was reduced to just a few resistance and propeller tests and some studies of submarine boats. The United States, by contrast, brought in female "laboratorians" to fill vacant technical posts at its naval experiment facilities, allowing the model basins to carry out and even expand their lines of research.[32]

The first wave of international technology transfers began in the interwar period, fueled by the rise of Adolf Hitler in Germany. Max Munk and Theodore von Kármán were Jewish, and growing apprehensive about the increasingly anti-Semitic tone of the National Socialist (Nazi) Party, they came to the United States in 1920 and 1936, respectively, to help lead research into frictional drag and lifting line theory. Other German advances in diesel engine technology and structural theory were brought to the EMB and other Navy laboratories, in many cases via German technical reports that were translated by Matthias C. Roemer at the EMB.[33]

In the Soviet Union, the path of technology transfer was fueled by the rise in 1924 of Stalin as the new national leader and successor to Vladimir Lenin. Stalin's proposed naval expansion was severely hampered by the lack of naval architects and marine engineers, many of whom fled the country during the Russian Revolution, and by his later Great Purge decimating a whole new generation of engineers. Russian naval leaders turned to Mussolini's Italian regime as the main source of naval technical and design expertise. Beginning in 1925 with visits by Italian naval delegates and port calls to Leningrad by Italian warships, by 1934 there was a robust and continuous exchange of naval architects between the two nations. The Ansaldo company in Genoa became the preferred design agent to help the Russians. Italian language classes were established in Leningrad, while Russian engineers took their families to Genoa. The pride of this exchange was the *Kirov*-class (Project 26) cruisers, begun in 1935 and based on Ansaldo's *Raimondo Montecuccoli* light cruisers but with considerable redesign of the structures to account for Russian research into longitudinal framing.[34]

After the outbreak of World War II in 1939, the mobilization of science to support the war effort dwarfed the efforts made in World War I. In Britain, the renamed William Froude Laboratory at Teddington was overwhelmed by special projects on top of the ceaseless demands for hull, propulsion, and seakeeping tests of new warships and submarines. Some of the more noteworthy model testing projects were the Habbakuk seadrome, a two-million-tonne ice-and-sawdust ocean landing platform, never built; amphibious tanks to ford rivers and invade beaches; and the Dambuster bouncing bombs, developed by Barnes Wallis to breach a series of dams in the industrial Ruhr valley, which they did in May 1943. The most famous of these special projects was the development and testing of the Mulberry harbors, floating breakwaters championed by Winston Churchill that protected the cargo offloading positions at Normandy after the June 1944 invasion.[35]

Even before the United States entered the war in December 1941, it was ramping up its scientific and design efforts to support what would become an explosion in shipbuilding. The nation would ultimately deliver 43,000 vessels of all types to maritime service, many of which were developed and tested at the David Taylor Model Basin. To accommodate this veritable eruption in duties, its workforce grew from 200 in 1940 to almost 1,000 by war's end, many of whom were women employed as technical assistants (see chapter 5), and all of whom worked on one of the two to three shifts per day, including on weekends. Critical projects included new torpedo developments, testing of landing craft both in open water and in beaching, and ship propeller and noise reduction to avoid the submarine threats. Like their British counterparts, American naval architects also worked on special projects, including the welding board of investigation (see chapter 4) and on tests to determine whether aircraft carriers could safely transit the Panama Canal.[36]

The Allied forces began intensive efforts to recruit and extract prominent German scientists and engineers even before the fall of the Nazi regime in April 1945. The best known of these efforts was Operation Paperclip, which brought more than 1,200 technical personnel to the United States, including Wernher von Braun, who would lead NASA's rocket programs that launched the Apollo vehicle to the moon in 1969. The U.S. Navy, already aware that future battles would likely take place underwater as well as on the surface, was particularly interested in Germany's expertise in hydrodynamics, submarine design, and acoustic quieting. It recruited many naval architects to work at the David Taylor Model Basin, including Georg Weinblum, an authority on wave-making resistance; Hermann Lerbs, an expert on the noise-producing phenomenon of cavitation; and Heinrich Heep, who had helped Hellmuth Walter develop his advanced high-speed submarines, which saw only limited service during the war (Walter himself was recruited by the British). Other U.S. and British naval technical missions were dispatched to postwar Germany, where they dismantled and brought back laboratory equipment (e.g., the Pelzerhaken Cavitation Tunnel, brought in 1947 to Newcastle University) and advanced weapons and machinery. This combination of talent and technology would give the United States and its allies a strategic edge over the Soviet Union during the early years of the Cold War.[37]

**International Cooperation**

The Cold War (1947–1991) both refuted and proved Joseph Banks's adage that "the sciences were never at war." On the one hand, the Western bloc (primarily NATO, the North Atlantic Treaty Organization) deployed and jealously guarded its technological superiority in areas like missile guidance, radar stealth, and submarine acoustics to offset the numerical superiority of Eastern bloc (Soviet Union and allies) forces. On the other hand, the Cold War period saw an unprecedented level of international collaboration and knowledge sharing in basic sciences at all levels of government, industry, and academia—such as the International Geophysical Year 1957–1958, which coordinated earth and space sciences among scores of nations on both sides of the Iron Curtain that separated the Eastern and Western blocs. For naval architects, that international cooperation was most famously embodied in the International Towing Tank Conference (ITTC).

The ITTC was a direct outgrowth of the aforementioned 1932 Hamburg conference, which brought together representatives from the major ship model testing basins around the world. The Hamburg conference was intended as a one-time event, similar to the many international conferences and symposia that had been occurring ever since the Great Exhibition of 1851. However, at an after-dinner speech at the conference, the consulting engineer Giovanni (also called Jan or John) de Meo, at the instigation of Dutch naval architect Lauren Troost, "pleaded strongly for international technical cooperation" in this domain and a continuing series of meetings to exchange information. In July of the following year, Troost hosted the first International Conference of Tank Superintendents, in which the heads of 22 European model basins met in The Hague to outline the framework for coordinating research and cooperation between them (figure 6.10). The goals, as de Meo explained at the conference opening, were "establishing the fundamental classification and nomenclature" and creating a standard system of publishing results. The conference met annually for three more years in various locations, presenting (for example) the results of testing the same model hull in different national tanks to establish correlations among them and comparing different procedures used by towing tanks to conduct model self-propulsion tests. The last meeting was held in Berlin in late May 1937, at the same time the Nazis openly began raids on the Jewish communities in the city, which the conference attendees undoubtedly would have been able to witness firsthand. The deteriorating political situation, followed by World War II, meant no subsequent conferences were held for another decade.[38]

| 1. G. S. Baker | 7. K. Nakamura | 13. J. F. Allan | 19. M. Legendre |
| 2. E. G. Barrillon | 8. G. Kempf | 14. E. Castagneto | 20. H. S. Howard |
| 3. N. Kal | 9. F. Horn | 15. J. M. Burgers | 21. W. P. v. Lammeren |
| 4. E. Vossnack | 10. F. Gebers | 16. H. Munday | 22. W. v. Beelen |
| 5. L. Troost | 11. A. W. Riddle | 17. A. van Driel | |
| 6. J. de Meo | 12. T. B. Abell | 18. H. M. Weitbrecht | |

**Figure 6.10**
First International Conference of Tank Superintendents attendees, The Hague, July 1933. Lauren Troost and John de Meo are in right front row.
Credit: Author's Collection.

The first postwar conference was held in London at the Teddington facility in September 1948, just weeks after the city hosted the first postwar Olympic Games, dubbed the Austerity Olympics because of the harsh economic climate gripping the nation. Despite the pervasive rationing, the participants were treated to sumptuous lunches and cruises on the Thames, while their wives visited Windsor Castle. Conference attendees began the much-needed work of standardizing research among the participating facilities, in much the same fashion that postwar industries were already standardizing products and services in a rapidly globalizing world economy. Instead of ad hoc work by different members, the conference established technical committees to rationalize procedures and reporting in ship resistance, propulsion, cavitation, and so on, across the different facilities. The Presentation Committee created standard nomenclature and symbols (and which this book follows). From 1951 to 1954 the conference was renamed the International Towing Tank Conference and held meetings in different nations every three years, including those within the Soviet bloc (Leningrad hosted the 1981 conference, during the Cold War). Today, the ITTC has over 200 members from over 70 organizations around the globe, maintaining standardized procedures on everything from cavitation experiments and ice model testing to computer-aided computational fluid dynamics.

The first ITTC gatherings occurred in the middle of the Great Depression (1929–1939). American naval architects and their sponsoring agencies simply could not afford to make the transatlantic crossings to attend the conferences. Ken Davidson of the Stevens Institute proposed establishing a parallel organization of U.S. and Canadian (and later, South American) model basins with the same purpose of coordinating and standardizing research. The first meeting of the American Towing Tank Conference (ATTC) took place at Stevens in 1938 and from then on met almost uninterrupted by war until the present day. Among its accomplishments was the adoption in 1947 of Karl E. Schoenherr's equation (given earlier) as the standard turbulent friction line to be used for all ship and model resistance calculations in U.S. and Canadian ship model basins; in 1957, the ITTC adopted a version of Schoenherr's equation, as modified by George Hughes at the British NPL (see figure 6.11 for the evolution of the turbulent friction line). Although ITTC and ATTC at first had similar goals of defining standard procedures, the ATTC has now largely ceded this activity to the ITTC and today focuses on reviewing new experimental and numerical methods that could be considered in future by the ITTC. Other nations like Japan and Korea have also established separate towing tank committees for research coordination.[39]

International cooperation in naval architecture took place in domains other than hydrodynamics. As noted in chapter 4, problems with welded ships during World War II led to the establishment of a welding board of investigation to identify and fix the underlying causes. In the board's final report in 1946, it recommended "that an organization be established to formulate and coordinate research in matters pertaining to ship structure" on a permanent basis. From that date, the Ship Structure Committee was formed with members from the U.S. Coast Guard, U.S. Navy, other U.S. government agencies, and the American Bureau of Shipping; later the Canadian government joined the organization. Today the committee sponsors research projects into many areas, such as aluminum fatigue and behavior of steel at low temperatures.[40] Other

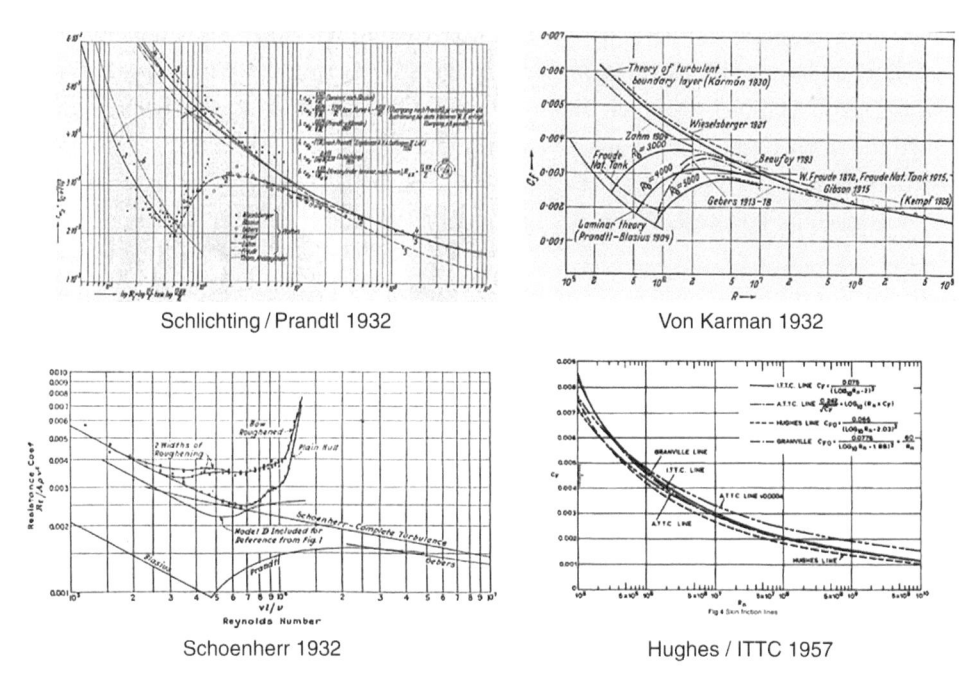

**Figure 6.11**
Evolution of the turbulent friction line, 1932–1957.
Credit: Archives Deutsches Museum, Munich.

international naval architecture research is sponsored through the NATO Naval Armaments Group (e.g., on seakeeping and hull loads in a seaway), which also coordinates and publishes formal NATO standards (STANAGs) and Allied Naval Engineering Publications (ANEPs) to codify procedures and results.

## The Endless Frontier

Although the early twentieth century saw several examples of government-funded research into science and engineering (notably the model basins described in this chapter), World War II proved to be the catalyst for the wholesale consolidation of national research goals and funding to come under governmental control. France established the Centre national de la recherche scientifique (National Center for Scientific Research) in 1939, at the beginning of the war, to coordinate laboratory activities for the war effort. Reduced in scope during the Nazi occupation, afterward it became the overarching body to set national research goals and created a series of joint laboratories in many different domains, including ship hydrodynamics. The United Kingdom, by contrast, had limited the centralization of research funding to the medical field until 1965, when it established the Science and Engineering Research Council, which was intended to rationalize the haphazard nature of setting research priorities and assigning funding for, among other fields, fluid dynamics and water engineering (this has since been subsumed under the broader Research Councils UK). The German postwar system combined federal and state (*Länder*) funding to academia and public and private research institutions such as the Max Planck Institutes and the Fraunhofer Institutes, both of which conduct research into fields such as fluid dynamics. In Japan, science policy is coordinated through a cabinet-level position within the government, while funding is provided through the Ministry of Education, Culture, Sports, Science, and Technology, which supports (among other things) crosscutting initiatives in computational fluid dynamics.

In the United States, NACA had been the coordinating body for aeronautics since 1915, but its scope was constrained to aircraft, while excluding the many systems and weapons (radar, bomb sights) they were already carrying. Even as the conflict raged in Europe, in 1940 the Americans established the National Defense Research Committee to coordinate and conduct all scientific research related to the war. This was soon replaced by the even more powerful Office of Scientific Research and Development, which under the leadership of the engineer Vannevar Bush pioneered the development of radar, proximity fuses, and the atomic bomb, among its many other activities. Historians often repeat the statement by Lee DuBridge of the MIT Radiation Laboratory that while the atomic bomb ended the war, radar won it. Both technological developments were due directly to the phenomenal success of the office's centralized approach to government-sponsored science and funding scientists and engineers to carry out both basic research (understanding fundamental scientific phenomena) and applied research (using these phenomena to solve specific problems).

In November 1944, with the Allies definitively on the road to victory, President Franklin D. Roosevelt looked past the war's end and asked Bush to outline how he could replicate the success of the Office of Scientific Research and Development in a peacetime economy. Bush's report, completed after Roosevelt's death, was released in July 1945 just days after the first atomic bomb test in New Mexico (and two weeks before the attack on Hiroshima). Evocatively titled *Science, the Endless Frontier*, it was short—just 30 pages—and to the point: science was vital to the nation, for "without scientific progress no amount of achievement in other directions can insure our health, prosperity, and security as a nation in the modern world." It was as much a call to arms as was Roosevelt's "date that will live in infamy" speech just four years earlier, but here, the enemies were not the Axis powers but disease, famine, and defense insecurity. Bush proposed that the federal government be the primary patron for this renewed focus on science and that Congress create a national research foundation to direct and coordinate all research in medicine, natural sciences, and national defense, which would be carried out primarily by colleges, universities, and research institutes.[41]

In the event, Congress delayed until 1950 the creation of a smaller National Science Foundation, pointedly limiting its scope to carrying out civilian-focused research in basic science, such as mathematics, biology, physics, and hydrodynamics. Meanwhile, the U.S. Navy stepped into the military research gap, establishing in 1946 the Office of Naval Research (ONR) under the direction of Harold Bowen, a naval officer who had been at odds with Vannevar Bush during the war. Even then, naval research was not completely centralized under ONR; most notably, the development of nuclear propulsion for naval ships was given to Hyman Rickover at the Bureau of Ships. However, ONR's research structure was (and remains) very close to Vannevar Bush's original model; the agency provides direction and coordination, while the actual research (both basic and applied) is carried out by colleges, universities, and research institutes in the United States and worldwide (the latter administered by a series of field offices around the globe). ONR even helped fund a deep submergence vessel, *Aluminaut*, originally built on speculation by the Reynolds Metals Company to demonstrate the undersea market potential of aluminum but which proved highly valuable in oceanographic research and deep-sea salvage.[42]

One of the most enduring legacies of ONR has been the biannual Symposium on Naval Hydrodynamics, jointly sponsored with naval, academic, and civilian institutions around the globe. Begun in 1956 and continuing to the present day, it was the brainchild of Phillip Eisenberg, who had recently moved from David Taylor Model Basin to become a project manager at ONR (and later went on to found Hydronautics). Within the decade that had passed since the end of World War II, the number of model basins and other research facilities (e.g., civil hydraulics laboratories such as at the one at University of Iowa) had mushroomed, and it was already becoming problematic for scientists and engineers to keep track of the many developments in naval hydrodynamics. Eisenberg's goal in holding the symposium was to highlight the status of the field to the scientific community and also to establish an overall agenda that could help guide researchers around the globe to address items of critical importance.[43]

Right from the start, Eisenberg planned to make the symposium an ongoing event in the same mold as the ITTC and planned accordingly. He established a continuing funding structure and partnered with the National Academy of Sciences to help pay travel expenses for scientists and publish the proceedings. He also ensured that the symposium would be held in a different venue each time, which would strengthen the basic research partnerships with other organizations and allow the investigators to explore new collaborations in areas of fundamental significance. The Symposium on Naval Hydrodynamics quickly became one of the most important and well-attended conferences for basic naval architecture research. The papers, often numbering more than 100, are arranged into a dozen topic areas ranging from propulsors to cavitation, to viscous flows, to maneuvering and ship motions, and conference proceedings can run over 1,000 pages. Unlike many other technical conferences that take just one or two days, these are weeklong events, giving the 200-plus participants ample time during and after each day's events to plan the next scientific breakthroughs in naval architecture.[44]

**The Social History of the Bulbous Bow**

I end this chapter with a study of naval architecture's most visible artifact, the bulbous bow. The ISO bulbous-bow symbol is seen on every type of vessel around the world, from passenger ships to crude oil carriers (figure 6.12). The modern underwater bulb that projects from the front of the ship was conceived in the model basin, evolving from laboratory life to workaday utility.[45] The bulb partially cancels out waves formed at the bow, thereby reducing resistance and improving efficiency. But the use of projecting bulbs on ships dates back to antiquity. Despite the continuity of the artifact itself, the purpose and the perception of the bulbous bow have changed dramatically over the years.

The earliest evidence for projecting bows, known as cutwaters, are from art works dating around 850 to 700 B.C. These show long, slender extensions that are clearly too fragile to serve as reinforced rams for combat (figure 6.13). Additionally, the textual evidence indicates that rams were not used in combat until 550 B.C., almost 300 years after cutwaters appeared. Model basin experiments at the Stevens Institute of Technology showed that these cutwaters significantly reduce bow waves and decrease wavemaking resistance (figure 6.14). This study lends strong support to the idea that early shipwrights initially developed cutwaters as a means of improving hydrodynamic efficiency and allowing galleys to go faster under oar and under sail.[46]

By the time the cutwater had evolved into the weapon known as the ram, the ship types had also evolved from single-row galleys to warships with multiple tiers of rowers. The most famous type of ram warship was the trireme, with three banks of oarsmen to provide the power necessary to punch a hole in an enemy ship with its massive bronze waterline ram (figure 6.15). At the battle of Salamis in 480 B.C., the effectiveness of the ram was demonstrated when roughly 380 Greek triremes defeated a flotilla of

**Figure 6.12**
A modern bulbous bow on a Chinese merchant ship, with the international symbol for bulbous bows (ISO standard 6050:1987) just to the left of the anchor.
Credit: Alexandre Sheldon-Duplaix.

about 1,000 Persian and Phoenician galleys. The act of ramming involved not speed but intricate maneuvering, requiring close coordination of the oarsmen to rapidly changing commands and a keen eye to judge distance and angle of attack. Oared warships were highly vulnerable to having their oars swept away by an attacking ship, so the typical fleet formation consisted of ships in a tight line abreast to protect one another's flanks. Opposing fleets attempted to outflank or break through the lines to attack individual vessels.

The end of the Roman Empire and the rise of Byzantium saw a marked change in the nature of naval warfare, from large oared galleys to smaller vessels with sails that could be quickly lowered for battle. The waterline ram was replaced by an above-water,

**Figure 6.13**
Cutwater on oared galley circa 700 B.C.
Credit: William M. Murray.

iron-tipped spur, which was no longer a ship-killing weapon but a means of locking one ship to another in preparation for boarding. With the introduction of gunpowder during the 1400s, even this vestige of the ancient ram bow began to disappear, and by the 1500s, the cannon-armed sailing warship became the means of projecting power at sea, putting an end to the ram for over two centuries.

As explained in chapters 2 and 3, the trireme was invoked in the original 1842 concept of the screw-propelled ironclad ram as a way to improve the French fleet's inferior position with respect to the larger and more heavily gunned British fleet through the absolute combat of ramming. The first French ironclads fitted with rams, *Solférino* and *Magenta* of 1861, were countered the same year in Britain with two ram-equipped *Defence*-class ironclads. A year later, in March 1862, the Confederate ironclad CSS *Virginia* drew first blood in its assault on Union warships when it rammed and sank USS *Cumberland*. Over the next decade, the bow ram (figure 6.16) was seen by many navies to be the weapon of the future.[47] The Battle of Lissa in 1866, where the Austrian flagship rammed and sank its Italian counterpart, and the 1879 Battle of Iquique, where a Peruvian ironclad rammed and sank a Chilean steamer, seemed to presage a new way of war. As with the triremes of old, agility and not speed was essential for this mode of combat, and navies continued to improve the steering and maneuverability of warships throughout the late nineteenth century.

Nevertheless, in this period naval gunnery was improving to the point that close-quarters action was becoming outmoded as battle tactic. At the same time, it was increasingly apparent that the ram was more dangerous to friends than foes: in 1875 HMS *Iron*

**Figure 6.14**
Attenuation of bow wave with cutwater (bottom) compared with standard merchant-hull bow (top).
Credit: Stevens Institute of Technology.

*Duke* accidentally sank HMS *Vanguard* in a fog, in 1878 SMS *König Wilhelm* holed and sank SMS *Grosser Kurfürst* during an emergency turn to avoid another collision, and in the Battle of Lissa, the Italian ironclad *Ancona* rammed and badly damaged its own escort *Varese*. In fact, from 1865 to 1905 more ships were sunk by accidental strikes from ram bows than by intentional ramming in battle. HMS *Dreadnought* (1906) was one of the last warships to be built with a ram bow.

Although the industrial-age ram started as a weapon, observant naval architects like Joseph Woolley and Robert E. Froude noted that the lengthened bow had distinct

**Figure 6.15**
Ram bow of *Olympias*, a modern replica of a Greek trireme.
Credit: Trireme Trust.

hydrodynamic benefits. In 1865, side-by-side trials of two near-identical vessels showed that HMS *Helicon*, with a ram-type bow, was a knot faster than HMS *Salamis* with a waveline bow, which Woolley attributed to an "unexplained cause." In the 1880s, when tank test results showed that models with ram bows were faster than models with conventional bows, Robert Froude explained it away to the well-known phenomenon that longer ships generate less wavemaking resistance. There the matter lay for 20 years.[48]

The development of the bulbous forefoot at the turn of the twentieth century was directly inspired by William Froude's earlier experiments with the *Swan* and *Raven* models and occurred almost simultaneously in the United States and Russia. On the U.S. side, the bulbous forefoot was invented by David W. Taylor at the EMB; in 1905 (just when the ram bow was fading from use as a weapon) he tested a bulb at the foot of the bow, which reduced wavemaking resistance at high speeds. Taylor gave credit to Froude's work as his inspiration:

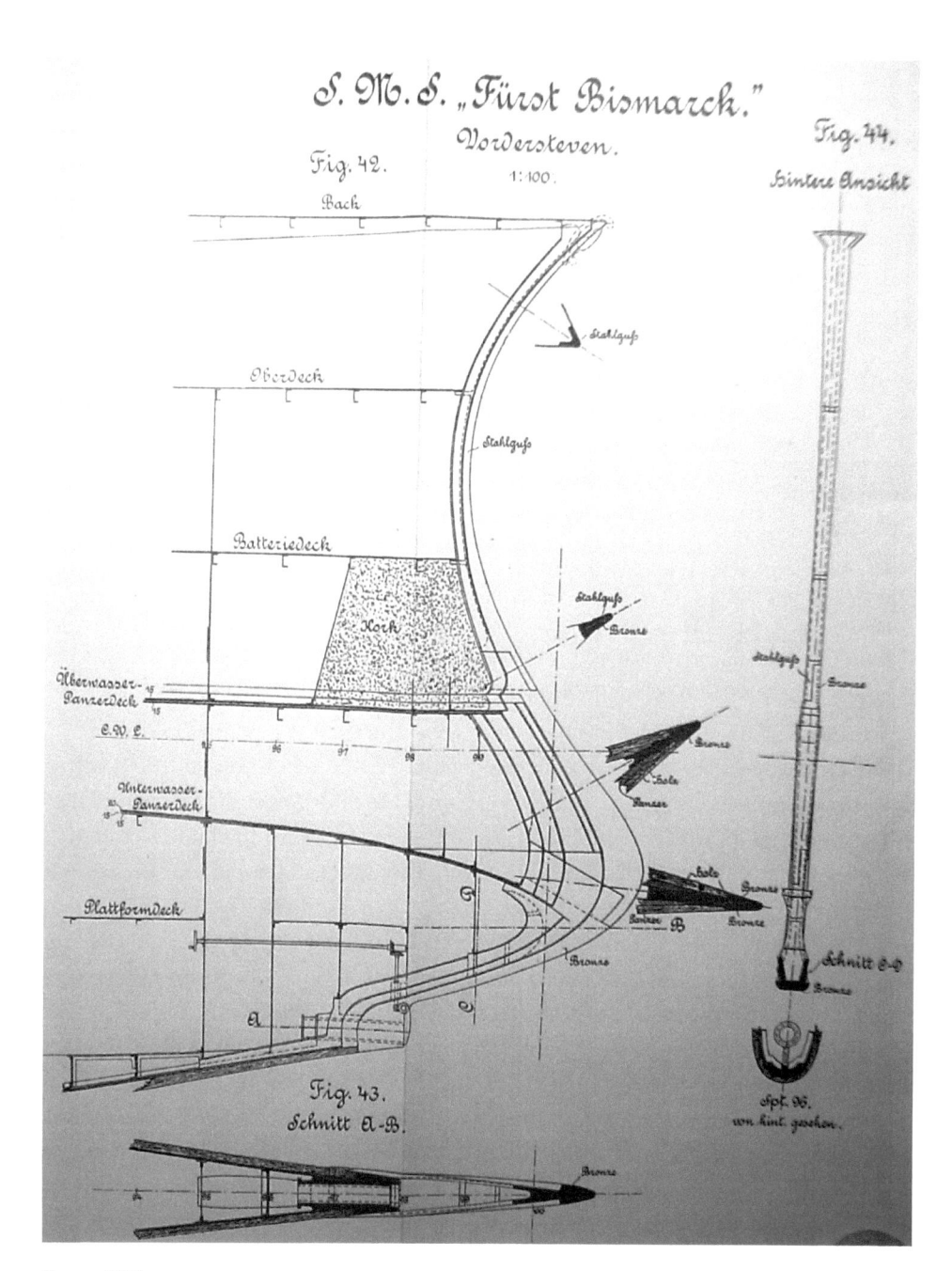

**Figure 6.16**
Cross section of the ram bow of armored cruiser SMS *Fürst Bismarck*, 1896.
Credit: Naval History and Heritage Command.

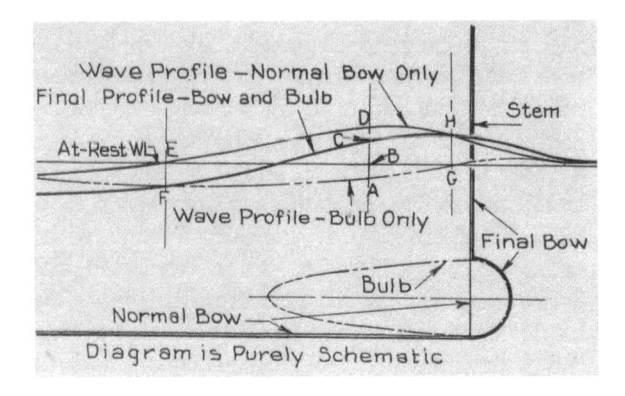

**Figure 6.17**
Mechanism of wave cancellation by an underwater bulb at the bow.
Credit: SNAME.

> The bulbous bow was not altogether the result of hit or miss methods. In the literature of ship resistance I had seen a reference (by Mr. William Froude) to a so-called swan model having full but narrow water lines close to the bow.... The theory seemed to be to have first a false bow as it were, corresponding to a small ship. This would create a small bow wave and in its hollow was located the second or true bow, making a second bow wave that would neutralize the first.[49]

Figure 6.17 shows the basic mechanism of wave cancellation described by Taylor.[50]

On the basis of these results, Taylor designed the BB 28 USS *Delaware* (1909) with a pear-shaped bulbous forefoot. From that time until the mid-1960s, all U.S. capital ships (battleships, aircraft carriers, and heavy cruisers) were built with the bulbous forefoot, sometimes referred to as the Taylor bulb. Taylor also designed the bulb for the passenger ship SS *Malolo* (1926) while working for the Gibbs Brothers. The Taylor bulb was soon a standard feature on passenger ships on both sides of the Atlantic; the German liners *Europa* and *Bremen* (1928) went on to capture the Blue Riband for fastest time across the Atlantic, owing in part to being fitted with *Taylorwulstbugs*.

Across the Atlantic, a pair of young Russian naval architects, unaware of Taylor's invention, developed their own version of the bulbous forefoot. In 1911 Nikolai Konstantinovich Artseulov and Vladimir Ivanovich Yourkevitch, also inspired by Froude's *Swan* and *Raven* experiments, tested a pair of models of a new class of battlecruiser *Borodino* (also called *Izmail*), one model with a bulbous forefoot and one without, at the Saint Petersburg model basin. However, they found only a slight improvement in powering, so the battlecruisers were built with a traditional cutaway stem.

Yourkevitch fled Russia in 1920 after the Russian Revolution of 1917. By 1930 he was in Paris as a lathe-turner at the Renault automobile factory while working as a consultant naval architect on the side. That year, Yourkevitch caught wind of a new transatlantic liner project and, through his connections with a former Russian admiral now living in France, proposed his bulbous forefoot to the Penhoët shipyard in Saint-Nazaire, which was building the ship. After tests at HSVA in 1931, which showed the Yourkevitch bulb gave a 5 percent improvement in resistance compared with a more traditional hull, Penhoët adopted the design. It became a prominent feature of the liner, now christened *Normandie* (figure 6.18), which captured the Blue Riband in 1935.

*Normandie*'s bulbous forefoot was seen in the popular culture as part of the streamlined aesthetic of the 1930s art deco era but given little thought as to what it actually represented. The popular press tripped over themselves attempting to (incorrectly) explain the workings of the bulbous forefoot, completely missing the wave-cancellation effect: the most repeated trope was that the bulbous forefoot "makes a hole in the water for the ship to travel in." As for Yourkevitch, he moved to New York in 1939 as war was looming over Europe. In 1942 he was on hand to watch his beloved *Normandie*, then being converted to a troopship, catch fire and capsize at the pier. Embittered and angry, Yourkevitch died in 1964 in Yonkers, New York, just as his famous bulbous forefoot was being eclipsed by an upstart invention from Japan.

The modern bulbous bow, shown earlier in figure 6.12, was developed by the Japanese naval architect Takao Inui not as a means to improve powering performance of ships but rather as an outgrowth of a careful, scientific study of how to control ship wavemaking. In 1943 Inui was a 23-year-old graduate researcher in naval architecture at the Tokyo Imperial University, just beginning his study of wave dynamics as World War II raged around him. He was following a long line of researchers who had sought to understand how ships produce waves, ever since John Scott Russell developed his wave-line theory a century earlier. In 1898 an Australian mathematician named John Henry Michell developed a remarkable, complex formula that explained the wave resistance of thin ship forms. In the 1920s and 1930s his work was expanded by scientists such as Thomas Havelock, Cyril Wigley, and Georg Weinblum to explain wavemaking, in particular how a deeply submerged sphere (representing a bulbous bow) could partially cancel hull waves.[51]

Japanese researchers had been closely following this work and were conducting hydrodynamics research on par with their European counterparts, although the information flow was strictly one way; they received the latest technical papers from Britain, France, and Germany, but little of their work was published outside Japan. Inui's research closely followed the work of Havelock, who had developed a purely analytic approach to understanding ship wavemaking, compared with Wigley, Weinblum, and others who concentrated more on experimental results for predicting ship resistance.

**Figure 6.18**
*Normandie* at its launching in 1932, showing its prominent bulbous forefoot.
Credit: Archive Deutsches Museum, Munich.

During his graduate research from 1943 to 1946, half of Tokyo was destroyed in bombing raids, but Tokyo Imperial University was largely spared and he was able to continue his work almost unhindered.

Inui was also spared from interruptions because, unlike the United States and Britain, which enlisted its research universities in the war effort, Japan kept academic and military R&D far apart. In fact, the National Experimental Tank at Meguro, just a few kilometers from Tokyo Imperial University, had been working completely independently on a very practical problem: how to ensure that the Imperial Japanese Navy's new super-battleship *Yamato*, the largest in the world, would achieve the required speed of 27 knots given its massive displacement and wide beam. After experimenting with over 50 variations of the Taylor bulb, the engineers hit upon a novel solution—a large bulb that, instead of being flush with bow stem, protruded forward three meters. This not only gave lower resistance than the regular Taylor bulb but also reduced the spray over the deck forward, which could affect gunnery in high seas. When launched in 1940, *Yamato* was the first vessel constructed with a protruding bulb, a fact that only later became known to Takao Inui.

After the war, Inui and his colleagues continued their research into wavemaking, making important contributions to the application of the Michell formula to ship forms and improving the theory for submerged spheres. In 1954 Inui traveled to Oslo for the seventh ITTC meeting to present a compilation of results from Japanese research. For the first time, European and American researchers took note of how advanced the Japanese science was, more astonishing still given the intensity of the war and the deprivations afterward. Flush with his success in Oslo, Inui now focused his attention on measuring wave patterns to derive the energy expended by the ship in wavemaking. In 1956, while still a doctoral candidate and not even a full professor (these would come in 1958), he authorized the University of Tokyo model basin to be equipped with a novel, and expensive, stereophotogrammetric system (funded with a grant from the Tōyō Rayon Company) to make three-dimensional images of the waves as the models traveled down the basin. To enhance the reflectivity of the water, aluminum powder had to be sprinkled on the surface, which was a mess to clean up afterward.

Inui's wave pattern experiments convinced him that it would be possible to make a "waveless" hull form (figure 6.19) by placing a protruding underwater bulb approximately 6 percent of the ship's length forward of the bow, with a corresponding bulb at the stern, whose wave pattern would cancel out the bow and stern wave patterns. Model tests done in 1960 confirmed that the bulbs almost completely eliminated wavemaking resistance, and trials of two passenger ferries of the Kansai Steamship Company, one retrofitted with a waveless bulb, showed it reduced the required power for top speed by 13 percent, or the equivalent of a half-knot increase (figure 6.20). However, when Inui

Fig. 19. First 'waveless' model C-201F2xA4 (L = 2.5 m)

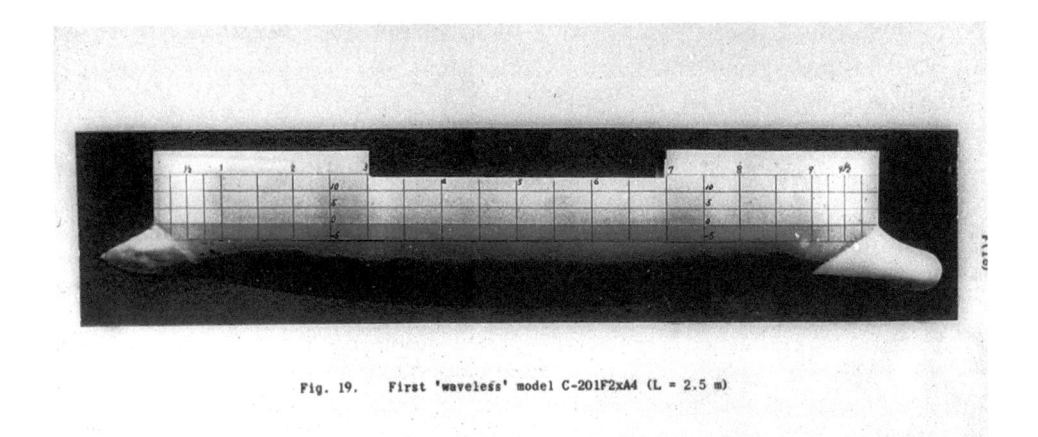

**Figure 6.19**
Takao Inui's first waveless hull form, in 1960.
Credit: Takao Inui.

presented these results to the 1962 Society of Naval Architects and Marine Engineers meeting in New York, he was greeted with open skepticism.

Naval architects were not the only skeptics. In 1961 Inui proposed the bow bulb for the new Japan National Railway ferry between Hokkaido and Honshu, but the ferry captains balked at the idea of a large bulb protruding in front of their ship, claiming that it would interfere with the ship's maneuverability. Another shipowner, Nedlloyd, also balked. It was only in 1963 that NYK Line agreed to have the Mitsubishi Nagasaki Shipyard construct its newest cargo ship *Yamashiro Maru* with a purpose-built bulb based on Inui's theory, which reduced fuel consumption almost 25 percent compared with a similar vessel.

By then, the initial skepticism had given way to serious interest. One of Inui's closest collaborators on the bulb, Tetsuo Takahei, spent two years at the University of Michigan researching its application to merchant ships, which led to the construction of a pair of Esso tankers and an NYK container ship fitted with bulbs. Within a few years, shipyards and shipping companies were developing their own bulbous-bow designs, often employing Inui's former students as their consultants. By the 1980s, bulbous bows were being designed and built around the world, and today the bulb has evolved into numerous shapes and configurations depending on usage, though naval architects still give the occasional nod to its inventor by referring to the "Inui bulb."[52] Although Inui took out patents on the bulbous bow, these were for protection only, and he never made any money from it, since public servants were not allowed to profit from research carried out for the government (University of Tokyo is a national institution).

**Figure 6.20**
Comparative tests on bulbous bow, Osaka Bay, 16 March 1961. *Kurenai Maru* with "waveless bulb" (top); *Murasaki Maru* without bulb (bottom).
Credit: Takao Inui.

Takao Inui (figure 6.21) received accolades worldwide as the "father of the bulbous bow." Yet throughout his life he remained surprisingly humble about his globally recognized invention. To me he expressed only a slight remorse, lamenting that today's ship designers seem to depend on the bulbous bow to make up for any errors, a cure for their lack of finesse in designing a proper hull.

In October 1964, *Yamashiro Maru*—the very first ship equipped, the year before, with an Inui bulb—collided with the East German cargo ship *Magdeburg*, which was carrying a load of British freight bound for Cuba. Like the rams at the battles of Salamis and Lissa, the bulbous bow of *Yamashiro Maru* punched a large underwater hole in *Magdeburg*, which quickly heeled over and sank in the middle of the Thames. This incident, occurring just two years after the Cuban Missile Crisis, caused an international uproar, with accusations of cloak-and-dagger sabotage by Western intelligence agencies. The accident was a harbinger of things to come. As with the iron rams of the nineteenth century, protruding bulbous bows are still a potential weapon because they pose the threat of turning a merely serious collision into a deadly catastrophe by puncturing the struck vessel below the waterline. The bulbous bow was implicated in the deadliest peacetime collision since World War II when in 1986 the passenger ship *Admiral Nakhimov* was holed by the bulk carrier *Petr Vasev*, resulting in 423 fatalities.

**Figure 6.21**
Takao Inui at the model basin of University of Tokyo in 2003.
Credit: Author's photo.

Such collisions are occurring with increasing frequency in highly trafficked sea lanes of the world, and even tightly subdivided warships can suffer serious damage, as happened in 2017 when in two separate incidents the American destroyers USS *Fitzgerald* and USS *John S. McCain* were holed and badly damaged (each with considerable loss of life) by merchant ships fitted with bulbous bows. The maritime industry continues to weigh the dangers of the bulbous bow against its fuel efficiency benefits and examine potential safeguards and protection against catastrophic collisions. For the foreseeable future, Takao Inui's invention shows no sign of going away.

# 7  The Ghost in the Machine

Conjugating the irregular verb "to design":
I create
You interfere
He gets in the way
We cooperate
You obstruct
They conspire
—David K. Brown, *Warrior to Dreadnought* (1997), p. 7

This book describes the rise of naval architecture from the beginnings of the industrial age (1800) to the dawn of the information age (2000). Yet in the same way that steam power and iron structures were present more than a half century before the start of the industrial age, the electronic computer and software-enabled design and analyses began to appear a half century before the dawn of the information age. The reason that businesses and governments adopted the electronic computer in the twentieth century was the same reason they adopted steam and iron in the nineteenth century—*predictability*. Horst Nowacki, one of the pioneers of computer-aided ship design, noted that the shift from mechanical to digital computing did not fundamentally alter the way naval architects worked but rather replicated what they did with mechanical tools and graph paper, using instead punch cards, graphical interfaces, computer printouts, and electronic files. This process was driven primarily by the need to create repeatability in design and manufacturing, replace tedious and error-prone human calculations, and carry out computationally intensive, time-consuming tasks. Only as computer tools matured did other organizational aspects of computer-aided design and analysis become apparent across all engineering fields; most notably, this involved the deskilling of complex processes, which allowed either the same work to be accomplished more quickly and more accurately by fewer employees, or more work to be done with the same number of personnel.[1]

Many textbooks of computer-aided design and engineering give potted histories of their specific domains, but they often disagree in chronology, events, and even who was responsible for which advancements. The field is still fairly young and, so far, has few synthetic histories like those for mechanical drawing, notably Peter Booker's *A History of Engineering Drawing*, the magisterial *The Art of the Engineer* by Ken Baynes and Francis Pugh, and the minor masterpiece *Drawing Instruments, 1580–1980* by Maya Hambly. I have therefore chosen to illustrate only a few aspects of computer-aided design and analysis in naval architecture, since the context of these developments within the larger sphere of engineering has yet to be established. Note also that I do not cover computer-aided developments in the Soviet Union and Russia, which were accomplished almost entirely independently from Western bloc influence; that field is still little studied, because for many years Russian computers were confined to classified military applications, while their wider, public use was enmeshed in a political tug-of-war between scientific utility and the perception that they embodied decadent Western values. Instead, I have chosen to focus on an aspect of computer-aided design that is unique to naval architecture: the ship synthesis model, which attempts to replicate, in a simplified, repeatable form, the work of the ship designer in developing the concept, overall dimensions, and top-level characteristics of the ship.[2]

This chapter describes the evolution of naval architecture tools, beginning with the traditional drafting and calculating instruments of the nineteenth and early twentieth centuries, developed as a means to more accurately control and predict the characteristics and performance of the emerging steam, iron, and steel technologies. It then describes the process, which began in the mid-twentieth century, to replace mechanical design and computing tools with computers, that is to say, changing from the machine to the "ghost in the machine" (i.e., software) in hull design and synthesis. Finally, it examines how industrialization has changed the nature of the ship design process across 200 years.[3]

**Splines, Curves, and Integrators**

By the mid-1700s, naval architects were designing ships on paper in a specific three-view format—waterlines, profile or side (buttock) lines, and body section. They used various forms of compasses and elliptical trammels to make curves of fixed or varying radii for hull frames, just as civil architects used these instruments to design domes and arches in buildings. However, shipbuilders also used devices such as the drafting bow and mechanical spline to create complex curves—for example, for longitudinal waterlines—which had no real civil architectural counterpart before the nineteenth century. Figure 7.1 shows examples of drafting bows, almost unchanged for over two centuries, that were used to draw irregular but fair (i.e., smooth) curves in ship design. The first drafting bows, using thumbscrews to create the complex curves, were described in shipbuilding treatises as early as the 1620s. The illustration from a Swedish author in 1691 precisely matches an actual bow manufactured in England in 1693 that was employed in an American shipyard until the early nineteenth century.[4]

The mechanical spline (sometimes referred to as a batten or penning batten) was a later invention for fairing lines on ship plans, making its first appearances in the mid-eighteenth century. The spline was the logical extension of the ribbands that earlier shipwrights had used to control the shape of the hull while building at full scale, and by the early nineteenth century they were well established in use. Figure 7.2 shows how the mechanical spline is used to create fair waterlines and avoid bumps and hollows, which cause fluid resistance. The naval architect establishes a series of points based on the body plan, then adjusts the spline along those points, holding it in place with heavy weights. Once the desired shape is achieved, the naval architect determines if it is fair by lifting weights (they have hooks that fit into a notch in the spline's edge) one at a time and observing the spline—if it jumps or moves, the line is not yet fair and must be readjusted (in mathematical terms, the transverse force of the spline is proportional to the third derivative of the deflection, so "fair" is defined as that derivative being zero). In the eighteenth century, cedar, pine, ebony, and holly were preferred as having the right "spring"; splines from the twentieth century to today (they are still used by shipbuilders) are plastic. The weights are typically one to two kilograms; called ducks, whales, newts, or pigs because of their shape; made from lead, iron, or bronze; and usually have a felt bottom so they can slide easily on the paper. During the twentieth century, splines and weights were common in the design bureaus of airplane and car manufacturers, who were themselves trying to develop fair curves to minimize air resistance.[5]

The drawing device that became most identified with naval architects was the ships' curve, for many would create their own individualized sets, which according to Colin Tipping, a British shipbuilder, were then "[passed down] from [a] retiring draughtsman.... A set of the most used curves was a valuable addition to the draughtsman's toolbox." Yet the ships' curve was in fact quite similar to the drawing curves used by civil architects as early as 1700. For civil architects, these were aids to creating the complex curves found in domes, onion-shaped roofs, and so on. For naval architects, ships' curves were used to draw body plans, buttock lines, and the intricate curves around the bow and stern. Because they were fixed curves (usually manufactured from

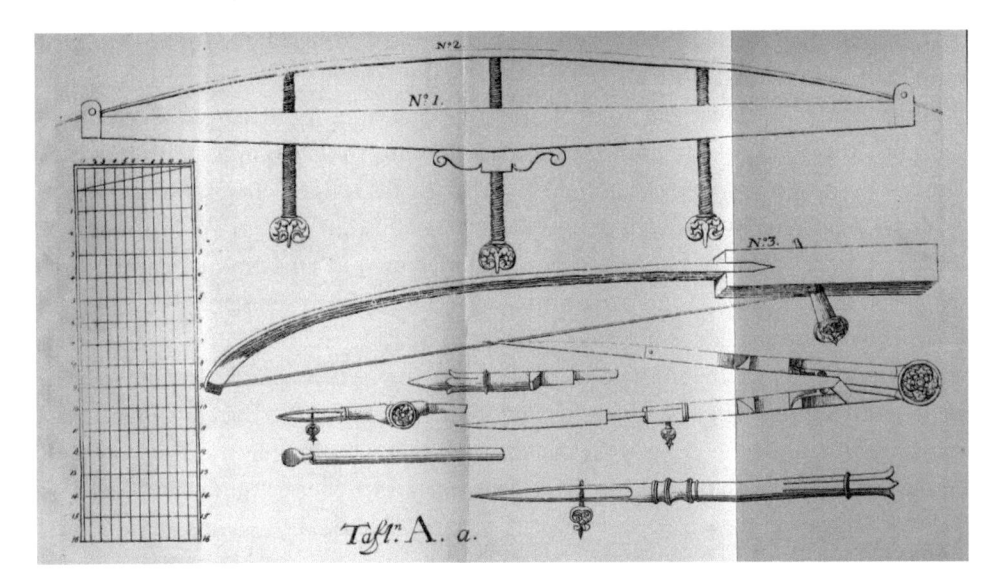

**Figure 7.1**
Adjustable drafting bows. Drafting bow (this page), illustrated in Rålamb, *Skeps Byggerij*, 1691. Drafting bow (opposite page) manufactured in England in 1693 by William Addisson and later used in the Barker and Magoun shipyard in Salem, Massachusetts, through 1836.
Credits: Cushing Library, Texas A&M University (this page); Peabody-Essex Museum (opposite page).

pearwood, hickory, or beech), they generally were used to establish the initial hull geometry, which was then completed with the use of flexible tools like the drafting bow and mechanical spline. Constructors like Frederik Henrik af Chapman in Sweden and Josiah Fox in the United States (who began his career in Britain) had their own set of curves that they employed from one design to the next (figure 7.3).[6]

During the age of sail, individual naval architects and draftsmen apparently fabricated their own curves, but by the industrial age they were being manufactured commercially as standard sets of 30 to 60 curves of varying sizes and radii (the longer, shallower curves were often called sweeps). The most famous set was called the Copenhagen Set, apparently named for the city of its manufacture, whose earliest examples date to 1817. Other, similar sets were referred to as English, Hamburg, and German curves. There was often considerable overlap with civil drawing curves, and naval architects frequently used those as well. A French naval constructor, Adrien d'Etroyat, noted in 1845 that such curves, called *pistolets* (other nations called them French curves), "were widely available on the market in varied shapes and sizes." The larger sweeps were almost indistinguishable from railroad curves and radius curves used by civil engineers for surveys and to trace rail lines, bridges, and roadways. By the late nineteenth century, ships curves were part of the catalog of large instrument makers such as the Eugene Dietzgen Company of Chicago, W. F. Stanley in London, and Keuffel and Esser Company in Hoboken, New Jersey. Despite the variety of manufacturers across different nations, most sets of ships' curves had almost identical forms and sizes (figure 7.4). Starting in the twentieth century, these manufacturers replaced wood with plastic, such as celluloid and acrylic (figure 7.5), which was both cheaper and easier to standardize.[7]

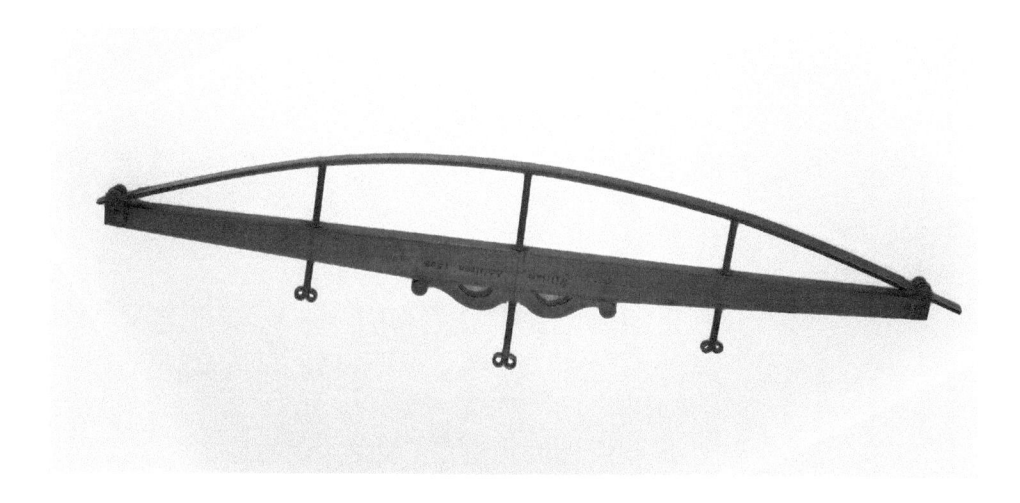

Despite the forms and types of ships' curves having survived almost unchanged over the course of two centuries and across many nations, it is not clear how their shapes were selected for manufacture; some were apparently fabricated at the instruction of naval architects and draftsmen for specific purposes. Josiah Fox's curves included a template for the stem of the frigate *Crescent*, which he designed in 1797 for the dey of Algiers. James W. Queen and Company, a Philadelphia instrument maker, claimed in its 1883 catalog that "the curves were made by us from drawings furnished by the chief draughtsman in the [Philadelphia] Navy Yard at League Island, and are the standard pattern used in the United States," though neither the drawings nor the name of the draftsman have come to light. By contrast, Howard Chapelle, former curator of maritime history at the Smithsonian Institution and a naval architect himself, thought that "ship-curves were made from arbitrary patterns.... I am doubtful that these curves were made by use of formula. Naval architects and marine draftsmen usually have a dozen or [so] curves which they use constantly; the rest of the Copenhagen set is rarely used."[8]

The commercial manufacture of ships' curves and mechanical splines provided a small degree of standardization of design practice but did not improve the accuracy or

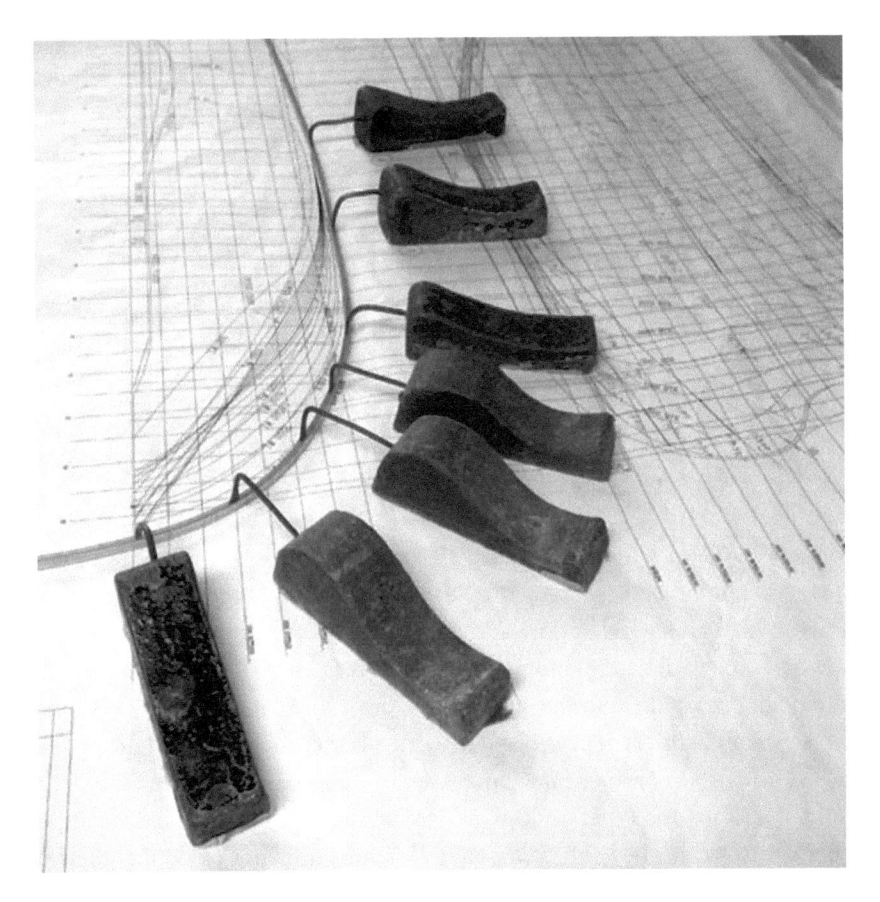

**Figure 7.2**
Mechanical spline with ducks.
Credit: Core77.

reduce the complexity of calculations. This was especially true for stability calculations, which as discussed previously, came under increasing scrutiny in the late nineteenth century after a series of stability accidents, notably the loss of HMS *Captain* and the capsize on launch of SS *Daphne*. Even before those accidents, techniques for determining stability under a range of loading conditions, drafts, and angles of heel, such as the method of wedges and locus of metacentric curves, had become available. The problem was, even these simplified methods required considerable time and effort; as the British constructor William H. White noted, "Naval architects had simply to decide whether or not it was worth the trouble to perform an elaborate calculation in order to ascertain the variations in the stability of any ship designed by them." Often that decision was

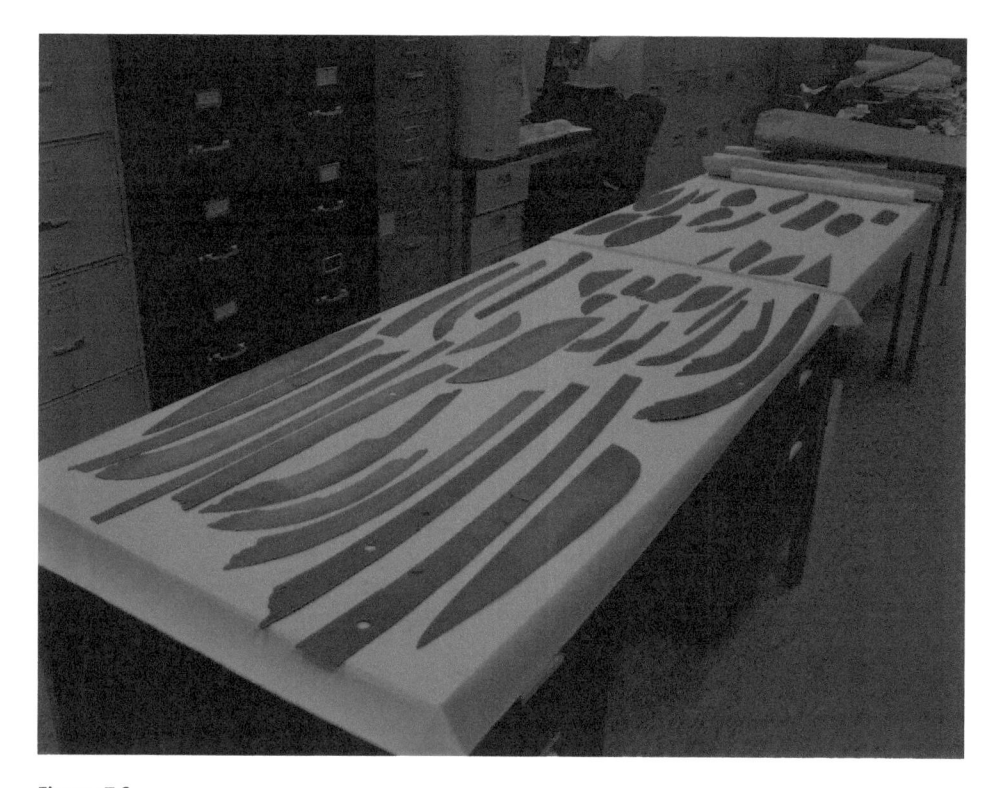

**Figure 7.3**
Ships' curves of Josiah Fox, active in Britain and United States, 1785–1809.
Credit: Office of the Curator of Models, Naval Surface Warfare Center Carderock Division.

not left to the individual naval architect; the shipbuilder might decide for him that the time needed for the calculations was better spent getting the ship's drawings complete and to the shipfitters and shopworkers.[9]

The first practical machine that could reduce the time and effort of stability calculations was developed for completely different engineering fields. At the beginning of the nineteenth century, several engineers simultaneously invented the planimeter to automatically calculate areas bounded by closed curves, which surveyors as well as civil, mechanical, and railroad engineers routinely needed in their design work. These first planimeters were cumbersome to use, but in 1854 a Swiss mathematician, Jakob Amsler-Laffon, developed the polar planimeter that substantially eased the task. Amsler's planimeter sold very well, and in 1867 he developed a new model, the Amsler Integrator (figure 7.6), which added the capability to automatically calculate the first and second moments of inertia of irregular areas. Although these calculations are

**Figure 7.4**
Ships' curves of Antonio Mas (pearwood), Academia de Ingenieros y Maquinistas de la Armada (El Ferrol) class of 1919.
Credit: José Maria de Juan-Garcia Aguado.

**Figure 7.5**
Copenhagen ships' curves (acrylic) manufactured by Eugene Dietzgen of Chicago, circa 1960.
Credit: Smithsonian Institution National Museum of American History.

precisely the type used by naval architects to calculate curves of stability, the integrator developed by Jakob Amsler and improved by his son Alfred was not initially marketed to that community.

The French navy was the first to take notice of Amsler's new device, in 1875, followed by the British navy in 1880, and by the Scottish shipbuilder William Denny in 1884. As the demand for standardized stability information increased among navies and commercial shipowners, the Amsler Integrator slowly gained in popularity by reducing these computationally intensive, time-consuming calculations to rote mechanical actions. According to the naval architect and historian David K. Brown, the Amsler Integrator, compared with hand calculations, sped up the time for calculating a single curve of stability from 8 hours to just 20 minutes, and did so with fewer errors. This was helped along by the introduction in 1898 of the Tchebycheff method for placing ordinates at uneven intervals as needed, which made numerical integration simpler

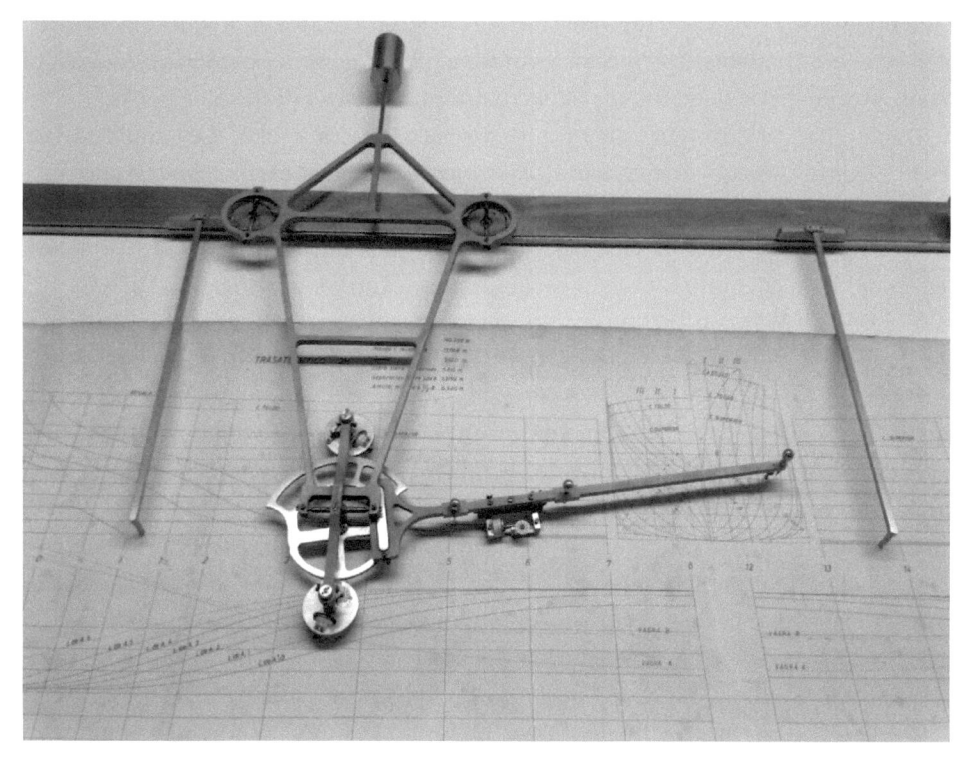

**Figure 7.6**
Amsler Integrator.
Credit: José Maria de Juan-Garcia Aguado.

and faster than the older Simpson's or trapezoidal methods. By the early twentieth century, the Amsler Integrator, coupled with various mechanical adding machines and slide rules, which were now commonplace in engineering and business offices, meant that naval architects—as well as draftsmen, engineering assistants, women tracers and computers—could routinely carry out the complex calculations required by an increasingly demanding marketplace.[10]

**Computer-Aided Design and Analysis**

Even with mechanical integrators, calculators, and slide rules, naval architects could and did continue to make mistakes. The cartoon in figure 7.7 is a humorous view of the failure to follow one of naval architecture's cardinal rules. Ships are symmetrical port and starboard, so to save time and calculating effort, the common practice was to measure areas, volumes, and moments on half the ship (say, the starboard side) and then multiply by two to get the value for the full ship. "You forgot to multiply by two" was a common correction by professors in students' homework assignments (at least in mine), written in bold red ink that stood out from the pages of the blue books.

Despite the cartoon portrayal of arithmetical errors as a joke, they could in fact result in grave consequences. When HMS *Atherstone*, the first of the *Hunt* class destroyers laid down in Britain at the beginning of World War II, had its inclining experiment performed in early 1940, its metacentric height turned out to be one foot (0.3 meter) less than intended, quite a lot for a small destroyer. Further investigation showed that one of the assistant constructors, under enormous time pressure to carry out calculations to get warships to sea, mistakenly assigned the center of gravity of the upper deck at 7 feet instead of 17 feet (2 meters instead of 5 meters). This meant that the weight and vertical moment estimates for the entire class of ship were faulty, and the ships were delayed several critical months while they were extensively modified before they could be placed into war service.[11]

The first electronic computers built during and immediately after World War II—Harvard Mark I for the U.S. Navy, ENIAC for the U.S. Army—were seen as giant calculators, capable of quickly carrying out complicated calculations and without the kind of arithmetical errors that plagued even the most capable humans. Within just a few years, that perception shifted to imagining computers as thinking machines; a 1950 *Time* magazine cover-page article, "The Thinking Machine," spotlighted the Navy's Harvard Mark III computer as the harbinger of electronic machines capable of automating a wide range of mental tasks, from ordering matériel to inspections, just as mechanical machines had automated physical labor in factories and on assembly lines;

**Figure 7.7**
I forgot to multiply by two.
From *The Binnacle* (Webb Institute, Glen Cove NY) 10, no. 1 (9 March 1945).
Credit: Office of the Curator of Models, Naval Surface Warfare Center Carderock Division.

the article even predicted that computers would "pay bills, blow the factory whistle and pay the help (if any)."[12]

The following year, the first commercial general-purpose computer, UNIVAC I, began production and became famous when it correctly predicted that Dwight D. Eisenhower would win the 1952 presidential election, despite pollsters favoring Adlai Stevenson. David Taylor Model Basin established the Applied Mathematics Laboratory and purchased the UNIVAC I (unit number six) in 1953, which as noted earlier, was placed under the charge of Betsy Holbertson. The laboratory was a one-stop location for all the model basin's computational needs, from pressure distribution on propeller blades to calculating the neutron distribution in a nuclear reactor, to analyzing the logistical demands of a major shipbuilding project. Naval architects could now tabulate and calculate, without arithmetical error, a ship's weight and center of gravity (a problem that vexed many more than just a single harried British constructor) and could perform a complete speed-power prediction for a ship with 20 minutes of data preparation and 2 minutes of computer time, instead of two days of desk calculation. By 1960, the lab had installed an IBM 7090, a high-speed transistorized computer, that coupled with the FORTRAN programming language, became the workhorse for scientific and engineering organizations around the world.[13]

The IBM 7090 (soon replaced by the IBM 360 family of mainframes) was just one example of the plethora of American-made computers that by the mid-1960s accounted for about half the world's market. A major factor in its success was that the Department of Defense (DoD) had invested early in electronic computers and their associated systems, both software and hardware, to improve military capabilities. In the 1950s and 1960s, the U.S. Air Force worked with MIT and several companies to develop SAGE (Semi-Automatic Ground Environment), a continent-wide air-defense system. The SAGE Whirlwind computers were the first to deploy features we take for granted today, notably real-time input and output and graphical user interfaces. The U.S. Navy developed the Navy Tactical Data System, with similar capabilities but smaller and lighter weight, to be placed aboard warships. In the 1970s, the DoD embarked on an "offset strategy" to counter the much larger Russian military with smarter weapons and sensors, almost single-handedly creating entire industries, such as microprocessors, advanced computing, and networking, that went on to become the foundation for commercial minicomputers and the Internet, marking the transition to the information age.[14]

The early forays into computing for ship engineering primarily involved batch processing, usually a combination of punched data cards and a line printer, which could handle a single bounded program. Modern computer-aided design and engineering (CAD/CAE), generally recognized as including interactive graphics on a screen controlled directly by the engineer, had its roots in computer-aided manufacturing (CAM). While MIT was developing the SAGE defense system in the early 1950s, it also embarked on another air force project, using the same Whirlwind computer to automate the production of airframes by numerically controlling the cutting and machining of individual parts. A decade later, shipyards began introducing numerically controlled manufacturing and production for steelwork. In the United Kingdom, the British Ship Research Association developed a lines-fairing and steel-cutting system called Britships that was employed in several shipyards. In France, the Lorient yard employed numerical nesting and cutting for steel frames. And in Norway, perhaps the best known of the shipyard CAM systems, Autokon, was developed for the highly automated steel assembly process at the Götaverken shipyard in Arendal (Gothenburg) in Sweden and subsequently was sold to shipyards around the world.[15]

The growth from CAM to CAD/CAE started with the CAD Project at MIT in 1963. The first and most visible product, called Sketchpad, was the forerunner of the three-dimensional computer models now ubiquitous in design offices worldwide. Unlike most batch-process machines, this had a graphical user interface that could be controlled interactively on a large cathode-ray-tube screen. One of the most challenging aspects of CAD was the mathematical representation of two-dimensional curves and three-dimensional surfaces, especially when fairing those curved surfaces as for a ship's hull. Early systems such as Autokon (which expanded into CAD/CAE after it was taken over by the Aker Group) employed cubic splines, a mathematical depiction of mechanical splines, constructed of separate segments of third-order polynomials and defined by specified control points. In the 1970s, other methods for developing fair lines, curves, and surfaces were introduced to CAD systems, such as B-spline, nonuniform rational B-spline (NURBS), and Bézier, the latter named for the French engineer, Pierre Bézier, who popularized them in his designs for automobile bodies at Renault. Even as these systems evolved from two-dimensional drawings to three-dimensional product models, in general they still were limited to outputting a set of numerical offsets that defined a hull surface, which then could be used by other programs to calculate stability, create structural drawings, and so on.[16]

For a quarter century, computer-aided calculations had made possible advances in naval architecture that solved previously intractable problems and allowed designs based on first principles of physics and mechanics instead of rules of thumb, such as using actual structural loads in a seaway instead of balancing on a hypothetical wave. For example, in 1952, Hermann Lerbs developed a rigorous lifting-line analysis for designing wake-adapted propellers that was solvable only by numerical means. In 1953, the work by Manley St. Denis and Willard J. Pierson Jr. on ship motions in random seas paved the way to probabilistic (i.e., statistical) treatment of sea states and ship motion response. This also gave rise to the use of strip theory, a means of determining the sea loads and dynamic reactions of the ship by dividing it into discrete sections, calculating the added mass and damping of each section (which can be thought of as entrained water moving along with the section as it oscillates in the seaway), and integrating the results in three dimensions. These sea loads were verified by placing strain gauges on vessels like the SL-7 high-speed container ships in the 1970s.[17]

Another development was the U.S. Navy's Ship Hull Characteristics Program (SHCP) created in 1966 to calculate intact and damaged stability of complex hulls from a given set of offsets. At the same time, numerous researchers, especially in the aircraft industry, were developing novel computer-aided analyses of structural stress and vibration through the use of meshes, which became known as finite element methods. Meanwhile, others were creating probabilistic reliability-based methods for predicting hull loading in a seaway and paving the way for probabilistic damaged stability assessments. However, with the limited computing power available at the time, there were no means for a single program to integrate them all.[18]

During the 1970s and 1980s, the concurrent developments of use of minicomputer and microcomputer workstations, increase in computing power, and rise of early networks allowed the realization of integrated suites of ship design and analysis programs, which had been imagined since the earliest days of batch processing. In Norway in 1970, the Aker Group and Det Norske Veritas expanded Autokon into a system called PRELIKON, which integrated programs for hull form development and calculations for intact stability and resistance, and it maintained the design information in a database that was shared among the modules. In the U.S. Navy, the ship design office NAVSEC (later renamed NAVSEA) developed the Computer Aided Ship Design and Construction system in 1976 to support a nuclear cruiser project (later canceled). It became the backbone for future design efforts. It combined SHCP with other modules to define hull geometry and calculate ship motions in a seaway, again using a centralized database to share and update information among modules. In 1988, the British navy's research and design establishments developed GODDESS (GOvernment Defence DEsign System for Ships), a suite of programs that combined hull form definition, hydrostatics, damaged stability, hydrodynamics, seakeeping, propeller, and structural design.[19]

By the 1990s, the advent of the Internet and powerful minicomputers and workstations allowed complex software suites such as FORAN and CATIA (the latter created by the French aerospace firm Dassault for its Mirage fighter jets) to be developed and adapted to fit the engineering needs of even small naval architecture offices. The change from machine to software over the course of 20 years—roughly 1980 to 2000—was physically noticeable in both drawing room and classroom, where the surfaces of drafting tables were taken up by computer printouts instead of Mylar sheets for plans and then the drafting tables themselves were removed and replaced altogether by computer workstations. As CAD/CAE became an integral part of shipbuilding, regular symposia and conferences kept participants up to date, such as the International Conference on Computer Applications in Shipbuilding (started in 1973) and the Conference on Computer Applications and Information Technology in the Maritime Industries (begun in 2000).[20]

CATIA was not the only system used by naval architects that had been originally developed by the aerospace industry. Computational fluid dynamics (CFD) is today one of the most powerful tools in engineering, often touted as being a numerical towing tank or numerical wind tunnel, a digital alternative to model testing. CFD mathematically simulates fluid flow using the Navier-Stokes equations referred to in chapter 3, a series of differential equations that describe fluid flow in terms of shear stresses, pressure, and energy. Although the first CFD computer models were developed in the 1950s to analyze implosion phenomena of thermonuclear warheads, it was only in 1967 that a method for simulating the potential flow around thee-dimensional bodies was developed by researchers at Douglas Aircraft (under contract, it should be noted, from the David Taylor Model Basin). The method divided the body into discrete panels—similar to what structural engineers were already doing with finite elements—and could calculate velocity and pressure distributions around both aircraft fuselages and ship hulls.[21]

The aerospace industry quickly saw the potential for using CFD to simulate a wide variety of designs and problems—for example, the analysis of transonic flows across wings and airfoils—at far less cost than building and testing models in wind tunnels. A wide range of industries, including automotive, civil engineering, meteorology, and even air conditioning, took note of the benefits of CFD methodologies and code—among them, the ability to color code graphics to help visualize fluid flow—and invested heavily in modeling techniques (figure 7.8). By comparison with these other industries, the investments in CFD by the naval architecture community have remained fairly modest, focusing, for example, on discrete problems such as approximating turbulent flow around hull forms using Reynolds-averaged Navier-Stokes (RANS) methods. Since 1975 the community has recognized the distinctiveness of this field by regularly holding the International Conference on Numerical Ship Hydrodynamics, separate from the much larger Symposium on Naval Hydrodynamics. For some time, the high price of CFD software and the supercomputers needed to run the programs led some engineers to dub CFD "color for dollars." However, starting in the 1990s, the lower costs for computation and the availability of commercial software code has made CFD available even to modest-sized design firms. The dream of creating a numerical towing tank, completely replacing physical models with digital ones, is still regularly touted even as model basins continue to operate worldwide.[22]

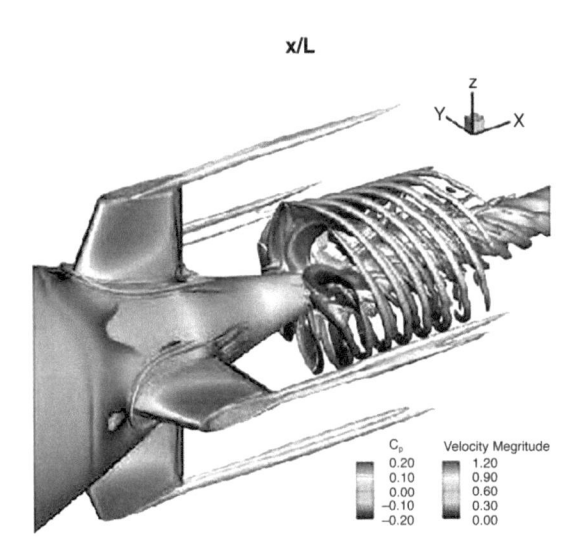

**Figure 7.8**
Representation of flow around stern and propeller of a submarine via CFD.
Credit: ITTC.

In the early days of computer-aided ship design, managers fully expected (as a 1970 *Naval Engineers Journal* article explained) that these improvements in software would result in shortened design times, shortened procurement cycles, decreased costs, and better utilization of scarce engineering resources. That was not the case. Instead, ship design and engineering has obeyed the famous Parkinson's law, articulated by British naval historian Cyril Northcote Parkinson, that "work expands so as to fill the time available for its completion." Another article, also written in the *Naval Engineers Journal* but more than 40 years later, showed that the number and types of studies demanded in a ship design project had risen dramatically compared with a generation earlier. The vast increase in computing power and available software programs had not decreased the design time, procurement time, or number of personnel involved in a project but rather paved the way for deepening and broadening the engineering effort for ship projects, each of which demands more studies, drawings, and data than ever before.[23]

**Ship Synthesis Models**

Horst Nowacki, as stated earlier, noted that computers did not fundamentally alter the way naval architects created drawings and engineering calculations but instead replicated in electronic form what they previously did mechanically. This was also true for a small but intriguing subcategory of computer-aided ship design tools—ship synthesis models, which were developed to reproduce the early-stage design process for feasible ship concepts but in a rapid, repeatable manner. These models were first developed in the 1960s, quite early in the CAD/CAE chronology, and though they started with commercial vessels, within a few years almost all development became focused on naval ships.

Naval architects traditionally conveyed the tacit knowledge of the ship design process through hands-on experience. The wave of professionalization in the late nineteenth and early twentieth centuries brought with it the codification of these processes so that they were standardized and repeatable. Before and during World War I, the U.S. Bureau of Construction and Repair developed a series of internal memoranda that detailed how preliminary designs, cost calculations, and specifications should be developed for naval vessels. In 1941, even as Britain was fighting Nazi Germany alone, a Swan Hunter Shipyard naval architect took pains to publish an article explaining the design process for commercial ships.[24]

In 1959, the MIT professor John Harvey Evans devised the now-ubiquitous design spiral (figure 7.9) to visualize this process, laying out step by step how an experienced naval architect would approach a new ship design. Perhaps the most important part of this visualization is the iterative nature of ship design, with a series of design cycles spiraling inward and each assumption and calculation represented by an individual spoke in the wheel. The designer begins with an initial set of design assumptions and requirements, which dictate overall dimensions and displacement, machinery type and power, hull form parameters, payload or cargo, and arrangements, and then carries out calculations to determine if these result in a stable, structurally sound ship that meets speed requirements. The designer then calculates the weights and centers of gravity of the overall vessel. Next, the designer compares the calculated values with the initial assumptions: Does the calculated weight equal the assumed displacement? Does the calculated engine power equal that of the assumed machinery plant? And so on. These comparisons are almost never equal on the first go-around, which means the designer readjusts the initial assumptions and goes through the cycle of calculations again. The spiral refers to the idea that the design starts on the outer circle, and after several turns around the cycle it will "converge on the ultimate, refined and balanced solution indicated by the inner closed circle," in which the calculated weights equal the assumed displacement, and so on. Although there have been many subsequent attempts at creating different ship design spirals, they all hearken back to Evans's original model.[25]

Evans's design spiral set off a spate of research into developing ship synthesis models using the spiral as their basis. In 1963, several naval architects at the U.S. Maritime Administration (MARAD) developed a Least Cost Ship design program that optimized

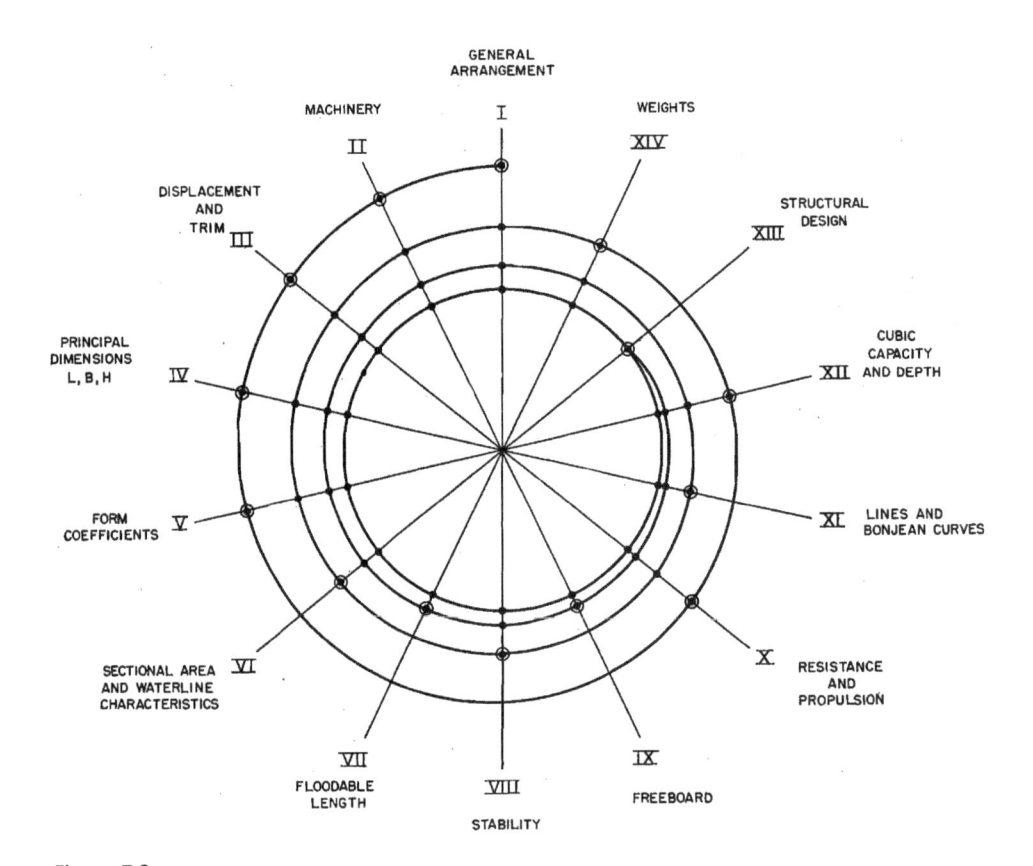

**Figure 7.9**
Ship design spiral, 1959.
Credit: Reproduced with permission from the American Society of Naval Engineers.

ship dimensions for speed, cargo load, stowage factor, and range and was run on the IBM 7090 at the David Taylor Model Basin Applied Mathematics Laboratory. It established three preliminary designs for U.S.-built merchant ships, which over the course of a decade evolved into the MARAD PD-214 project, which was never built. The same year, the Center for Naval Analyses (CNA) one of the U.S. Navy's think tanks, was asked to study new aircraft carriers and as an interim step developed a ship synthesis program to evaluate cargo ship alternatives. About the same time, Litton Industries (a large defense contractor) created a ship optimization program based on linear programming methods first developed at the University of Michigan. MIT also began work on a ship feasibility program. There was heady optimism in those early days of ship synthesis models. One advocate in 1964 predicted that soon a naval leader would come into a single room that had six engineering groups (each with its own computer handling a different aspect of the ship design), hand them a few top-level operational requirements (sometimes called staff requirements), and over the course of a single "hard day's work," the design would be electronically passed around the room until the "best compromise is reached ... the [leader] presses a button, and all the necessary data for construction is outputted at the proper shipyard."[26] The one-day-one-push-of-a-button paradigm never would come to pass, for creating a feasible design is but one part of any shipbuilding project, which also must be planned, approved, and financed. As noted earlier, computer automation of ship design simply increased the demand for studies, and ship synthesis models did nothing to automate the project itself, which still required the same number and level of technical reviews; tradeoffs among cost, schedule, and system performance; management oversight; and budgetary controls and audits.

The increasing capabilities of ship synthesis models engendered their own feedback spiral in the demand for more studies. The first major push for large-scale analyses of alternative design concepts was the DX/DXG project, intended to propose replacements for World War II–era cruisers and destroyers that were reaching the end of their service lives. In 1965, John Schmidt of CNA teamed up with Jim Mills of NAVSEC (with help from many others in both organizations) to develop the first destroyer synthesis program, dubbed CODESHIP by CNA but DD01 by NAVSEC. The programmers divided it into a main program and 15 subroutines that calculated displacement, engine power, hull resistance, weights, and more, following the Evans design spiral. The main program compared initial assumed values with final calculated values and adjusted those for each spiral, until a converged, balanced design was achieved. Resistance and power were calculated using David Taylor's original hydrodynamics data (which subsequently had been modified by Morton Gertler); weight relationships (e.g., tonnes of steel per unit length) were estimated from historical return data of many comparable ships. The resulting output had no graphics but instead was numbers indicating displacement, weights, power, fuel requirements, and so on, which could allow rough estimates of costs and provide naval architects a starting point to begin more detailed hull design work for the selected concepts, using graphics-based CAD/CAE tools.

Within a few years CODESHIP diverged from DD01, which NAVSEC soon upgraded to DD07. The process of generating a design concept alternative in the 1960s was quite cumbersome. The naval architect at NAVSEC in Hyattsville, Maryland, would determine the main parameters and create a stack of computer punch cards. A courier would take the cards to David Taylor Model Basin in Carderock—a 20-mile journey each way—to be run in a batch process on the IBM 7090 where the DD07 program was housed. The line-printer output would be couriered back. The whole process could take two to three days for a single run, assuming there were no mistakes. Even with this time-consuming procedure, by 1971 the DD01 through DD07 programs had generated 1,300 separate design concepts, which were narrowed to become the DD 963 *Spruance* class destroyers and FFG 7 *Oliver Hazard Perry* class frigates (by contrast, previous studies done by hand generated 20 to 50 design concepts each). NAVSEC also attempted to develop more complex synthesis programs for aircraft carriers, amphibious ships, and submarines, but these proved beyond the capabilities of the computers at the time and were abandoned.[27]

Synthesis models arrived at the time when the U.S. Navy was closely examining the Soviet Union's navy, whose capabilities were being expanded under the direction of its commander-in-chief Sergey Gorshkov. One means to assess those capabilities was to conduct comparative naval architecture analyses of their ships. The British and U.S. navies had done such an analysis in World War II on the German battleship *Bismarck*, after plans of the ship were recovered from a sailor who survived its sinking by the British in May 1941. Taking this information and that from other intelligence sources, both navies were able to work backward from the ship dimensions to characterize its armor, armament, and propulsion. That study showed that the surviving sister ship *Tirpitz* was larger and more powerful than Germany had advertised, which informed the Allied navies on how best to defeat it (it was sunk at anchor by bombers in late 1944). The Cold War problem was how to analyze Russian ships without a readily available set of technical data.[28]

The key to solving this problem lay in the fact that the postwar Soviet navy continued to build its ships based on Italian design practices learned before World War II, such as on the Ansaldo-inspired *Kirov*-class cruisers mentioned in chapter 6. The Americans were of course NATO allies with Italy, under which auspices the two nations shared technical information, and NAVSEC had already performed comparative analyses of the two navies' warships. Starting in 1971 and through much of the rest of the Cold War, NAVSEC (and later, NAVSEA) created a "reverse ship synthesis model" using DD07 as its foundation. Starting with known dimensions and other characteristics (from both open and intelligence sources) and using estimating relationships largely based on Italian design practices, the performance and operational characteristics of Russian ships were assessed. Among other findings, the U.S. navy noted that Russian ships were generally smaller for a given payload than American ships, yet their hull forms were more seakindly, that is, able to operate in higher sea states, than American ships. This led to studies and model basin tests in the 1970s and 1980s that resulted in improved seakeeping hulls for U.S. warships, like the DDG 51 *Arleigh Burke* class.[29]

By the mid-1970s, advances in microprocessors (helped along by the Department of Defense's offset strategy) had dramatically increased the computing power available to engineers, allowing programs to be run in real time instead of batch mode. At that time, the Boeing Corporation, which was building patrol hydrofoil missile craft, was given a contract by the David Taylor laboratory to develop the synthesis model HANDE (Hydrofoil ANalysis and DEsign) to run on the new real-time computer architecture. The completed version (1977) not only ran faster than the DD07 model but had the beginnings of graphical outputs that enabled engineers to visualize important characteristics like speed-power curves. In 1980, the David Taylor laboratory again contracted with Boeing to extend the HANDE model into a new ship synthesis program, ASSET (Advanced Ship and Submarine Evaluation Tool), for evaluating a range of ship types, including hydrofoils, destroyers, and advanced hulls. By 1986, the upgraded DD09 model was merged with ASSET and brought directly under the control of the David Taylor laboratory.[30]

Similar advances in synthesis models were made in other nations. In Canada, SHOP5 (SHip OPtimization version 5) was used during the NATO Frigate Replacement project of the 1980s. At the same time in Britain, the Forward Design Group opted to develop a stand-alone synthesis program, CONDES (COncept DEsign System), which was used as a precursor to GODDESS. By the 1990s and first years of the 2000s, improvements in computing power and networking allowed the development of synthesis models for additional ship types like aircraft carriers, amphibious ships, and submarines and increasingly blurred the line between concept evaluation and CAD/CAE for design and production. For example, ASSET today processes a wide range of graphical data (see figure 7.11), and in Britain, both synthesis and engineering design tools have been brought together by the QinetiQ company into a single commercial design system, Paramarine.[31]

**Ship Design Process and Tools across the Ages**

David K. Brown's puckish conjugation of the verb "to design" at the beginning of this chapter underlines the often contentious nature of creating a complex system like a ship. The different parties within a shipbuilding project—naval architects, project managers, budget directors—struggle to control the famous iron triangle of cost, schedule, and system performance and are continually negotiating compromises to keep the project on track. Some of this friction can be attributed to the idealized notion that the design spiral begins with a universally-agreed-to set of design assumptions and requirements and that the ship flows directly from these requirements. This has never been the case. Each party visualizes the final ship in different ways, based on her or his operational and institutional experience, which can be quite distinct from what the other parties imagine or even what the requirements spell out. Rowland Baker, another experienced British naval constructor who was a mentor to David Brown, noted, "As the chicken comes before the egg, so does the warship [come] before the Staff Requirement."[32]

This visualization of the final product is highly influenced by the design tools available, and as these tools evolved across the two centuries that this book covers, so did the nature of the ship design process. Ship design circa 1800, at the dawn of the industrial age, differed greatly between the naval and commercial worlds. Merchant ship design was often at full scale without the intermediate use of scaled drawings, and only rarely did shipbuilders perform any naval architecture calculations. By contrast, most navies were using scale drawings as a means of controlling the design process and creating a medium of communication between the shipbuilder and the naval engineer. Wooden ships' curves and mechanical splines were the primary tools of the trade. The few naval architectural calculations—metacenter, ratio of bow resistance, and the *point vélique*—were carried out by hand. Plans were drawn and calculations laboriously carried out over several months by a single person and then submitted just once for approval, without any revisions or suggestions for improvements—in other words, there were no indications of an iterative design spiral. One exception to this rule was a series of designs for the Spanish 74-gun ship *San Iledefonso* carried out by several teams of naval constructors and reviewed by naval officers until in 1785 the final design was agreed to and constructed.[33]

By 1900, industrialization was in full swing. Mechanical tools and scientific research gave naval architects more control over the ship design process, which allowed a process resembling the design spiral to emerge. Both commercial and naval ships went through broadly similar design processes, although commercial ships tended to have just a few initial studies before the company settled on a design that was completed according to classification society rules and standards. For naval ships, the bureaucratic process was much longer, and technical standards were set by the naval constructors themselves. The director of naval construction (called the chief constructor in the United States) would receive direction from the naval administration for a new project and confer with his constructors to study a few (perhaps a dozen) design alternatives. Each design study would begin with a parent ship—that is, an existing ship comparable to the one under consideration—and data collected from that ship. The constructor would create a rough drawing indicating armor, armament, and machinery layout and use parent ship data to estimate weights, stability, and costs. Robert Froude's constant system of notation would allow a rapid calculation of speed and power, which also gave fuel load. Structural strength was only rarely calculated. This process might take a single day, and within a short time a range of design alternatives was available for consideration by the naval administration. This was the first "turn" in the design spiral.

The next step was to take one or two alternatives (or some amalgam thereof) and make more detailed studies, which the U.S. Navy's Bureau of Construction and Repair nicknamed "spring styles" after ladies' fashion catalogs, which typically came out in the spring (figure 7.10). Once the final approvals were given—the second turn in the spiral—the process of developing a final design could begin. Velum was stretched over the drafting table, and the hull lines were drawn to scale (scales of 1:100 to 1:200 were common) using mechanical splines and ships' curves. Stability calculations were carried out with the aid of the Amsler Integrator, which was also used for the balance-on-a-wave calculations for hull strength. More detailed weight and center of gravity estimates were made on the basis of calculated scantlings, speed-power calculations were refined, and the whole was checked against the initial assumptions to determine whether more turns around the design spiral were needed. Constructors maintained design notebooks (in the United Kingdom called ships' covers) to document all decisions. After many months or even years of design and calculations, model basin tests confirmed the final results, and the design was turned over for shipyards to bid on construction.[34]

By 2000, ship design had become differentiated more by vessel complexity and market than by whether the ships are commercial or military. For example, the majority of commercial tonnage is break-bulk, container, or tanker ships. Most are built in Asia by just a handful of large companies, each of which maintains a series of off-the-shelf designs that can be rapidly modified to suit customer needs. High-value, custom-designed vessels like passenger ships and research vessels are closer to naval ships in terms of bureaucratic and decision-making processes, approvals, and technical complexity. The U.S. amphibious assault ship LPD 17 (*San Antonio* class) is one example of the ship design process at the transition from the industrial to the information age.

The project for a new amphibious assault ship began in 1988, when the Cold War dictated that the likeliest assaults would be focused on nations of the Warsaw Pact. The fall of the Berlin Wall in 1989 created requirements for a more versatile ship, dubbed LX, which were explored with dozens of concepts using ASSET and other ship synthesis models (figure 7.11). After establishing several alternatives in 1991, engineering teams spent two years developing more detailed designs. A final configuration was selected in 1993 and the shipbuilding contract was awarded in 1996 under the LPD 17 designation. To create viable designs, teams went through one turn of the design spiral every 6 to 12 weeks until a converged solution was arrived at. All design work was done by computer, but that did not mean that the different tools were integrated or were always compatible. Hull and structural three-dimensional models were made on commercial workstations, but the graphics were sometimes clumsy, and it was quite common to print out drawings so that engineers could make pen-and-ink changes for input. Transfer of data between the different software programs was often problematic, notably for the U.S. Navy's Ship Hull Characteristics Program stability program, which had to account for highly complex tankage and well deck spaces. Shipyard design, construction, and lead ship delivery was

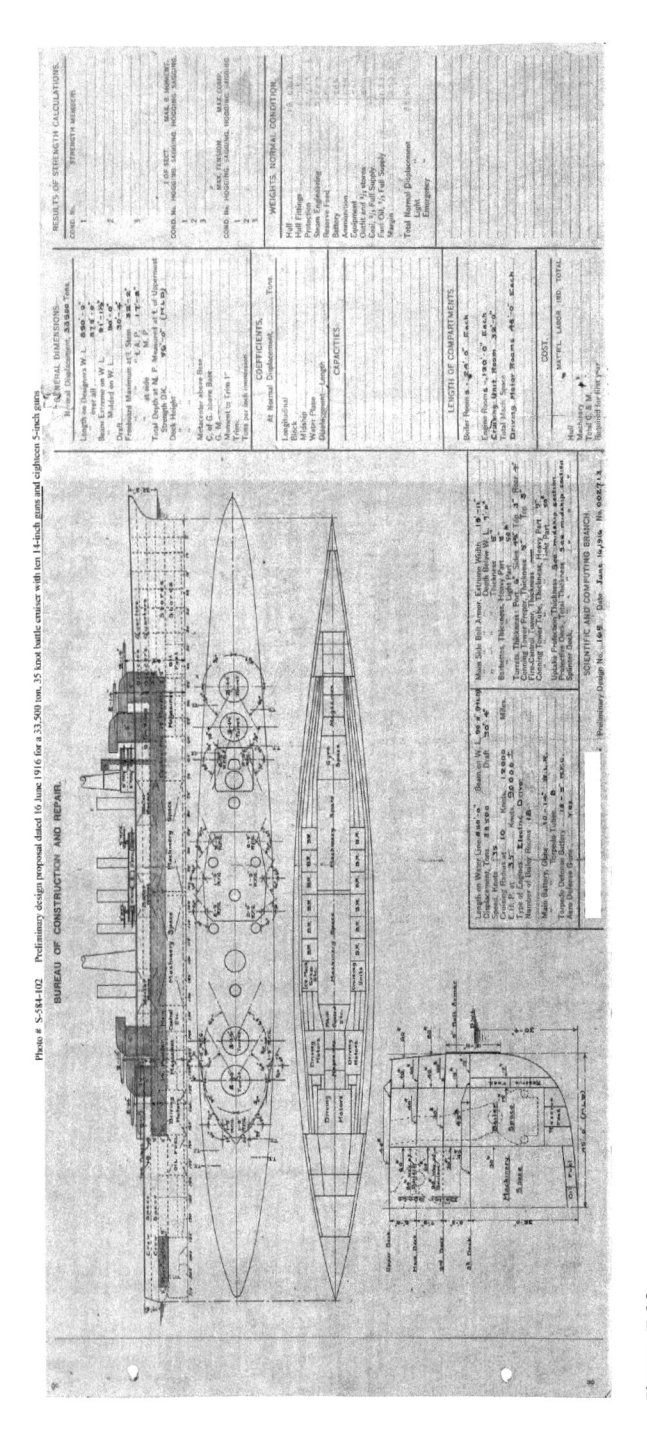

**Figure 7.10**
Battlecruiser spring-style 1916 sketch.
Credit: U.S. Naval History and Heritage Command.

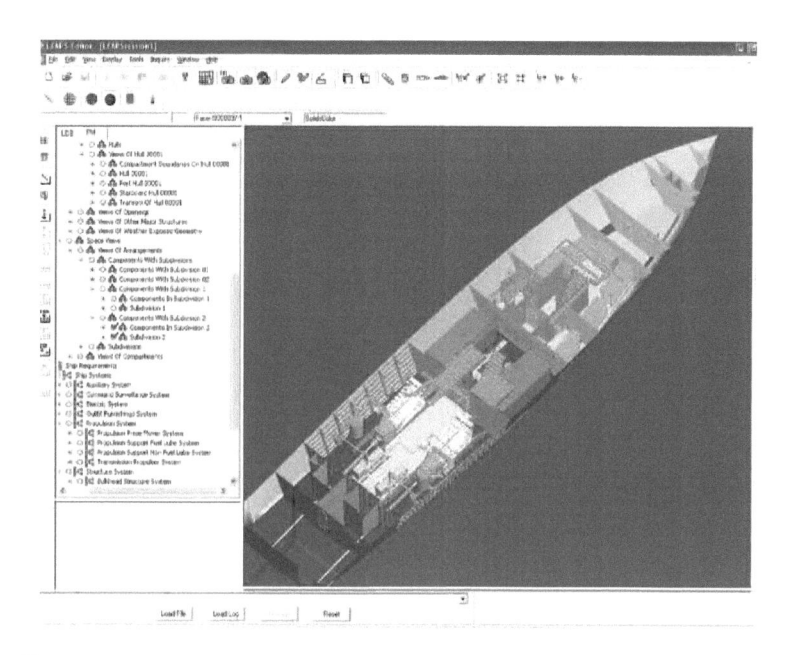

**Figure 7.11**
Screenshot of a ship synthesis model.
Credit: Naval Surface Warfare Center Carderock Division.

completed in 2006, almost two decades after the project began.[35] Integrated CAD/CAE tools have evolved considerably since then, in almost every field of engineering; they are installed on personal computers and linked with each other and with remote databases via the Internet. Paper has almost entirely disappeared from most design offices.

The availability of CAD/CAE tools for ship synthesis has increased the demand for more studies and analysis rather than decreased project time. These tools shape not only how we perceive the final ship but also the process for getting there. The momentum for wider use of computers in ship design and construction is increasing. That is not because CAD/CAE is cheaper than more traditional engineering but rather because the workforce is inexorably moving in that direction. The US Bureau of Labor Statistics indicates that while careers in naval architecture are increasing faster than average, those for drafters and technicians are flat, and for machinists—who fabricate the mechanical tools and equipment used in model basins—the career growth is negative. Meanwhile, software developers are growing at a rate far faster than average. The dream of the numerical towing tank may one day be realized, not through improvements in software, but through the lack of anyone left to keep physical towing tanks in operation.[36]

# Epilogue: From Metacenter to Metasystem

The computer and the Internet have given naval architects the tools to create and analyze ships, not as collections of individual domains, but as fully integrated systems. At the same time, ships and vessels have become part of a complex metasystem of transportation, energy production, and maritime security. This final chapter examines these trends as the industrial age gives way to the information age, in which connected teams of naval architects are now able to bring more intellectual and computing power to bear than at any time in history to solve problems and meet challenges on a global scale.

### Naval Architecture in the System

As I discuss in the preface, naval architecture, like all engineering disciplines, has always been about prediction. The first successful effort at developing a means of predicting ship performance was the invention of the metacenter in 1746 by Pierre Bouguer for assessing stability. From that time, naval architecture advanced in different domains—stability, hydrodynamics, structures—with predictive theories and design practices developed more or less independently from one another. The design spiral, articulated in 1959, attempted to describe visually how those independent domains came together in a ship project.

At the same time, a discipline called systems engineering was being created in the aerospace and military armaments industries. It identifies the overall objectives of the system and organizes the design process accordingly to predict how that design process will produce a system to meet specified characteristics. Systems engineering is marked by multidisciplinary teams and design efforts that advance concurrently instead of in a stepwise fashion and is often marked by extensive prototyping and testing before production. Naval architects, while well aware of the systems engineering approach, were often hampered by the lack of integrated engineering tools to work in such a concurrent manner. By the turn of the twenty-first century, however, powerful computers gave rise to the integrated design and engineering software suites mentioned in chapter 7. At the same time, robust communications networks also have allowed multidisciplinary teams to work concurrently on complex designs even when separated geographically, as was the case when the Electric Boat and Newport News shipyards together developed the SSN 774 *Virginia* class submarines and when BAE Systems led a consortium of seven shipyards to design and build the HMS *Queen Elizabeth* aircraft carriers.[1]

An essential part of systems engineering is identifying the overall objectives of the system. For naval architecture, these objectives were embodied in the prescriptive standards developed in the nineteenth and twentieth centuries, like the Rahola criteria for stability. Ship designers always had the nagging feeling that this one-size-fits-all approach was inadequate for the wide variation of ships and vessels in operation. As remarked in chapter 7, research on probabilistic seakeeping in the 1950s opened the doors to a wider view of systems objectives in other domains. However, it was not until the advent of effective software tools, and more importantly, the ability to accumulate and process large amounts of data, that probabilistic methods could be introduced to replace deterministic standards based upon prescribed events (e.g., two-compartment flooding for damaged stability). In the 1970s, Safety of Life at Sea regulations were rewritten to allow the use of probabilistic damaged stability assessments, in which an overall level of acceptable risk is determined by international authorities and statistical analyses of prior accident data are used to evaluate ship designs (probabilistic assessments became mandatory in the 1990s for certain types of cargo ships). The probabilistic approach is now extensively employed in structural design and fire safety, and this trend is being extended to other domains. The European Union continues to sponsor much of this development, with the National Technical University of Athens and the Universities of Glasgow and Strathclyde particularly involved in the research.[2]

Another trend in the systems engineering approach to naval architecture has been the convergence of the design practices and standards used for commercial and military ships, as the overall systems objectives for both types of vessels often overlap and the resources behind each are stretched thinner. Some of this movement has been from military to commercial; for example, many of the same noise reduction techniques employed in warships to avoid underwater detection are now used for modern passenger and cruise ships to protect sensitive marine life and improve passenger comfort. But most of the movement has been the other way, from commercial to military and is particularly visible in the increasing employment of classification societies to develop and enforce rules for naval vessels. This process was not initiated by improvements in computers or design techniques but rather by the rapid diminution of naval budgets after the end of the Cold War. The 1990s marked a significant decrease in the number of ships and their design infrastructure, to the point that support staff were barely able to maintain current ships, let alone continue developing standards for new ones.

Among the first to recognize this trend was Guy Gibbons, a naval architect with the UK Ministry of Defence ship structures group, who saw that the British Naval Engineering Standards were becoming "outdated and irrelevant." At the same time, more of the design work was being shifted to industry instead of being done in-house at the ministry, which was also losing the staff needed to update and maintain its engineering standards. Allowing shipyards to set and self-certify their own standards posed an unacceptable risk, very much like allowing a bank to audit its own books without regulatory oversight. In 1997, Gibbons convinced the ministry and Lloyd's Register to have Lloyd's update and maintain these engineering standards as a set of naval classification rules that would be endorsed by the ministry and that would then be used to certify the work done by shipyards. Once warships were built, they would be entered into Class, as with commercial vessels and inspected regularly. This arrangement ensured that navies could continue to hold shipyards to required design, construction, and maintenance standards, even while facing staffing cuts themselves. Within a short period, many other navies and classification societies entered into similar arrangements, which are becoming more common today; Det Norske Veritas and the Norwegian navy, Germanischer Lloyd and Germany, Bureau Veritas and France, and so on.[3]

Not all navies have followed this paradigm, and in the case of the U.S. Navy it had very serious consequences. By 1995, NAVSEA was cutting its staff, had eliminated its general specifications for ships, and was shifting design work to the commercial industry. Following on from the example of Lloyd's Register and the Ministry of Defence, the American Bureau of Shipping and NAVSEA agreed to pursue Naval Vessel Rules classification. However, in a meeting in October 1999 with the bureau and the commercial shipyards that built warships, the technical director of NAVSEA, Gregg Hagedorn, made it clear to those shipyards that they could "create" and "self-certify" their own standards, that is, they were permitted to design and build warships for the navy without oversight from the American Bureau of Shipping. At the same time, NAVSEA withdrew support for developing Naval Vessel Rules, which then stalled. Since NAVSEA had lost most of its ability to maintain and enforce specifications, this meant that shipyards were allowed to design and build U.S. Navy warships without being required to use any specific standards. When the Littoral Combat Ship (LCS) project was awarded to two shipyards in 2001, there were no mandatory standards in place. At that time, the navy and industry estimated that each warship, both of which were based on civilian ferry designs, would cost $220 million. When the LCS project office belatedly restarted the

Naval Vessel Rules effort in 2004 and demanded that the LCS ships be built to those standards, the costs tripled to $630 million each, which led to major congressional investigations and cuts in the number of ships. As a 2010 report to Congress noted, "The belated application of the [Naval Vessel Rules] to both LCS designs was a major factor in the cost growth on those ships." Since that time, NAVSEA has once again dropped the Naval Vessel Rules effort.[4]

The systems engineering approach for naval architecture will have its limits. Systems engineering grew out of the twentieth century aerospace and military armaments industry, which from the start employed extensive full-scale prototyping and testing of aircraft and systems as the way to assess design objectives before going into full production. By contrast, naval architecture remains still a product of nineteenth-century engineering, a heritage it shares with civil engineering. Despite advances in software that allow greater use of first principles of physics and mechanics in design, ships and skyscrapers are still built to unique sets of rules, with only small-scale testing prior to production. For this reason, ships are much less costly to develop than aircraft, even when their production costs are equivalent. For example, both a large passenger jet and a midcapacity cruise passenger ship are highly complex engineered systems having comparable complexity and safety requirements to protect their passengers and with each costing on the order of a half-billion dollars to build a single unit (destroyers are on the order of $1 billion). Yet the development cost—research, engineering, and design—for the passenger jet is 200 times greater (two orders of magnitude) than for the cruise ship. The trend holds even when comparing military ships and aircraft, as shown in figure E.1. The main reason for this extraordinary difference is the extensive full-scale prototyping and testing for aircraft prior to production, compared with the rules-and-small-scale-testing paradigm for ships. The likelihood is very small that maritime and naval industries would embrace full-scale prototyping, as those costs would have to be weighed against any potential improvements to safety and reliability.[5]

**Naval Architecture in the Metasystem**

While systems engineering has been generally successful at predicting how a design process will produce a system, it cannot predict how multiple systems, developed separately by differing organizations, will be developed and work together. Yet that is the environment of most major engineering projects in the twenty-first century. The standard model for this is the Internet, which today links everything from communications devices like cellphones and maritime radios to services like banking and ship registries, to physical objects like refrigerators and offshore wind turbine farms. These are

|  | UK Type 23 frigate | UK Typhoon fighter | US DDG 51 destroyer | US F22 fighter |
|---|---|---|---|---|
| Units | 16 | 620 | 62 | 187 |
| Development | **$0.7B** 34x | **$24B** | **$ 3B** 10x | **$28B** |
| Procurement | $4.3B | $23B | $60B | $34B |
| Total | $5.0B | $47B | $63B | $62B |

|  | T-AKE Cargo ship | C-17 Cargo plane | Cruise Passenger Ship | Airbus A380 Passenger plane |
|---|---|---|---|---|
| Units | 12 | 190 | 10 | 65+ |
| Development | **$0.1B** 50x | **$ 7B** | **$0.06B** 200x | **$13B** |
| Procurement | $4.6B | $59B | $ 6B | $22B+ |
| Total | $4.7B | $66B | $ 6B | $35B+ |

**Figure E.1**
Ship versus aircraft development costs (as of 2005).
Credit: Author.

variously called systems-of-systems, megasystems, and metasystems and are all marked by the underlying concept that none of the pieces are developed under a single point of control but rather must work together through interfaces that are developed and agreed to by all parties. At the same time, advancements in networks and in software and hardware that allow greater autonomy (often described as artificial intelligence) are changing the way societies interact with technology. To predict how ships and vessels will integrate into these metasystems, naval architects will have to pay far more attention to the interfaces between a ship and its operating environment so that these vessels are not left trailing in the wake of rapid developments in technology and changes in markets and missions.[6]

The intermodal ISO shipping container is the paradigm for what has come to be known as "open architecture," which relies on widely available industrial standards to dictate the interfaces between a platform (e.g., the ship) and its environment. Developed in the 1950s by the American trucking magnate Malcom McLean, the container is a standard size (usually 20 or 40 feet long, designated as twenty-foot equivalents or TEUs) that can carry anything from electronics to tires to bicycles and can be carried seamlessly by ships, trains, and trucks as part of an operating environment called the intermodal transportation system. Container ship design is no longer optimized around the ship itself but rather as one part of a complex metasystem that includes the port loading and unloading infrastructure. In fact, the ship is not even the most expensive bit; an 18,000 TEU container ship might cost $200 million, while its cargo for a single voyage might be worth $300 million. The same is now true for warships. A modern destroyer's hull and machinery, the parts that naval architects are concerned with, now represent less than half the cost of the ship; the combat and mission systems (radars, sonars, guns, and the like) are where the money is spent, and of that, software is the single most expensive item. Because software and combat systems evolve so quickly compared with the platform, warship designers are embracing open architecture through the use of modularity, in which mission systems can be updated through a warship's life without ripping apart the hull. The LCS project, for example, uses containerized mission systems, while in Germany and Denmark, MEKO and Stanflex systems allow for plug and play of different weapons and sensors within standardized locations in the ship.[7]

The powerful computers and networked systems of the information age require far greater amounts of electricity than previous ages. At the same time, grave concerns about the global environment are pulling nations away from traditional fossil fuels and toward generating electricity in cleaner ways with, for example, offshore wind turbine farms and tidal power plants. Careers for naval architects used to be focused exclusively on ships and shipping; today's graduates are just as likely to see positions opening in offshore industries, energy companies, and even the aerospace industry (which designs wind turbine blades, among other things) that help to stem the tide of environmental change. The traditional university degree in naval architecture is also evolving beyond ships and into the global economy, in many cases not even using the phrase "naval architecture." For example, at Virginia Tech in the United States, naval architecture studies are part of Aerospace and Ocean Engineering, while at University of Tokyo naval architecture was subsumed in 1998 into Environmental and Ocean Engineering. At the same time, the global nature of the discipline had led to many universities pooling resources, faculty, and students into global degree programs, for instance the EMSHIP European Master's Course in Advanced Design in Ship and Offshore Structures.[8]

**Naval Architecture as Legacy**

> There does not exist in the world a more thankless career and profession, for those who dedicate themselves to it, than that of naval architecture: whose works, although the most sumptuous, most carefully constructed, most costly to a nation and the most enthusiastically pursued, are also the most heartbreaking to the men who have assembled and directed those works, only to watch them disappear from one moment to the next like smoke, without leaving the least vestige of proof of the little-appreciated concern, care and sleepless nights of those unfortunate men who executed them.[9]
> 
> —Honorato Bouyón, former engineer-director of Spain's Cuerpo de Ingenieros de Marina, 1833

Bouyón's gloomy comment must be put into some context. He had come of age in the Spanish navy at its apogee; when allied with the French navy, it had defeated its most implacable adversary, Great Britain, in the War of American Independence in 1783. He helped build the Spanish fleet to its greatest power toward the end of the eighteenth century, only to watch its near-complete destruction at the Battle of Trafalgar in 1805. By 1833, when he penned these words, Spain was on the cusp of civil war, it had lost almost its entire empire in the Americas, and its navy was adrift and without purpose. He was watching not only his ships but his entire legacy "disappear ... like smoke." Yet Bouyón got one thing wrong. The ships may have disappeared, but their names, at least, live on in history. Not so those of their designers, even when they have put their stamp on entire fleets across several nations, or have given to their own nation their last full measure of devotion.

Jacques Stosskopf was a highly decorated French artillery officer during World War I, after which, with his close friend Jean Dieudonné, he attended "X" and became a naval architect in the Génie Maritime. He was soon designing torpedo-boats and the *Mogador*-class fast destroyer. In 1939, just as World War II broke out, he was placed in charge of new construction at the naval shipyard in Lorient. When Germany defeated France in June 1940, it occupied Lorient and built a submarine base there. Stosskopf— who was of Alsatian descent and spoke fluent German—appeared to be the perfect go-between linking the French with their German occupiers. He was roundly despised by his compatriots as a traitor. In reality, he had become one of the most successful agents of the Alliance network of the French resistance, under the legendary Marie-Madeleine Fourcade. For four years he relayed critical intelligence on German submarine technology and operations, until he was discovered in 1944 and executed in a concentration camp. I first learned of Stosskopf while I was working at the Direction des Constructions Navales, where his name was given to the principal auditorium; his name has also been honored by the French navy in many other ways.[10]

Rogério Silva Duarte Geral D'Oliveira was a naval cadet during World War II in neutral Portugal, after which he was sent to Britain to train as a naval constructor, returning to Portugal in 1950. He designed passenger and cargo vessels before he was tasked in 1961 with adapting an American warship design that became the *Almirante Pereira da Silva* class of frigates. In 1965 he led the design of a robust multipurpose corvette, which became the *João Coutinho* class. Built in both Germany and Spain for the Portuguese navy from 1970 to 1971, the design proved so effective that it was replicated for several new classes of corvettes across five nations: Portugal's *Baptista de Andrade*, Spain's *Descubierta*, France's *d'Estienne d'Orves*, and Germany's MEKO 140, which was built in Argentina as the *Espora* class. Many of the vessels continue in service today in dozens of nations around the world. D'Oliveira himself rose to the rank of *contra-almirante* (rear admiral) and, though retired from military service, continues even today to consult, teach, and write on the subject of naval architecture.[11]

Andrew B. Summers was educated as a naval architect at MIT and began work in 1973 at NAVSEC (soon to become NAVSEA). Like many of his colleagues, he worked on several futuristic designs that were never built. This changed in 1982 when he became the chief naval architect for the DDG 51 *Arleigh Burke* class of destroyers. He led the design through its commissioning into service in 1991. Although designed during the Cold War, the DDG 51 and its variants are well adapted to a changing world. A projected 77 ships will be built—perhaps the largest class of warships ever constructed in the modern era—and are likely to sail until the end of the twenty-first century, or for almost 100 years of service. Meanwhile, another five nations are constructing some variant of the DDG 51: Japan's *Kongō* and *Atago* classes, Korea's *Sejong-the-Great* class, Spain's *Álvaro de Bazán* class, Norway's *Fridtjof Nansen* class, and Australia's *Hobart* class. Summers went on to become chief naval architect for the high-technology destroyer DDG 1000 *Zumwalt* class and retired shortly before the ship was launched in 2013. He continues to consult on naval architecture and design.[12]

As with building designers like Frank Lloyd Wright and Frank Gehry, the designers of ships merit their place in history. With *Bridging the Seas* and its predecessor *Ships and Science*, I now have given this "thankless career and profession" of naval architecture its very first history and inscribed the legacies of its developers and practitioners so that they will not, as Bouyón had once feared, "disappear like smoke."

# Notes

**Preface**

1. Benford and Mathes, *Your Future in Naval Architecture* (1968).

2. See, for example, Laurence Dunn, "Merchant Ship Design: Some Aesthetic Considerations," *TINA* 100 (1957), pp. 1–11; John Charles Roach and Herbert A. Meier, "Visual Effectiveness in Modern Warship Design," *JASNE/NEJ* 91, no. 6 (1979), pp. 19–32; NAVSEA, *Warship Appearance* (1981); Quartermaine, *Building on the Sea* (1996); David J. Andrews, "The Art and Science of Ship Design," *IJME* 85 (2007), pp. 9–26; Nowacki and Lefèvre, *Creating Shapes in Civil and Naval Architecture* (2009); Sheridan, "Synthesis of Aesthetics for Ship Design" (2013).

3. Pollard and Robertson, *The British Shipbuilding Industry, 1870–1914* (1979), p. 45; Pugh, *The Cost of Seapower* (1986), pp. 160–161; Headrick, *The Tentacles of Progress* (1988), pp. 42–44; Modelski and Thompson, *Seapower in Global Politics, 1494–1993* (1988), pp. 80–81; Davies, *Belief in the Sea* (1992), pp. 65–66; Todd and Lindberg, *Navies and Shipbuilding Industries* (1996), pp. 131–138; Johnman and Murphy, *British Shipbuilding and the State since 1918* (2002), pp. 101–103.

**Terms, Symbols, Units of Measure, and Money**

1. *PNA3* (1988), vol. 1, pp. 304–306.

2. Measuring Worth website https://www.measuringworth.com/, accessed February 2015.

**Prologue**

1. Among the hundreds of books and scholarly articles on the Industrial Revolution, several useful primers are Musson and Robinson, *Science and Technology in the Industrial Revolution* (1968); Hudson, *The Industrial Revolution* (1992); and Allen, *The British Industrial Revolution in Global Perspective* (2009).

2. Several notable biographies of Brunel are Brunel, *The Life of Isambard Kingdom Brunel* (1870); Rolt, *Isambard Kingdom Brunel* (1957); Pugsley, *The Works of Isambard Kingdom Brunel* (1976); Vaughan, *Isambard Kingdom Brunel* (1991); Buchanan, *Brunel* (2002); and Brindle, *Brunel* (2005).

3. "Engineering: In Need of Heroes," *Economist*, May 16, 1998, pp. 91–93; Christine MacLeod, "The Nineteenth Century Engineer as Culture Hero," in Kelly and Kelly, *Brunel* (2006), pp. 61–79.

4. Hyman, *Charles Babbage* (1982), pp. 157–163.

5. For the Britannia Bridge, see Fairbairn, *The Life of Sir William Fairbairn* (1877); and Rosenberg and Vincenti, *The Britannia Bridge* (1978).

6. Brown, *The Way of the Ship in the Midst of the Sea* (2006), pp. 23–25.

7. Brunel, *The Life of Isambard Kingdom Brunel* (1870), p. 233.

8. See John Armstrong and David M. Williams, "The Perception and Understanding of New Technology: A Failed Attempt to Establish Transatlantic Steamship Liner Services 1824–1828," *TNM/LMN* 7, no. 3 (2007), pp. 41–56.

9. The story of *Great Western* is recounted in Griffiths, Lambert, and Walker, *Brunel's Ships* (1999), pp. 13–26, 91–104.

10. *The Times* (London), "Steam Communication with Distant Parts," 27 August 1836, p. 5.

11. "Atlantic Steam Navigation," *Edinburgh Review* 25 (1837), pp. 118–146; Paul Quinn, "IK Brunel's Ships—First Among Equals?," *TNS/IJHET* 80, no. 1 (2010), pp. 80–99, at pp. 84, 96.

12. Burgess, *Engines of Empire* (2016), pp. 35–38.

13. The competition between Cunard and Brunel is described in Fox, *Transatlantic* (2003).

14. Details of the development of *Great Britain* are from Corlett, *The Iron Ship* (1975); and Griffiths, Lambert, and Walker, *Brunel's Ships* (1999), pp. 27–36, 76–88, 127–136.

15. "The 'Great Eastern,'" *Times* (London), 30 April 1857, p. 11.

16. Biographical information on Russell is from Emmerson, *John Scott Russell* (1977); and Andrew Lambert, "John Scott Russell—Ships, Science and Scandal in the Age of Transition," *TNS/IJHET* 81 (2011), pp. 100–118.

17. For Russell and the wave line, see Larrie D. Ferreiro and Alex Pollara, "Contested Waterlines: The Wave-Line Theory and Shipbuilding in the Nineteenth Century," *T&C* 57, no. 2 (2016), pp. 414–444 (Russell quote, p. 417).

18. Accounts of the design and construction of *Great Eastern* are from Brunel, *The Life of Isambard Kingdom Brunel* (1870), pp. 289–339; Rolt, *Isambard Kingdom Brunel* (1957), pp. 235–264 (Brunel's quote, p. 236); Emmerson, *John Scott Russell* (1977), pp. 65–157; and Griffiths, Lambert, and Walker, *Brunel's Ships* (1999), pp. 137–153. Brunel's calculations are from the Brunel Collection, General Calculation Books 1850–1858; and Pugsley, *The Works of Isambard Kingdom Brunel* (1976), pp. 142–146. See also George S. Emmerson and R. Angus Buchanan, "L.T.C. Rolt and the Great Eastern Affair of Brunel versus Scott Russell," *T&C* 20 (1980), pp. 553–569 and *T&C* 24 (1983), pp. 98–106, 107–113.

19. Brunel's notebooks, sketchbooks, and calculations regarding *Great Eastern* are dispersed throughout the Brunel Collection, University of Bristol Library, Special Collections, Bristol, UK.

Russell's notebooks are cataloged as MS 516, books 1–7 (circa 1859), Science Museum Library, London.

20. Wright, "Ship Hydrodynamics" (1983), pp. 80–81.

21. Brunel, *The Life of Isambard Kingdom Brunel* (1870), p. 306.

22. Brunel, *The Life of Isambard Kingdom Brunel*, p. 341.

23. Rolt, *Isambard Kingdom Brunel* (1957), p. 243.

24. Emmerson, *John Scott Russell* (1977), p. 95.

25. Brown, *The Way of the Ship in the Midst of the Sea* (2006), pp. 36–37.

26. Brown, *The Way of the Ship in the Midst of the Sea*, p. 40.

## Chapter 1

1. Hugo, *Les Misérables* (1862), vol. 1, p. 531; Ruskin, *The Harbours of England* (1856), in *The Works of John Ruskin* (1904), vol. 13, p. 28. Hugo wrote *Les Misérables* between 1845 and 1848, before Ruskin wrote *Harbours of England*, but it remained unpublished for over 14 years. Therefore, although the two authors expressed almost identical sentiments, Hugo was not influenced by Ruskin nor vice versa.

2. North, "Sources of Productivity Change in Ocean Shipping" (1968).

3. Ferreiro, *Ships and Science* (2007), pp. 158–179, 303–304.

4. Maindron, *Les Fondations de prix à l'Académie des Sciences* (1881), pp. 22, 46, 56–60.

5. George Atwood and Honoré-Sébastien Vial du Clairbois, "A Disquisition on the Stability of Ships," *PTRS* (1798), pp. 201–310; Bonjean, *Nouvelles Echelles de Déplacement* (1810); Ferreiro, *Ships and Science* (2007), pp. vii–viii, 256–257; *PNA3* (1988), pp. 31–34.

6. Dupin, "Mémoire sur la Stabilité des Corps Flottants," in *Applications de géométrie et de méchanique* (1822). Biographical information on Dupin is from Christen and Vatin, *Charles Dupin* (2009); and Bradley, *Charles Dupin* (2011). Later works built on Dupin's efforts: Moreau, *Principes fondamentaux de l'équilibre et du mouvement des corps flottans dans deux milieux résistans* (1830); Poisson, *Traité de Mecanique* (1833); and Bravais, *Sur l'Equilibre des corps flottants* (1840). The most complete treatment of three-dimensional stability in terms of energy was given by Émile Guyou, "Théorie nouvelle de la stabilité de l'équilibre des corps flottants" (New theory of stability for floating bodies), *RMC* 60 (1879), pp. 682–705. For a full analysis of the development of stability theory, see Vallée-Poussin, "Histoire des théories de stabilité des corps flottants 1727–1879" (2015).

7. Lemmers, *Techniek op schaal* (1996), pp. 152–153; Gatti, *Un raggio di convenienza* (2008), p. 25; Fratta, *La fabbrica delle navi* (1990), pp. 100–103; Tupinier, *Mémoires du Baron Tupinier* (1994), pp. 103–133; Marzari, *Progetti per l'Imperatore* (1990), pp. 21–43; Staccioli, *Mare Scienza e Tecnica* (1999), p. 10.

8. Bopp-Vigne, "Emigrés francais" (2004), pp. 422–427; Chatenet, "Jean-Baptiste de Traversay" (2004), pp. 495–497; Zorlu, *Innovation and Empire in Turkey* (2008), pp. 78–90.

9. Davin, "Un toulonnais, Jean-Louis Barrallier"; Lutun, *1814–1817 ou l'Epuration dans la Marine Française* (2005), pp. 190–196; Gardiner, *Frigates of the Napoleonic Wars* (2000), pp. 43–46, 93–95.

10. Steel, *The Elements and Practice of Naval Architecture* (1805); Barrallier to Admiralty, 9 Sept. 1812, NAUK ADM 1/4382.

11. Quote from Sewell, *A Collection of Papers on Naval Architecture* (1791), p. iii. The original papers on naval architecture printed on the blue wrappers of the *European Magazine* were not preserved in library collections. For the history of the society, see Andrew W. Johns, "An Account of the Society for the Improvement of Naval Architecture," *TINA* 52 (1910), pp. 28–40; Schaffer, "Fish and Ships" (2004); and Schaffer, "The Charter'd Thames" (2007).

12. Hutchison, *A Treatise on Naval Architecture* (1794), pp. 63–66; Gore, *Result of Two Series of Experiments* (1799). Gore used Mark Beaufoy's Greenland Dock apparatus to conduct his model tests on resistance, and Mark Beaufoy later repeated Gore's stability experiments. See Beaufoy, "On the Stability of Vessels" (1816).

13. Beaufoy and Beaufoy, *Nautical and Hydraulic Experiments* (1834). Explanations of Beaufoy's experiments are given in Wright, "Ship Hydrodynamics 1710–1880" (1983), pp. 39–67; and Thomas Wright, "Mark Beaufoy's Nautical and Hydraulic Experiments," *MM* 75 (1989), pp. 313–327. For Bossut's model experiments, see Ferreiro, *Ships and Science* (2007), pp. 164–167.

14. For the closing of the society, see NMM manuscript SOC/17, 1800–1801. For Sewell's obituary, see *Gentleman's Magazine* 72, no. 2 (November 1802), p. 1078.

15. Among many works on wooden ship construction, two excellent starting points are Steffy, *Wooden Ship Building and the Interpretation of Shipwrecks* (1994); and Hocker and Ward, *The Philosophy of Shipbuilding* (2004).

16. For wooden ship structures under loads, see John F. Coates, "Hogging or 'Breaking' of Frame-Built Wooden Ships: A Field for Investigation?," *MM* 71, no. 4 (1985), pp. 437–442; Francisco Fernández González, "Ship Structures under Sail and under Gunfire," in Fernández González et al., *Technology of the Ships of Trafalgar* (2006); Bernard Luttmer and Bart Boon, "Aspects of Longitudinal Strength of Historic Ships," *Historic Ships Conference Proceedings*, RINA, London, 2007.

17. Goodwin, *The Construction and Fitting of the English Man of War* (1987), pp. 95–110; Gardiner, *Frigates of the Napoleonic Wars* (2000), pp. 75–81. For British constructors in Spain, see Larrie D. Ferreiro, "Spies versus Prize: Technology Transfer between Navies in the Age of Trafalgar," *MM* 93, no. 1 (2007), pp. 16–27.

18. Wadia, *The Bombay Dockyard and the Wadia Master Builders* (1957); Sutton, *Lords of the East* (2000); Bulley, *The Bombay Country Ships* (2000); Lambert, *Trincomalee* (2002).

19. Larrie D. Ferreiro, "Paul Hoste et Les Premiers Développements de L'Architecture Navale, 1685–1700" (Paul Hoste and the first developments of naval architecture, 1685–1700), *BATMA* 111/2626 (2012); Biaggio Pangalo, "Memoire consernant [sic] la conservation des vaisseaux du Roy" (Memoir concerning the preservation of the King's ships), *ANF* Marine D/1/9 f. 10, Brest (1699); Alexandre Gobert (sometimes Goubert), "Examen de la nouvelle construction de

l'architecture navale" (Examination of the new construction of naval architecture), *ANF* Marine D/1/10 f. 83–84, Brest (1700); Gobert, "Memoire sur la construction des Vaisseaux" (Memoir on the construction of vessels), *ANF* Marine D/1/10 f. 94–100, Versailles (1703).

20. Beauchesne, *Historique de la construction navale à Lorient* (1980), pp. 30–40; Peter, *Le port et l'arsenal de Toulon* (1995), pp. 130–139. See also Boudriot and Delacroix, *Le Fleuron* (1995).

21. Stibolt, *Afhandling om Skibes Kiølbrækkelighed* (1784); For plans for Stiboldt's (and other Danish) ships, see Holmen dockyard (Denmark) database, accessed April 2017.

22. Toll, *Six Frigates* (2006), gives a broad view of the early development of the U.S. Navy. For Humphreys's use of diagonal riders, see especially, Eddy, *Joshua Humphreys* (2005), pp. 60–75. Thanks to Robert Wasalaski for his information on *Constitution*'s 1993–1995 overhaul.

23. Alan Lemmers, "Shipworm, Hogbacks and Ducks' Arses: The Influence of William May on Sir Robert Seppings," *MM* 99, no. 4 (2013), pp. 410–428.

24. Glete, *Navies and Nations* (1993), vol. 1, p. 396.

25. Among the many works on this subject, two classic texts on the timber problem are Albion, *Forests and Sea Power* (1926/1999); and Bamford, *Forests and French Sea Power* (1956).

26. Alan Lemmers, "Shipworm, Hogbacks and Ducks' Arses," *MM* 99, no. 4 (2013), pp. 410–428.

27. Although Seppings was familiar with diagonal systems such as *Leiden*'s, he credited the wooden truss bridge at Schaffhausen, Switzerland, as his inspiration for the X-shaped truss system. Robert Seppings, "On a New Principle of Constructing His Majesty's Ships of War," *PTRS* 104 (1814), pp. 285–302; Thomas Young, "Remarks on the Employment of Oblique Riders," *PTRS* 104 (1814), pp. 303–336. For Dupin's analysis, see Charles Dupin, "De la Structure des Vaisseaux Anglais, Considerée dans ses Derniers Perfectionnements" (On the structure of English vessels, considered in their latest improvements), *PTRS* 107 (1817), pp. 86–135. Modern assessments include Thomas Wright, "Thomas Young and Robert Seppings: Science and Ship Construction in the Early Nineteenth Century, *TNS/IJHET* 53 (1981), pp. 55–72; Gardiner, *Frigates of the Napoleonic Wars* (2000), pp. 81–86; Andrew Lambert, "Sir Robert Seppings and the British Naval Response to the Lessons of Trafalgar," in Fernández González et al., *Technology of the Ships of Trafalgar* (2006); and Lambert, "Science and Seapower" (2006). My thanks to Andrew Lambert for his additional insights into Seppings's work.

28. Lemmers, *Techniek op schaal* (1996), pp. 148–172; Tredrea and Sozaev, *Russian Warships in the Age of Sail* (2010), pp. 220–226; Independence Seaport Museum, John Lenthall Collection, F1 L90.43, USS *Pennsylvania* (1837).

29. Robert Seppings, "On a New Principle of Constructing Ships in the Mercantile Navy," *PTRS* 110 (1820), pp. 133–143. Overviews of other diagonal framing systems are in Linda Maloney, "A Naval Experiment," *AN* 34 (1974), pp. 188–196; and Alisa Steere, "Mid-17th to Mid-19th Century Design Innovations That Strengthened Ship Hulls," *NRJ* 49, no. 2 (2004), pp. 84–89.

30. UNESCO, "The Watertight-Bulkhead Technology of Chinese Junks" (2010); Xi Longfei and David W. Chalmers, "The Rise and Decline of Chinese Shipbuilding in the Middle Ages," *IJME*

146, no. A2 (2004), pp. 59–70; Franklin, "Sundry Maritime Observations" (1785), p. 301; Bentham, *The Life of Brigadier General Sir Samuel Bentham* (1862), p. 110.

31. Dirk Böndel, "Hydrodynamic Design and Performance Comparisons of British, French, and American Frigates of the Trafalgar Era," in Fernández-González et al., *Technology of the Ships of Trafalgar* (2006). Böndel derived speed and resistance data for each navy from normalized sets of hulls.

32. For the development of concepts in ship resistance and hydrodynamics, see Ferreiro, *Ships and Science* (2007), pp. 113–185.

33. Calkoen, *Wiskundige scheeps-bouw en bestuur* (1805); Loon, *Beschouwing van den Nederlandschen scheepsbouw* (1820); Loon, *Handleiding tot den Burgerlijken Scheepsbouw* (1838); Vandersmissen, "Folkert van Loon" (2004). My thanks to Bart Boon and Gerbrand Moeyes for their clarifications of these works. On the Swedish experiments, see Lagerhjelm et al., *Hydrauliska forsook* (1818–1822); Beaufoy and Beaufoy, *Nautical and Hydraulic Experiments* (1834), pp. xxix–xxxvi; and Frängsmyr et al., *The Quantifying Spirit in the 18th Century* (1990), pp. 295–300.

34. Thibault, *Recherches expérimentales* (1826). For an analysis of this work, see Rank, *Die Theorie des Segelns* (1984), pp. 271–277. For early sail and masting theory, see Ferreiro, *Ships and Science* (2007), pp. 101–112. For sail design, see Harland, *Seamanship in the Age of Sail* (1984).

35. For Schank's sliding keels, see Steel, *The Elements and Practice of Naval Architecture* (1805), pp. 158–174. From 1800 to 1819 Richard Gower built (and wrote about) a series of experimental ships, each named *Transit*, to test his hull and rigging ideas, but these were not widely adopted. See Gower, *Original Observations* (1833); and MacGregor, *Fast Sailing Ships* (1988), pp. 50–52.

36. Ferreiro, "Développements et avantages tactiques du doublage en cuivre des coques des navires français, britanniques et espagnols" (2018); Solar, "Opening to the East" (2013).

37. Humphry Davy, "On the Corrosion of Copper Sheeting by Seawater, and on Methods of Preventing This Effect, and on Their application to Ships of War and Other Ships," *PTRS* 114 (1824), pp. 151–246 and *PTRS* 115 (1825), pp. 328–346; Humphry Davy, "Additional Experiments and Observations on the Application of Electrical Combinations to the Preservation of the Copper Sheathing of Ships, and to Other Purposes," *PTRS* 114 (1824), pp. 242–246; McCarthy, *Ships' Fastenings* (2005), pp. 101–113.

38. Andrew Lambert, "Captain Sir William Symonds and the Ship of the Line 1832–1847," *MM* 73 (1987), pp. 167–178; Lambert, *The Last Sailing Battlefleet* (1991), pp. 67–87; Morriss, *Cockburn and the British Navy in Transition* (1997), pp. 205–211, 237–242; Chapelle, *The History of the American Sailing Navy* (1949), pp. 448–450.

39. Ferreiro, *Ships and Science* (2007), pp. 113–185; John Scott Russell, "On the Solid of Least Resistance," *BAAS Report*, 1835, pp. 107–108.

40. Hyman, *Charles Babbage* (1982), pp. 88–90.

41. Larrie D. Ferreiro and Alex Pollara, "Contested Waterlines: The Wave-Line Theory and Shipbuilding in the Nineteenth Century," *T&C* 57, no. 2 (2016), pp. 414–444 ("excavating," p. 426);

Wright, "Ship Hydrodynamics" (1983), pp. 68–81 ("to fill out the lines," p. 80); Darrigol, *Worlds of Flow* (2005), pp. 47–56.

42. *Mechanics' Magazine* 32 (1840), pp. 130–136; *Literary Gazette*, 1840, pp. 566–568; Fishbourne, *Lectures on Naval Architecture* (1846), pp. 112–138; Longfellow, *The Seaside and the Fireside* (1849), pp. 7–29; Verne, *Twenty Thousand Leagues Under the Sea* (1998), pp. 82–83. Longfellow was a friend of American shipbuilder Donald McKay and undoubtedly learned of the wave line through him; Verne was always well versed in the latest technological and scientific advancements.

43. James R. Napier, "On Sections of Least Resistance in Ships of Limited Breadth and Draft of Water," *PPSG* 5 (1864), pp. 217–221; Martin Bellamy, "A Ludicrous Travesty? James R. Napier and the *Lancefield*," *MM* 98, no. 2 (2012), pp. 161–177.

44. Larrie D. Ferreiro, "A Biographical Sketch of John Willis Griffiths from Primary and Archival Sources," *NRJ* 52, no. 4 (2007), pp. 221–228.

45. Some excellent references are Albion, *Square Riggers on Schedule* (1965); Crothers, *The American Clipper-Built Ship* (1997); Ujifusa, *Barons of the Sea and Their Race to Build the World's Fastest Clipper Ship* (2018).

46. Quote from *Literary Gazette*, 1857, p. 980. See also McKay, *Some Famous Sailing Ships and Their Builder Donald McKay* (1928); and Chapelle, *The Search for Speed under Sail* (1967).

47. MacGregor, *Fast Sailing Ships* (1988), pp. 199–125, 167–168 (quote p. 168); MacGregor, *British and American Clippers* (1993), pp. 36–40; Vos, *Nederlandse Clippers* (2003), pp. 70–76; Briot and Briot, *Les Clippers Français* (1993), pp. 128–132.

48. J. Phipps, "On the Sailing Powers of Two Yachts, Built on the Wave Principle," *BAAS Report*, 1846, pp. 112–113.

49. Quotes: Griffiths, *A Treatise on Marine and Naval Architecture* (1856), p. 196; and Russell, *The Modern System of Naval Architecture* (1865), vol. 1, p. 613. For the yacht *America*, see Rousmaniere, *The Low Black Schooner* (1986). For the relationship between Steers and Griffiths, see George Steers Papers, John W. Griffiths to George Steers, June 28, 1856.

50. For the use of wave lines in yacht design, see Kemp, *Yacht Architecture* (1897), pp. 118–162. For *Joe Lane*, see Chapelle, *The History of American Sailing Ships* (1935), pp. 213–215. For similar principles, see Eardley-Wilmot, *Reminisces of the Late Thomas Assheton Smith* (1902), pp. 140–160; Charles Dawson, "Thomas Assheton Smith's Steam Yachts," *MM* 92, no. 3 (2006), pp. 331–338; Montagu, *Naval Architecture* (1852), pp. 31–40; Colin Archer, "The Wave-Principle Applied to the Longitudinal Disposition of Immersed Volume," *TINA* 19 (1878), pp. 218–231.

51. Herreshoff, *Capt. Nat Herreshoff, the Wizard of Bristol* (1953), pp. 128–140; Pastore, *Temple to the Wind* (2005), pp. 8–9.

52. Howells, "The Response of Old Technology Incumbents to Technological Competition" (2002); and Mendonça, "The Sailing Ship Effect" (2013). A multivariate statistical analysis supports the finding that older sailing ship technologies shaped newer steamship development and

not vice versa (Damásio and Mendonça, "Modeling Insurgent-Incumbent Dynamics" [2018]). For a comprehensive view of how wooden warships powered by sail maintained relevance in an increasingly industrialized navy, see Brown, *Before the Ironclad* (1990); and Lambert, *The Last Sailing Battlefleet* (1991).

53. Masefield, "Ships," in *Salt-Water Poems and Ballads* (1916), pp. 131–132.

**Chapter 2**

1. Hugo, *Les Misérables* (1862), vol. 1, p. 335; Ruskin, *The Works of John Ruskin* (1903), vol. 8, p. 67, and (1907), vol. 28, p. 452.

2. See Sutcliffe, *Steam* (2004).

3. See Albion, *The Rise of New York Port, 1815–1860* (1939/1984).

4. O'Har, *Shipbuilding, Markets and Technological Change in East Boston* (1995), pp. 12–13.

5. Paine, *The Sea and Civilization* (2013), p. 11.

6. Ferreiro, "The Aristotelian Heritage in Early Naval Architecture" (2010); Leonhard Euler, "Mémoire sur la force des rames" (Memoir on the force of oars), 1747, in *Leonhardi Euleri Opera Omnia*, 2nd Series, vol. 20 (1974), pp. 101–129.

7. Paddle wheel boats are described in several parts of Needham, *Science and Civilisation in China*, vol. 4, *Physics and Physical Technology: Part 2, Mechanical Engineering* (1965), and *Part 3, Civil Engineering and Nautics* (1971).

8. For general works, see Joseph Eliav, "Paddle Wheels for Sailing Men-of-War," *MM* 102, no. 3 (2016), pp. 275–289. For Spanish paddle wheelers, see Molina, *El ingenio de Blasco de Garay* (1996); Ricardo Molina, "Mechanical Inventions for Sailing Without Sails or Oars," *JASNE/NEJ* 116, no. 2 (2004), pp. 15–18; and the following chapters in Achútegui Rodríguez, *I Simposio de historía de las técnicas* (1996): Alejandro Mira Monerris, "La propulsión mecánica en los siglos XVI, XVII y XVIII" (Mechanical propulsion in the 16th, 17th and 18th centuries), pp. 61–80; and Nicolás García Tapia, "Barcos impulsados por ruedas de paletas en España en el siglo XVI" (Paddle wheel boats of 16th-century Spain), pp. 159–164. For French paddle wheelers, see Jean-Mathieu de Chazelles and François Du Quet (also spelled Duquet, Du Guet, and Duguet), "A Method for Rowing Men of War in a Calm," *PTRS* 31 (1720), pp. 239–250.

9. Leonhard Euler, "De promotione navium sine vi venti" (On the movement of ships without the force of wind), in *Leonhardi Euleri Opera Omnia* (1974), vol. 20, pp. 196–228; Daniel Bernoulli (with commentary by Frans A. Cerulus), "Recherches sur la manière la plus advantageuse de supliéer l'action du vent sur les grands vaisseaux" (Research on the most advantageous manner to supplement the action of wind on large vessels), 1753, in *Die Werke von Daniel Bernoulli* (2004), vol. 8, pp. 35–251; Cerulus, "Daniel Bernoulli and Leonhard Euler on the Jetski" (2006).

10. Manstan and Frese, *Turtle* (2010).

11. Cerulus, "Daniel Bernoulli and Leonhard Euler on the Jetski" (2006), p. 87.

12. The following sections on the early development of steam power are derived from several general histories, notably Flexner, *Steamboats Come True* (1944); Gardiner, *The Advent of Steam* (1993); Sutcliffe, *Steam* (2004); and Clark, *Steamboat Evolution* (2006).

13. Although Fulton does not mention a visit to *Charlotte Dundas*, Symington clearly recalls it. See Dickinson, *Robert Fulton, Engineer and Artist* (1913), pp. 179–181.

14. John Armstrong and David M. Williams, "Technological Advance and Innovation: The Diffusion of the Early Steamship in the United Kingdom, 1812–34," *MM* 96, no. 1 (2010), pp. 42–61.

15. For SS *Savannah*, see Busch, *Steam Coffin* (2010).

16. Labaree et al., *America and the Sea* (1998), pp. 235–261; Roland et al. *The Way of the Ship* (2008), pp. 99–147; Krisman and Cohn. *When Horses Walked on Water* (1998).

17. Prinsep, *An Account of Steam Vessels* (1830); Sutton, *Lords of the East* (2000), pp. 116–126. For steamship percentages, see Mendonça, "The Sailing Ship Effect" (2013), p. 1727.

18. Tomblin, *From Sail to Steam*, 1988; Christine Macleod, Jeremy Stein, Jennifer Tann, and James Andrew, "Making Waves: The Royal Navy's Management of Invention and Innovation in Steam Shipping, 1815–1832," *H&T* 16 (2000), pp. 307–333.

19. Sinclair, *Early Research at the Franklin Institute* (1966); John Armstrong and David Williams, "The Steamboat, Safety and the State: Government Reaction to New Technology in a Period of Laissez-Faire," *MM* 89, no. 2 (2003), pp. 167–184.

20. For the Cunard and Collins competition, see Fox, *Transatlantic* (2003); Butler, *Atlantic Kingdom* (2001); and Fowler, *Steam Titans* (2017).

21. Marestier, *Mémoire sur les bateaux à vapeur* (1824). For the introduction of steamships in France, see Mollat, *Les Origines de la Navigation à Vapeur* (1970); Roberts, "The Introduction of Steam Technology in the French Navy" (1976); and Brisou, *Accueil, Introduction et Développement de l'Énergie Vapeur* (2001).

22. For the introduction of steam marine engineering in the Netherlands, see Jan M. Dirkzwager, "A Case of Transfer of Technology: Ship Design and Construction in 19th-Century Netherlands," in Jackson and Williams, *Shipping, Technology, and Imperialism* (1996), pp. 189–210; in Denmark, see Rasmussen, "Statslig eller Privat?" (1993); in Spain, see Casanueva and Fraidias, "El *Real Fernando*" (1990); in Italy, see Antonio Formicola and Claudio Romano, "La propulsione a vapore nella reale marina borbonica; Studi, progetti e realizzazioni" (Steam propulsion in the Royal Bourbon Navy: Plans, projects and achievements), *RM*, 1996, pp. 91–108; in the Ottoman Empire, see Yener, *From the Sail to the Steam* (2010); in Russia, see V. S. Virginsky, "The Birth of Steam Navigation in Russia and Robert Fulton," *T&C* 9 (1968), pp. 562–569; and Kipp, "The Russian Navy and the Problem of Technological Transfer" (1994); in Japan, see Nakaoka, "From Shipbuilding to Automobile Manufacturing" (1998); and in China, see Hsien-Chun Wang, "Discovering Steam Power in China, 1840s–1860s," *T&C* 51 (2010), pp. 31–54.

23. Harley, "The Shift from Sailing Ships to Steamships, 1850–1890" (1971); Smith, "The 'Crinoline' of Our Steam Engineers" (2011); Crosbie Smith, "Witnessing Power: John Elder and the

Making of the Marine Compound Engine, 1850–1858," *T&C* 55, no. 1 (2014), pp. 76–106. For more general histories of marine steam engineering in the mid-nineteenth century, see Smith, *A Short History of Naval and Marine Engineering* (1938); Rippon, *Evolution of Engineering in the Royal Navy* (1988); Griffiths, *Steam at Sea* (1997); Smith and Marsden, *Engineering Empires* (2005), pp. 88–128; and Smith, *Coal, Steam and Ships* (2018).

24. Nabor Soliani, "Liquid Fuel on Steam Vessels," in Melville, *Proceedings of the International Engineering Congress* (1894), vol. 1; Denis Griffiths, "British Marine Industry and the Diesel Engine," *TNM/LMN* 7, no. 3 (1997), pp. 11–40; Dahl, "Naval Innovation: From Coal to Oil" (2000–2001).

25. Allington and Greenhill, *The First Atlantic Liners* (1997), pp. 64–81.

26. Alberi Auber, "Historical Developments in Naval Propulsion" (2003); Reech, *Mémoire sur les machines à vapeur* (1844), pp. 154–160; Peter W. Barlow, "An Investigation of the Laws Which Govern the Motion of Steam Vessels, Deduced from Experiments," *PTRS* 124 (1834), pp. 308–332; Foster, *The Search for Speed under Steam* (1991), pp. 88–89. Many other inventors, too numerous to mention, also laid claim to the feathering paddle wheel.

27. James Hamilton, "On Waves Raised by Paddle Steamers and Their Positions Relatively to the Wheels," *TINA* 22 (1881), pp. 73–86.

28. Clark G. Reynolds, "The Great Experiment: Hunter's Horizontal Wheel," *AN* 24 (1964), pp. 5–24; Seaton, *The Screw Propeller* (1909), pp. 91–95.

29. The following paragraphs are based on several works that describe the early development of the screw propeller: Labrousse, *Des Propulseurs Sous-Marins* (1845); Bourne, *A Treatise on the Screw Propeller* (1852); Smith, *On the Introduction and Progress of the Screw Propeller* (1856); Seaton, *The Screw Propeller* (1909); and Augustin Normand, *La Genèse de l'Hélice Propulsive* (1962).

30. Stevens, "The First Steam Screw Propeller Boats," 1893, quote p. 106. Additional thanks to Carl Kriegeskotte.

31. Augustin Normand, *La Genèse de l'Hélice Propulsive* (1962), pp. 31–50.

32. Gardiner, *The Advent of Steam* (1993), pp. 85–87.

33. Andrew Lambert, "The Royal Navy, John Ericsson, and the Challenges of New Technology," *IJNH* 2, no. 3 (2002).

34. Brown, *Before the Ironclad* (1990), pp. 99–114.

35. This and the following paragraphs are derived from Varende, *Les Augustin-Normand* (1960), pp. 80–83; Roberts, "The Introduction of Steam Technology in the French Navy" (1976), pp. 351–378; Michel Lagrée, "Religion and Technological Innovation: The Steamboat in 1840s France," *H&T* 12 (1995), pp. 327–359; Brisou, *Accueil, Introduction et Développement de l'Énergie Vapeur dans la Marine Militaire Française* (2001), pp. 577–620; and Borde, "The Madness of Frédéric Sauvage of Boulogne" (2003).

36. Seaton, *The Screw Propeller* (1909), p. 105; Carlton, *Marine Propellers and Propulsion* (2007), pp. 7–10.

37. Smith and Marsden, *Engineering Empires* (2005), pp. 88–128.

38. Gobert, "Memoire sur la construction des Vaisseaux" (Memoir on the construction of vessels), *ANF* Marine D/1/10 f.94–100, Versailles (1703); Peter Goodwin, "The Influence of Iron in Ship Construction: 1660 to 1830," *MM* 84, no. 1 (1998), pp. 26–40; Dominique Brisou, "La transition bois-fer dans la construction navale militaire en France au XIX$^e$ siècle" (The wood-iron transition in nineteenth-century French military naval construction), in Villain-Gandossi, *Deux siècles de constructions et chantiers navals* (2002), pp. 183–197; García-Torralba Pérez, *Navíos de la Real Armada 1700–1860* (2016), pp. 534, 541.

39. Barker, "Iron Ships in Green Fields" (2008).

40. Walker, *Song of the Clyde* (1984), pp. 31–32; Walker, *Ships and Shipbuilders* (2010), pp. 63–65; William H. Chaloner and William O. Henderson, "Aaron Manby, Builder of the First Iron Steamship," *TNS/IJHET* 29, no. 1 (1953), pp. 77–91.

41. Abell, *The Shipwright's Trade* (1948), pp. 108–110. The book's title was taken from Rudyard Kipling's 1910 poem "A Truthful Song," which claimed, "How little things have altered in the shipwright's trade."

42. MacDougall, *Chatham Dockyard*, p. 41.

43. Richard Biddle, "The Health of Workers in the Royal Dockyard, Portsmouth," in Leggett and Dunn, *Re-inventing the Ship* (2012), pp. 93–112.

44. Clarke, *The Changeover from Wood to Iron Shipbuilding* (1986), pp. 65–67, 78–85; O'Har, "Shipbuilding, Markets and Technological Change in East Boston" (1995), pp. 156–164, 172–176; Thiesen, *Industrializing American Shipbuilding* (2006), pp. 90–91,

45. MacGregor, *British and American Clippers* (1993), pp. 124–127, 166; *RHM* (2007), vol. 7, "Les constructions navales dans l'histoire" (Naval construction in history), p. 201; Cathérineau, *Construction navale* (1854).

46. *Marine et Technique au XIX$^e$ Siècle* (1988), p. 354; Rosenberg and Vincenti, *The Britannia Bridge* (1978), p. 4.

47. Brisou, "La transition bois-fer"; Jeffrey Remling, "Patterns of Procurement and Politics: Building Ships in the Civil War," *TNM/LMN* 17/1 (2008), pp. 16–29.

48. Hughes, *Networks of Power* (1983).

49. Paul Quinn, "Wrought Iron's Suitability for Shipbuilding," *MM* 89, no. 4 (2003), pp. 437–461; Hamilton, *Anglo-French Naval Rivalry* (1993), pp. 85–95; Brown, *Before the Ironclad* (1990), pp. 88–98, quote p. 89; Lambert, *Warrior* (1987), pp. 9–20.

50. Paul Quinn, "The Early Development of Magnetic Compass Correction," *MM* 87, no. 3 (2001), pp. 303–315; Smith and Marsden, *Engineering Empires* (2005), pp. 32–35; True, *History of the First Half-Century of the National Academy of Sciences* (1913), pp. 213–217, quote p. 214.

51. See, for example, Russell, *The Fleet of the Future: Iron or Wood?* (1861); and Wiard, *Iron vs Wood for Great Ships* (1867).

52. Clarke, *The Changeover from Wood to Iron Shipbuilding* (1986), p. 1.

53. Carlos Alfaro Zaforteza, "Medium Powers and Ironclad Construction: The Spanish Case, 1861–1868," *NINH* 16 (2009), pp. 11–21; Bargoni et al. *Navi a Vela e navi Miste Italiane* (2001), pp. 433–440.

54. Sechrest, "American Shipbuilders in the Heyday of Sail" (1998); Fernández-González, *Aportación de Cataluña a la arquitectura naval* (2009); Delis, *Mediterranean Wooden Shipbuilding* (2016).

55. Tredgold, *A Practical Essay on the Strength of Cast Iron* (1842); Robert A. Jewett, "Structural Antecedent of the I-Beam, 1800–1850," *T&C* 8, no. 3 (1967), pp. 346–362; Sutherland, *Structural Iron, 1750–1850* (1997), pp. 65–75; Addis, *Structural and Civil Engineering Design* (1999), pp. 209–219.

56. Jeremy, "Damming the Flood" (1977); Cotte, *De l'espionnage industriel à la veille technologique* (2005), pp. 131–133; Montgéry, *Mémoire sur les navires en fer* (1824).

57. Joseph Bertrand, "Éloge de M. Dupuy de Lôme" (Eulogy of M. Dupuy de Lôme), *MASIF* 44 (1888), pp. 186–203; SHDM *dossier individuel* (individual file), Dupuy de Lôme, CC7 A791.

58. Laird Family Papers, Worcestershire Record Office. Worcester, UK, .ref. X970.51:4, BA 8699, pp. 89–92, quote p. 89.

59. Dupuy de Lôme, *Atlas joint au Rapport sur les Bâtiments en Fer* (1843), and *Mémoire sur la Construction des Bâtiments en Fer* (1844), quotes pp. 1, 4; Brown, *Before the Ironclad* (1990), pp. 80–81.

60. Battesti, *La Marine de Napoléon III* (1997), vol. 1, pp. 42–57; Hamilton, *Anglo-French Naval Rivalry* (1993), pp. 64–105.

61. Battesti, *La Marine de Napoléon III* (1997), vol. 1, pp. 161–247.

62. Lambert, *Warrior* (1987).

63. Larrie D. Ferreiro, "The Social History of the Bulbous Bow," *T&C* 52, no. 2 (2011), pp. 335–359, at p. 341.

64. For iron warship developments in this period, see Ropp, *The Development of a Modern Navy* (1987); Brown, *Warrior to Dreadnought* (1997), pp. 12–73; and Greene and Massignani, *Ironclads at War* (1998).

65. Clarke and Storr, *The Introduction of the Use of Mild Steel into the Shipbuilding and Marine Engine Industries* (1983).

66. Breemer, "The Great Race" (2011).

67. Louis-Marc-Antoine-Émile Willotte [Villotte], "Etudes sur la fabrication, l'emploi et les propriétés de l'acier" (Studies on the fabrication, usage and properties of steel), *MGM*, 2nd series, no. 1 (1860), pp. 245–295, quote p. 245.

68. Gardiner and Lambert, *Steam, Steel and Shellfire* (1992), pp. 95–111; Brown, *Warrior to Dreadnought* (1997), pp. 74–90; Ropp, *The Development of a Modern Navy* (1987), pp. 36, 64; Bargoni et

al., *Navi a Vela e navi Miste Italiane* (2001), p. 461; Sondhaus, *The Naval Policy of Austria-Hungary* (1994), p. 53; Wend, *Recovery and Restoration* (2001), pp. 4–5.

69. Cooling, *Gray Steel and Blue Water Navy* (1979), pp. 15–32; Edward W. Sloan, "Progress and Paradox: Benjamin Isherwood and the Debate over Iron vs. Steel in American Warship Design," *JASNE/NEJ* 94, no. 4 (1982), pp. 59–63.

70. Kipling, "The Ship That Found Herself," in *The Day's Work* (1898), pp. 95–122.

## Chapter 3

1. See Cardwell, *From Watt to Clausius* (1971).

2. Brown, *The Way of the Ship in the Midst of the Sea* (2006); Peter Froude (a relative of William Froude), interview with author, November 2012.

3. Ferreiro, *Ships and Science* (2007), pp. 250–256.

4. Sobel, *Longitude* (1995).

5. Frank Horner, "Baudin's Oscillometer," *MM* 69 (1983), pp. 63–64; Chatfield, *On the Advantages of Observing a Ship's Inclination at Sea* (1831).

6. Henry Moseley, "On Dynamical Stability, and on the Oscillations of Floating Bodies," *PTRS* 140 (1850), pp. 609–643; John King, "Origins of the Theory of Ship Stability," *TRINA* 140 (1998), pp. 222–238. Moseley did not know about previous theoretical research on dynamic stability conducted 20 years earlier by the chief engineer of the *Génie Maritime*, Philippe-Jacques Moreau, who applied the principle of least work to small oscillations of ships in calm waters. See Moreau, *Principes fondamentaux de l'équilibre et du mouvement des corps flottans* (1830).

7. Rolt, *Isambard Kingdom Brunel* (1957), p. 252.

8. Froude's correspondence with Brunel on his rolling experiments are in the Brunel Collection, Eastern Steam Navigation letterbooks, vols. 4 (1856–1857) and 5 (1857). He described his experiments in an 1873 paper delivered before the Royal United Services Institute, "Apparatus for Automatically Recording the Rolling of a Ship in a Sea-Way, and the Contemporaneous Wave-Slopes," in Froude, *The Papers of William Froude* (1955), pp. 192–205.

9. William Froude, "On the Rolling of Ships," *TINA* 2 (1861), pp. 180–229.

10. For the history of water wave theory, see Darrigol, *Worlds of Flow* (2005), pp. 31–100.

11. *PNA2* (1967), pp. 672–673.

12. Charles W. Merrifield, "Report of a Committee ... on the State of Existing Knowledge on the Stability, Propulsion, and Sea-Going Qualities of Ships," *BAAS Report*, 1869, pp. 10–47.

13. For HMS *Captain*, see McGee, "Floating Bodies, Naval Science" (1994); Brown, *Warrior to Dreadnought* (1997), pp. 40–52; Hawkey, *Black Night off Finisterre* (1999); and Leggett, *Shaping the Royal Navy* (2015), pp. 126–164.

14. Brown, *Warrior to Dreadnought* (1997), pp. 53–54; Leggett, *Shaping the Royal Navy* (2015), pp. 169–196.

15. Brown, *The Way of the Ship in the Midst of the Sea* (2006), pp. 55–69, 97–103, 127. My family and I stayed at Chelston Cross in 2002 when it was the Manor House Hotel and swam in Froude's model basin, which was the hotel swimming pool. The building is now the Manor House Apartments.

16. The results of full-scale trials by the French naval constructors Louis-Émile Bertin, Octave-René-Auguste Duhil de Bénazé, and Arthur-Paul Risbec, using motion recording devices very similar to those later developed by Froude, were translated in Britain soon after they were published in France. Bertin, "On Waves and Rolling," *Naval Science* 2 (1873), pp. 344–358, 476–497, and *Naval Science* 3 (1874), pp. 43–68, 198–228, 331–347, 486–500; and Duhil de Bénazé and Risbec, "On the Complete Motion of a Ship Oscillating in Still Water, with Experiments Made upon the *Elorn*," *Naval Science* 3 (1874), pp. 229–239, 348–358, and *Naval Science* 4 (1875), pp. 80–88, 224–230. See also La Vallée-Poussin, "Froude parle français" (2012).

17. William Froude, "Description of an Instrument for Automatically Recording the Rolling of Ships," *TINA* 14 (1873), pp. 179–190.

18. Ferreiro, *Ships and Science* (2007), p. vii.

19. Gardiner, *The Advent of Steam* (1993), p. 10; Flachat, *Navigation à vapeur transocéanienne* (1866) (for accident rates, see vol. 1, p. 376); Porter, *Trust in Numbers* (1996), pp. 114–147; Kaukiainen, "Shrinking the World" (2001).

20. Rankine, *A Manual of the Steam Engine and Other Prime Movers* (1859); Isherwood, *Experimental Researches in Steam Engineering* (1863–1865).

21. Ferreiro, *Ships and Science* (2007), p. 183.

22. Paul Quinn, "Performance of Early Powered Vessels," *MM* 89, no. 3 (2003), pp. 348–350, and 92, no. 1 (2006), pp. 95–97.

23. "Description des Dynamomètres Taurines employés pour mesurer la puissance des machines marines" (Description of Taurines's dynamometers used to measure marine engine power), *MGM* (1860), pp. 247–265, 354–383; Seaton, *A Manual of Marine Engineering* (1907), pp. 141–151.

24. Wright, *"Ship Hydrodynamics 1710–1880"* (1983), p. 58.

25. Bourne, *A Catechism of the Steam Engine* (1847), pp. 83–86; Jennifer Tann and Christine MacLeod, "Empiricism Afloat—Testing Steamboat Efficiency: Boulton Watt & Co. 1804–1830," *TNS/IJHET* 86, no. 1 (2016), pp. 228–243; John Farey Jr., "An Approximative Rule for Calculating the Velocity with Which a Steam Vessel Will Be Impelled through Still Water," *TICE/MPICE* 1 (1836), pp. 111–116; Wright, *"Ship Hydrodynamics 1710–1880"* (1983), p. 58.

26. Atherton, *The Capability of Steam Ships* (1853), p. 22; Marsden, "The Administration of the 'Engineering Science' of Naval Architecture at the British Association for the Advancement of Science" (2008); Brown, *Warrior to Dreadnought* (1997), p. 206. David K. Brown erroneously attributed the invention of the Admiralty coefficient to Chief Engineer Thomas Lloyd.

27. Wright, *"Ship Hydrodynamics 1710–1880"* (1983), p. 221.

28. William John Macquorn Rankine, "On Plane Water-Lines in Two Dimensions," *PTRS* 154 (1864), pp. 369–391; Rankine, "On the Mathematical Theory of Stream-Lines, Especially Those with Four Foci and Upwards," *PTRS* 161 (1871), pp. 267–306; Wright, "Ship Hydrodynamics 1710–1880" (1983), pp. 105–120; Darrigol, *Worlds of Flow* (2005), pp. 274–277; Marsden, "The Administration of the 'Engineering Science' of Naval Architecture at the British Association for the Advancement of Science" (2008), p. 83; Eckert, "Approaching Reality by Idealization" (2017), pp. 199–200.

29. Ferreiro, *Ships and Science* (2007), pp. 164–170; Reech, *Mémoire sur les machines à vapeur* (1844), pp. 164–176; Bourgois, *Memoire sur la resistance de l'eau* (1857), pp. iii–vi; Flachat, *Navigation à vapeur transocéanienne* (1866), vol. 2, pp. 311–316; Brisou, *Accueil, Introduction et Développement de l'Énergie Vapeur* (2001), vol. 2, pp. 614–618.

30. Dupuy de Lôme, *Notice sur les travaux scientifiques de M. Dupuy de Lôme* (1866), pp. 7–15; Bourgois, *Memoire sur la resistance de l'eau* (1857), pp. 244–246; Brisou, *Accueil, Introduction et Développement de l'Énergie Vapeur* (2001), vol. 2, pp. 634–638, 676–678.

31. Reech, *Mémoire sur les machines à vapeur* (1844), p. 166; Reech, *Cours de mécanique* (1852), pp. 265–275; Darrigol, *Worlds of Flow* (2005), pp. 278–279, Dupuy de Lôme quote p. 279.

32. Wright, "Ship Hydrodynamics 1710–1880" (1983), pp. 93–97, quote p. 95; Marsden, "The Administration of the 'Engineering Science' of Naval Architecture at the British Association for the Advancement of Science" (2008).

33. Charles Merrifield, "Report on the State of Existing Knowledge on the Stability, Propulsion, and Sea-Going Qualities of Ships," *BAAS Report* 39 (1870), pp. 10–47, quote pp. 24–25.

34. Froude, *The Papers of William Froude* (1955), pp. 129–133; Robert Mansel, "Letter of Reclamation," in William Denny, "Progressive Speed Trials," *TIESS* 28 (1885), pp. 65–198; Wright, "Ship Hydrodynamics 1710–1880" (1983), pp. 129–137; Brown, *The Way of the Ship in the Midst of the Sea* (2006), pp. 117–126; Bovis, *Hydrodynamique navale* (2009), p. 416.

35. Froude, *The Papers of William Froude* (1955), pp. 120–128; Brown, *The Way of the Ship in the Midst of the Sea* (2006), pp. 127–141, quotes p. 139.

36. Brown, *The Way of the Ship in the Midst of the Sea* (2006), pp. 143–176. I visited the site of the original tank in 2002, which is now a private residence across Seaway Lane from the rear of the Manor House Apartments.

37. Froude, *The Papers of William Froude* (1955), pp. 120–128.

38. Ferreiro, *Ships and Science* (2007), pp. 163–176; Darrigol, *Worlds of Flow* (2005), pp. 101–144.

39. Froude, *The Papers of William Froude* (1955), pp. 138–146, 232–253, quotes pp. 248–249; Wright, "Ship Hydrodynamics 1710–1880" (1983), pp. 151–157; Brown, *The Way of the Ship in the Midst of the Sea* (2006), pp. 169–182.

40. Leggett, *Shaping the Royal Navy* (2015), pp. 175–196, quote p. 184.

41. Brown, *The Way of the Ship in the Midst of the Sea* (2006), pp. 158–167, 200–206; Saldanha, *Memoir on the Novel Formation of the Bottom of Ships and Vessels, Proposed by the Brazilian Naval Architect Trajano Augusto de Carvalho* (1876).

42. Brown, *The Way of the Ship in the Midst of the Sea* (2006), pp. 208–210; Guard Books (Admiralty Civil Engineer's correspondence books), Simon's Town Museum, Simon's Town, South Africa, letter from Edward Adeane to Edward J. Stone, Admiralty House, 1 May 1879; "Death of Mr. Wm. Froude," *Cape Argus* (Cape Town, South Africa), 6 May 1879. Special thanks to Bill Rice for showing me Froude's gravesite and locating archival materials on Froude's illness and death.

43. Tideman, *Memoriaal van de marine* (1876–1880), pp. 75–90; Robert E. Froude, "On the 'Constant' System of Notation of Results of Experiments on Models Used at the Admiralty Experiments Works," *TINA* 29 (1888), pp. 304–318; Edmund V. Telfer, "The Design Presentation of Ship Model Resistance Data," *TNECIES* 79 (1963), pp. 357–390; Dirkzwager, *Dr. B.J. Tideman, 1834–1883* (1970), pp. 130–149; *PNA3* (1988), vol. 2, pp. 5–9.

44. Brown, *The Way of the Ship in the Midst of the Sea* (2006), pp. 215–223, William H. White quote p. 219; Don Leggett, "Replication, Replacing and Naval Science in Comparative Context, c. 1868–1904," *BJHS* 46, no. 1 (2013), pp. 1–21, Denny quote p. 9.

45. Horst Nowacki, "Leonhard Euler and the Theory of Ships," *JSR* 52, no. 4 (2008), pp. 274–290; Ferreiro, *Ships and Science* (2007), pp. 98–99.

46. Seguin, *Mémoire sur la Navigation à Vapeur* (1828); Reynolds, *Stronger than a Hundred Men* (1983).

47. Brown, *Before the Ironclad* (1990), pp. 108–120; Griffiths, Lambert, and Walker, *Brunel's Ships* (1999), pp. 27–52.

48. Auguste Taurines, "Théorie de la vis d'Archimède" (Theory of the Archimedes screw), *AMC* 27, no. 2 (1842), pp. 434–474; Bourgois, *Recherches Théoriques et Expérimentales sur les Propulseurs Helicoïdes* (1845), pp. 29–60.

49. The following sections are from William John Macquorn Rankine, "On the Mechanical Principles of the Action of the Propeller," *TINA* 6 (1865), pp. 13–35, quote p. 13; William Froude, "On the Elementary Relation between Pitch, Slip, and Propulsive Efficiency," *TINA* 19 (1878), pp. 47–63, quotes p. 48; Robert E. Froude, "A Description of a Method of Investigation of Screw-Propeller Efficiency," *TINA* 24 (1883), pp. 250–283; Robert E. Froude, "The Determination of the Most Suitable Dimensions for Screw Propellers," *TINA* 27 (1886), pp. 250–283; Alfred G. Greenhill, "A Theory of the Screw Propeller," *TINA* 29 (1888), pp. 319–345; Robert E. Froude, "On the Part Played in Propulsion by Differences of Fluid Pressure," *TINA* 30 (1889), pp. 390–405; Taylor, *Resistance of Ships and Screw Propulsion* (1893), pp. 62–85; Lennard C. Burrill, "On Propeller Theory," *TIESS* 90 (1946–1947), pp. 449–477; Carlton, *Marine Propellers and Propulsion* (2007), pp. 169–172.

50. Frederick W. Lanchester, "A Contribution to the Theory of Propulsion and the Screw Propeller," *TINA* 56 (1915), pp. 98–116; Ferreiro, "The Mutual Influence of Aircraft Aerodynamics and Ship Hydrodynamics" (2014).

51. Robert E. Froude, "A Description of a Method of Investigation of Screw-Propeller Efficiency," *TINA* 24 (1883), pp. 231–255; Robert E. Froude, "The Determination of the Most Suitable Dimensions for Screw Propellers," *TINA* 27 (1886), pp. 250–283; Brown, *The Way of the Ship in the Midst of the Sea* (2006), pp. 192–195, 225–227.

52. Larrie D. Ferreiro, "The Social History of the Bulbous Bow," *T&C* 52, no. 2 (2011), pp. 335–359; Depeyre, "Éperon et Bélier" (1999).

53. Røksund, *The Jeune École* (2007); Olivier, *German Naval Strategy, 1856–1888* (2004), pp. 130–152; Gabriele, *Benedetto Brin* (1998), pp. 43–58; Sondhaus, *The Naval Policy of Austria-Hungary* (1994), pp. 95–103; *Marine et Technique au XIXe Siècle* (1988), pp. 417–536.

54. Siméon Bourgois, "Théorie du gouvernail et de ses applications aux mouvements giratoires des navires à vapeur" (Theory of the rudder and its applications to rotational movements of steamships), *RMC* 27 (1869) March 1869, pp. 537–570, June 1869, pp. 255–293, September 1869, pp. 65–105; Bourgois, *Études sur les Manœuvres des Combats sur Mer* (1876); Giuseppe Gavotti, "Sulla Potenza di rotazione delle navi" (On the rotating power of ships), *RM* 12 (1879), pp. 241–265; Pollard and Dudebout, *Architecture navale* (1890–1894), vol. 4, pp. 64–102.

55. Nathaniel Barnaby, "On the Steering of Ships," *TINA* 4 (1863), pp. 56–78; Joseph Joëssel, "Rapport sur les Expériences Relatives aux Gouvernails" (Report on experiments relative to rudders), *MGM* (1873), pp. 213–232; Lammeren, *Resistance, Propulsion and Steering of Ships* (1948), pp. 322–323; Bennett, *A History of Control Engineering* (1979), pp. 98–100; Roy L. Harrington, "Rudder Torque Prediction," *TSNAME* 89 (1981), pp. 23–90; Brisou, *Accueil, Introduction et Développement de l'Énergie Vapeur* (2001), vol. 2, pp. 754–756.

56. Osborne Reynolds, "On the Steering of Screw-Steamers," *BAAS Report* 45 (1875), pp. 141–146; Clarke, "The Foundations of Steering and Maneuvering" (2004), pp. 2–3; Ferreiro, "The Mutual Influence of Aircraft Aerodynamics and Ship Hydrodynamics" (2014).

57. Todhunter, *A History of the Theory of Elasticity and of the Strength of Materials* (1886), pp. 1–7.

58. Renn and Valleriani, *Galileo and the Challenge of the Arsenal* (2001); Larrie D. Ferreiro, "Paul Hoste et les premiers développements de l'architecture navale, 1685–1700" (Paul Hoste and the first developments of naval architecture, 1685–1700), *BATMA* 111, no. 2628 (2012).

59. Bouguer, *Traité du Navire* (1746), pp. 151–161; Juan y Santacilia, *Examen Marítimo* (1771), pp. 174–185.

60. For a general overview of the history of structures and materials, see Timoshenko, *History of Strength of Materials* (1953); and Addis, *Structural Engineering* (1990). See the preface for definitions of stress, strain, elasticity, moment of inertia, and so on.

61. Sutherland, *Structural Iron, 1750–1850* (1997), pp. 65–75, 139–140, 293–310; Addis, *Structural and Civil Engineering Design* (1999), pp. 209–219; Kranakis, *Constructing a Bridge* (1997), pp. 85–94, 119–164.

62. Rosenberg and Vincenti, *The Britannia Bridge* (1978), pp. 13–38, 48.

63. Brunel, *The Life of Isambard Kingdom Brunel* (1870), p. 306. A British inventor, Richard Roberts, claimed that Fairbairn and Brunel had appropriated his ideas for double-bottom ship construction, but the claim was never substantiated, and Roberts did not receive any recognition. Hills, *Life and Inventions of Richard Roberts* (2002), pp. 46–47, 217–227.

64. John Vernon, "On the Construction of Iron Ships," *PIMechE* 14 (1863), pp. 115–149; William John Macquorn Rankine, "On Some of the Strains of Ships," *BAAS Report* 34 (1864), pp. 187–188. Note that the words "strain" (physical deformation of material) and "stress" (force per unit area) were still used interchangeably in 1864; it was not until the 1880s that the two terms were rigorously separated in scientific and engineering literature.

65. Fairbairn, *Treatise on Iron Ship Building* (1865), pp. 8–12; Rankine, *Shipbuilding, Theoretical and Practical* (1866), pp. 152–155 (Fairbairn was the author of chap. 2, "The Strength of Ships as a Whole").

66. Edward J. Reed, "The Strains of Ships in Still Water," *Naval Science* 1 (1872), pp. 351–365; Reed, "The Strains of Ships at Sea," *Naval Science* 2 (1873), pp. 12–23; Brown, *Warrior to Dreadnought* (1997), pp. 31–32.

67. White, *A Manual of Naval Architecture* (1877), pp. 214, 223, 313.

68. Attwood, *Text-Book of Theoretical Naval Architecture* (1899), pp. 206–260; Rawson and Tupper, *Basic Ship Theory* (1983), vol. 1, pp. 180–189.

69. William E. Smith, "Hogging and Sagging Strains in a Seaway, as Influenced by Wave Structure," *TINA* 24 (1883), pp. 135–153; Attwood, *Text-Book of Theoretical Naval Architecture* (1899), pp. 206–260; Rawson and Tupper, *Basic Ship Theory* (1983), vol. 1, pp. 180–189.

70. *SDC3* (1980), p. 233.

71. Brown, *Warrior to Dreadnought* (1997), pp. 184–186; Lehmann, "The Historical Development of the Strength of Ships" (2014), pp. 279–281. The reason for the breakup of HMS *Cobra*—buckling of thin-walled structures, poorly understood at the time—would not be established for another 30 years.

72. Brown, *Warrior to Dreadnought* (1997), pp. 30–31.

73. *PNA2* (1967), p. 198.

**Chapter 4**

1. Blind, *The Economics of Standards* (2004), pp. 18–21.

2. Ferreiro, *Ships and Science* (2007), pp. 35–38, 66, 286–287, 303.

3. Headrick, *The Tentacles of Progress* (1988), pp. 19–48.

4. Boisson, *Safety at Sea* (1999), pp. 46–52.

5. Of the dozen or so histories of Lloyd's Register, I have made particular use of Watson, *Lloyd's Register: 250 Years of Service* (2010). Additional thanks to Louise Bloomfield and Barbara Jones of the Lloyd's Register Library for their assistance with specific topics.

6. Jane Coates and John Coates, "Bernard Waymouth and the Change from Wood to Steel Ships," *TNS/IJHET* 71, no. 1 (1999), pp. 257–268, quote p. 264.

7. *Bureau Veritas 1828/1978* (1978), quote p. 5; Andersen and Collett, *Anchor and Balance* (1989), quote p. 25.

8. American Bureau of Shipping, *The History of the American Bureau of Shipping* (2013).

9. Brown, "The Development of Subdivision in Merchant Ships" (1996), quote p. 3.

10. Plimsoll, *Our Seaman: An Appeal* (1873); Jones, *The Plimsoll Sensation* (2006); Meek, *There Go the Ships* (2003), p. 101.

11. James Dunn, "Bulkheads," *TINA* 24 (1883), pp. 21–41, quotes pp. 22, 27; Émile Bertin, "The Use of Small Models for the Determination of Curves of Stability," *TSNAME* 2 (1894), pp. 27–36; Émile Bertin, "Stabilité d'un paquebot après un abordage en mer" (Stability of a passenger ship after a collision at sea), *BATM* 11 (1900), pp. 7–24; John J. Welch, "The Watertight Subdivision of Ships," *TINA* 56 (1915), pp. 1–30.

12. Lux, *Classification Societies* (1993), p. 37; Lindsay, *History of Merchant Shipping* (1876), vol. 3, pp. 632–633.

13. Philip J. Sims, "Comparative Naval Architecture of Passenger Ships," *TSNAME* 111 (2003), pp. 233–258.

14. Of the more than 200 books on RMS *Titanic*, I have a particular fondness for Garzke and Woodward, *Titanic Ships, Titanic Disasters* (2002). I was an original "plankholder" member of the SNAME Marine Forensic panel formed in 1995 under Bill Garzke's leadership, and I have since then participated in many forensic analyses of *Titanic* and other ships.

15. Larrie D. Ferreiro, "The Social History of the Bulbous Bow," *T&C* 52, no. 2 (2011), pp. 335–359, at p. 347. The article cites Captain Smith's comment in the *Washington Times* newspaper (16 April 1912, p. 4) by a former *Olympic* passenger, the day after *Titanic* sank.

16. Hilton, *Eastland* (1995).

17. Wescott Abell, "The Story of Safety at Sea," *TNECIES* 45 (1929), pp. 325–346; Boisson, *Safety at Sea* (1999), pp. 52–55.

18. IMO website, accessed July 2017.

19. *Final Report of a Board of Investigation* (1946); Robert H. Peebles, "Navy Shipbuilders 'Discover' Welding," *NINH* 6 (1987), pp. 157–166; Christopher J. Tassava, "Weak Seams: Controversy over Welding Theory and Practice in American Shipyards, 1938–1946," *H&T* 19, no. 2 (2003), pp. 87–108; Ship Structure Committee website, accessed August 2017.

20. Johnman and Murphy, "Welding and the British Shipbuilding Industry" (2005); NAUK, ADM 284 Admiralty Ship Welding Committee (1948–1954).

21. Chandler, *The Visible Hand* (1977).

22. For French dockyards, see Jouin, *600 ans de constructions navales* (1974). For British dockyards, see Haas, *A Management Odyssey* (1994). For the Amsterdam dockyard, see Lemmers, *Van werf tot facilitair complex* (2011). For American dockyards, see Hepburn, *History of American Naval Dry Docks* (2003).

23. Hutchins, *The American Maritime Industries and Public Policy* (1941), pp. xx–xxi.

24. This section is largely derived from Larrie D. Ferreiro and Stuart A. McKenna, "The Scientific and Management Revolution in Shipbuilding on the Two Clydes, 1880–1900," *NRJ* 58, no. 2 (2013), pp. 105–128, which includes the citations for the business records of Clydebank shipyards (William Denny and Brothers, Robert Napier and Sons, and J. & G. Thomson) at the Glasgow University Archive Services and the business records at the Hagley Museum Library in Wilmington, Delaware, and the Independence Seaport Museum in Philadelphia for the Delaware shipyards William Cramp and Sons, Harlan and Hollingsworth, and John Roach and Sons. It also has references to important secondary literature on the subject, notably Fassett, *The Shipbuilding Business in the United States of America* (1948); Tyler, *The American Clyde* (1958); Cooling, *Gray Steel and Blue Water Navy* (1979); Thiesen, *Industrializing American Shipbuilding* (2006); Pollard and Robertson, *The British Shipbuilding Industry, 1870–1914* (1979); Walker, *Song of the Clyde* (1984); and Johnston and Buxton, *The Battleship Builders* (2013).

25. Quotes from Ferreiro and McKenna, "The Scientific and Management Revolution in Shipbuilding on the Two Clydes (2013), p. 119.

26. Quote: ibid, p. 121.

27. Ferreiro, *Ships and Science* (2007), pp. 240–241.

28. For an overview of shipbuilding industrialization in Italy, Germany, Japan, and other nations, see O'Brien, *Technology and Naval Combat in the Twentieth Century and Beyond* (2001).

29. Witthöft, *Meyer Werft* (2005), pp. 33–54; Witthöft, *Tradition und Fortschritt* (2002), pp. 36–45.

30. Staccioli, *Mare Scienza e Tecnica* (1990), pp. 34–46; Fragiacomo, *L'industria come continuazione della politica* (2012), pp. 63–65; Schuster, *A Workforce Divided* (2002), pp. 39–114, quote p. 49.

31. Kanigel, *The One Best Way: Frederick Winslow Taylor and the Enigma of Efficiency* (1997), pp. 271–276; Petersen, "Fighting for a Better Navy" (1990); Thiesen, *Industrializing American Shipbuilding* (2006), pp. 168–212.

32. Hovgaard, *Structural Design of Warships* (1915), pp. x, 11–12. Hovgaard was not quite accurate when he claimed there were "no rules … to guide the designer." In 1898 and again in 1923, the Bureau of Construction and Repair published *Instructions for Displacement and Stability Calculations under the Bureau of Construction and Repair,* which laid out in detail all the stability calculations required for naval ships. From 1908 to 1999, the *General Specifications for Building Vessels of the United States Navy* was regularly issued and also contained formalized rules for designers.

33. Frieze and Shenoi, *Proceedings of the 16th International Ship and Offshore Structures Congress* (2006), vol. 2, pp. 221–222.

34. *SDC3* (1980), pp. 300–301.

35. ABS, *Rules for the Classification and Construction of Steel Ships* (1922), p. 29; American Railway Engineering, *General Specifications for Steel Railway Bridges* (1915), p. 12.

36. Hovgaard, *Structural Design of Warships* (1915), pp. 66–67; Brown, *The Grand Fleet* (1997), p. 71.

37. Larrie D. Ferreiro, "Goodall in America: The Exchange Engineer as Vector in International Technology Transfer," *CTTS* 4, no. 2 (2006), pp. 172–193; John K. Brown, "Design Plans, Working Drawings, National Styles: Engineering Practice in Great Britain and the United States, 1775–1945," *T&C* 41, no. 2 (2000), pp. 195–238.

38. James W. Kehoe, Clark Graham, Kenneth S. Brower, and Herbert A. Meier, "Comparative Naval Architecture Analysis of NATO and Soviet Frigates," *JASNE/NEJ* 92, no. 5 (1980), pp. 87–99, 92, no. 6, pp. 84–93; Larrie D. Ferreiro and Mark Stonehouse, "A Comparative Study of US and UK Frigate Design," *TSNAME* (1991), pp. 147–175, *TRINA* 136A (1994), pp. 1–55.

39. Frieze and Shenoi, *Proceedings of the 16th International Ship and Offshore Structures Congress* (2006), vol. 2, pp. 222–223.

40. William Denny, "On the Cross-Curves of Stability," *TINA* 25 (1884), pp. 45–56; Walker, "The Capsize of the *Daphne* in 1883" (2002).

41. Philip Jenkins, "On the Stability, &c, of Oil-Carrying Steamers," *TIESS* 32 (1889), pp. 225–255; John King, "Origins of the Theory of Ship Stability," *TRINA* 140 (1998), pp. 222–238; *PNA2* (1967), pp. 78–88; William H. White, "On the Designs for the New Battleships," *TINA* 30 (1889), pp. 150–215.

42. Committee on Safety of Construction, *Report and Minutes of Proceedings* (1914), p. 162.

43. Charles Frodsham Holt, "Stability and Seaworthiness," *TINA* 66 (1925), pp. 307–330; Scribanti, *La statica della nave* (1928); Ludwig Benjamin, "Über das Maß der Stabilität der Schiffe (On the degree of stability of ships), *JSTG* 15 (1914), pp. 594–614; American Marine Standards Committee, *Stability and Loading of Ships* (1929).

44. *Report of the Board of Trade Inquiry into the Loss of the Vestris* (1929); George H. Rock, "The International Conference on Safety of Life at Sea, 1929, with Special Reference to Ship Construction," *TSNAME* 37 (1929), pp. 89–128, quote p. 98; Wescott Abell and Alfred J. Daniel, "Safety of Life at Sea (1929 Conference)," *TINA* 71 (1930), pp. 1–24.

45. Ernesto Pierrottet, "A Standard of Stability for Ships," *TINA* 76 (1935), pp. 208–222; John C. Niedermair, "Stability of Ships after Damage," *TSNAME* 40 (1932), pp. 216–246; John C. Niedermair, "Further Developments in the Stability and Rolling of Ships," *TSNAME* 44 (1936), pp. 418–442; Cochrane and Niedermair, *Subdivision, Stability, and Construction of Merchant Ships* (1935), quotes pp. 1, 78; Lennard C. Burrill, "Seaworthiness of Collier Types," *TINA* 73 (1931), pp. 75–107, quote p. 78.

46. Arjava, *Raholan Kriteeri* (2002). I thank Jouni Arjava for additional information.

47. Rahola, "The Judging of the Stability of Ships" (1939), quotes pp. 93, 118.

48. Herd, "Rahola—40 Years On" (1979); *PNA3* (1988), vol. 1, pp. 106–114, 180–194; Kobylinski et al., *Stability and Safety of Ships* (2003), vol. 1, pp. 25–39.

49. Lehmann, *100 Jahre Schiffbautechnische Gesellschaft* (1999), vol. 1, pp. 76–77.

50. Aristophanes, *Birds* (414 B.C.), lines 375–382, in *The Comedies of Aristophanes* (1913), vol. 3, p. 47.

51. Eustace T. d'Eyncourt, "Notes on Some Features of German Warship Construction," *TINA* 62 (1921), pp. 1–12; Stanley V. Goodall, "The Ex-German Battleship *Baden*," *TINA* 62 (1921), pp. 13–48; U.S. Navy Bureau of Construction and Repair, *The Stability of Ships and Damage Control* (1931), quote p. 8; Schumacher, *Stability and Compartmentation of Ships* (1938); Brown, *Nelson to Vanguard* (2012), p. 19–20; Schaub, "U.S. Navy Shipboard Damage Control" (2014).

52. *Stability of Ships* (1942); David K. Brown, "Stability of RN Destroyers during World War II," *WT* 2 (1989), pp. 108–111; Niedermair, *Reminiscences of John C. Niedermair* (1978), pp. 272–275; Niedermair, John C. Niedermair Files, Box 10; Theodore H. Sarchin, interviews with author, November 2003.

53. "Ship Design Division Stability Criteria" (1957); Niedermair, John C. Niedermair Files, Box 10; Sarchin, interviews with author, November 2003.

54. Theodore H. Sarchin and Lawrence L. Goldberg, "Stability and Buoyancy Criteria for U. S. Naval Surface Ships," *TSNAME* 70 (1962), pp. 418–458; *Stability and Buoyancy of U.S. Naval Surface Ships* (1975); Steven W. Surko, "An Assessment of Current Warship Damaged Stability Criteria," *JASNE/NEJ* 106, no. 3 (1994), pp. 120–131; Deybach, "Intact Stability Criteria for Naval Ships" (1997); Alan J. Brown and Frédéric Deybach, "Towards a Rational Intact Stability Criteria for Naval Ships," *JASNE/NEJ* 110, no. 1 (1998), pp. 65–77; Sarchin, interviews with author, November 2003; Lawrence L. Goldberg, interviews with author, December 2003; Biran and López-Pulido, *Ship Hydrostatics and Stability* (2003), pp. 276–283.

55. Vincenti, *What Engineers Know and How They Know It* (1990), pp. 51–108; Alexander Z. Ibsen, "The Politics of Airplane Production: The Emergence of Two Technological Frames in the Competition between Boeing and Airbus," *TIS* 31 (2009), pp. 342–349.

56. Alphonse-Louis Marbec, "Sur une erreur commise de tout temps dans le calcul du déplacement d'un navire" (On an error committed all the time in the calculation of displacement of a ship), *BATM* 20 (1909), pp. 179–184.

57. Matthew G. Forrest, "The Society's Emblem," *MT* 2 (1965), pp. 2–3; *Stabilité des Bâtiments de Surface* (1998), p. 19.

## Chapter 5

1. Adams, *Familiar letters of John Adams* (1876), p. 381.

2. Ferreiro, *Ships and Science* (2007), pp. xiii–xvii; Larrie D. Ferreiro, "A Biographical Sketch of John Willis Griffiths from Primary and Archival Sources," *NRJ* 52, no. 4 (2007), pp. 221–228; Hamilton, *The Making of the Modern Admiralty* (2011), p. 162; Stuart A. McKenna and Larrie D. Ferreiro, "The Scientific and Management Revolution in Shipbuilding on the Two Clydes, 1880–1900," *NRJ* 58, no. 2 (2013), pp. 105–128.

3. Carr-Saunders and Wilson, *The Professions* (1933), pp. 155–165, 298–317; Millerson, *The Qualifying Associations* (1964), pp. 1–25; Wilensky, "The Professionalization of Everyone?" (1964); Layton, *The Revolt of the Engineers* (1986), pp. 25–52; Ferreiro, "Organizational Trust in Naval Ship Design Bureaus" (1998).

4. Ferreiro, *Brothers at Arms* (2016), pp. 18–20.

5. Ferreiro, *Ships and Science* (2007), pp. 279–293.

6. See Dedet, *Les Fleurs d'acier du Mikado* (1993).

7. SPEI, *Bi-centenaire du Génie Maritime 1765–1965* (1965); Bruno Belhoste and Konstantinos Chatzis, "From Technical Corps to Technocratic Power: French State Engineers and Their Professional and Cultural Universe in the First Half of the 19th Century," *H&T* 23, no. 3 (2007), pp. 209–225. See also Lutun, *Liste générale des élèves et des ingénieurs du corps des ingénieurs-constructeurs de la marine* (2013). When I worked as an exchange engineer at the Directorate of Naval Construction (DCN) in Paris in the 1990s, I was classed as an ingénieur d'armament.

8. Stewart, *Monturiol's Dream* (2003); Rodríguez González, *Isaac Peral* (2007).

9. Sánchez Carrión, "Los Ingenieros de Marina" (2009); Silva Suárez, *Técnica e ingeniería en España*. (2013), vol. 4, pp. 695–754, vol. 7, book 2, pp. 351–352. Additional thanks to José María Sánchez Carrión and Francisco Fernández González.

10. Ferreiro, *Ships and Science* (2007), p. 298; Dirkzwager, *Dr. B.J. Tideman, 1834–1883* (1970), pp. 81–85; Lemmers, *Techniek op schaal* (1996), pp. 152–153, 225–226.

11. Tredrea and Sozaev, *Russian Warships in the Age of Sail, 1696–1860* (2010), pp. 30–31, 406–407; Westwood, *Russian Naval Construction, 1905–45* (1994), pp. 15–18; McGlaughlin, *Russian and Soviet Battleships* (2003), pp. 356–357, 436–438. Additional thanks to Paul Wlodkowski.

12. Cosentino and Stanglini, *Il Corpo del Genio Naval* (2006); Gabriele, *Benedetto Brin* (1998).

13. Massie, *Dreadnought* (1991), pp. 468–497, quote p. 475.

14. Edward J. Reed, "Three Recent Critics of Naval Architecture," *Naval Science* 3 (1874), pp. 429–438; Hamilton, *The Making of the Modern Admiralty* (2011), pp. 137–144, 174; Stanley Sandler, "In Deference to Public Opinion: The Loss of HMS *Captain*," *MM* 59, no. 1 (1973), pp. 57–68, bunglers quote p. 160, court-martial quote p. 65; Leggett, *Shaping the Royal Navy* (2015), pp. 62–67, 146–154, Froude quote p. 146.

15. Parkes, *British Battleships 1860–1950* (1966), p. 230; Parkinson, *The Late Victorian Navy* (2008), pp. 118–160.

16. The documentation pertaining to the creation of the RCNC is in NAUK, ADM 116/31, case 212, "Correspondence Relative to the Formation of the Royal Corps of Naval Construction [*sic*], 1879–1884; Report of the Committee," which contains White's quote. A comprehensive biography of the corps is given in Brown, *A Century of Naval Construction* (1983), and Betts (ed.), *Vanguard to Dreadnought: 30 Years of Naval Construction 1983–2013* (2018). See also Manning,

*The Life of Sir William White* (1923), pp. 89–100; and Leggett, *Shaping the Royal Navy* (2015), pp. 210–219. Thanks to Andrew Lambert for his additional insights into the period.

17. Brown, *A Century of Naval Construction* (1983), p. 10. See also d'Eyncourt, *A Shipbuilder's Yarn* (1948).

18. Eskew, *Our Navy's Ships and Their Builders* (1962), pp. 253–254.

19. Roberts, *History of the Construction Corps of the United States Navy* (1937); Sprout, *The Rise of American Naval Power 1776–1918* (quote p. 212); Eskew, *Our Navy's Ships and Their Builders* (1962); Pedisich, *Congress Buys a Navy* (2016), pp. 17, 44, 52.

20. McBride, *Technological Change and the United States Navy* (2000), pp. 36–37; Thiesen, *Industrializing American Shipbuilding* (2006), pp. 140–168.

21. Bowen, *Ships, Machinery and Mossbacks* (1954), pp. 116–126; Larrie D. Ferreiro, "Genius and Engineering: The Naval Constructors of France, Great Britain, and the United States," *JASNE/NEJ* 110 (1998), pp. 99–132; Firebaugh, *Naval Engineering and American Sea Power* (2000), pp. 178–179.

22. Ferreiro, *Ships and Science* (2007), pp. 279–293; Unger, "The Technology and Teaching of Shipbuilding 1300–1800" (2013).

23. For overviews of engineering education in the nineteenth and twentieth centuries, see Wickenden, *A Comparative Study of Engineering Education in the United States and in Europe* (1929); Ahlström, *Engineers and Industrial Growth* (1982).

24. Among many histories of the École Polytechnique, I find Terry Shinn's *L'école polytechnique, 1794–1914* (1980) the most insightful.

25. Jules Lafont, "L'École Nationale Supérieure du Génie Maritime," *Neptunia* 10 (1948), pp. 12–13; SPEI, *Bi-centenaire du Génie Maritime 1765–1965* (1965), pp. 183–222; Sicard, *Saint-Nazaire et la construction navale* (1991); Berthiau, "Des maîtres entretenus aux ingénieurs, 1819–1971" (2005); Guedj, "L'ingénieur, le Génie Maritime et l'idée de Progrès Théoriques de l'Architecture Navale" (2016).

26. *II centenario de las enseñanzas de ingenieria naval* (1975); José María Sánchez Carrión, "La evolución de los planes de estudios necesarios para la obtención de las distintas titulaciones de Ingeniero Naval" (The evolution of the plans of study necessary to obtain the various titles of naval engineer), special issue, *IN* 67, no. 785 (1999), pp. 91–99; Ramón Blecua Fraga, "La Escuela de Ingenieros Navales de Ferrol, única en España desde 1860 a 1932" (The Naval Engineering School of Ferrol, the only one in Spain from 1860 to 1932), *IN* 72, no. 813 (2004), pp. 85–87; José María Sánchez Carrión, "Las distintas ubicaciones de las Escuelas de Ingenieros Navales hasta llegar a la Ciudad Universitaria de Madrid en 1948" (The specific locations of the Schools of Naval Engineers until the arrival at the City University of Madrid in 1948), *RM* 77, no. 857 (2008), pp. 82–86; Fernández González, *Aportación de Cataluña a la arquitectura naval* (2009), appendix. Thanks to José María Sánchez Carrión, José María de Juan-García Aguado, and Francisco Fernández González for additional information.

27. Kamp, *De Technische Hogeschool te Delft* (1955), pp. 284–299; Dirkzwager, "De voor geschiedenis van de opleiding tot scheepsbouwkundig ingenieur aan de Technische Universiteit Delft (1994); Davids and Schippers, "Innovations in Dutch Shipbuilding in the First Half of the Twentieth Century" (2008). Thanks also to Alan Lemmers.

28. Timoshenko, *Engineering Education in Russia* (1959), pp. 5–9; Westwood, *Russian Naval Construction* (1994), pp. 13–17; Rostovsev, *Stolichnyy universitet Rossiyskoy imperii* (2017).

29. Ostenc, *La Marine Italienne de l'Unité à Nos Jours* (2005), pp. 132–165; Massimo Figari and Arcangelo Menna, "L'università di Genova e la Marina Militare Italiana: La formazione degli ingegneri navali" (The University of Genoa and the Italian navy: The formation of naval engineers), *RM* 146 (June 2013), pp. 71–79.

30. Watson, *The German Genius* (2010), pp. 358–362, 834–835.

31. Günther Brenken, "The Education of Professional Engineers in West Germany," *JASNE/NEJ* 78, no. 6 (1966), pp. 1028–1034; Meiksens and Smith, *Engineering Labour* (1996), pp. 132–195; Lehmann, *Schiffbautechnische Ausbildung in Deutschland* (2001–2002).

32. Christensen, *European Historiography of Technology* (1993), pp. 111–162; Olsson, *Technology Carriers* (2000), pp. 35–56.

33. Robertson, "Technical Education in the British Shipbuilding and Marine Engineering Industries, 1863–1914" (1974), p. 227.

34. Chatfield, *An Apology for English Ship-Builders* (1833); Leggett, *Shaping the Royal Navy* (2015), pp. 27–58, 62–67.

35. Coles, "The Contribution of British Defence Departments to Technical Education and Instruction from about 1700" (1984), vol. 1, pp. 198–351; Haas, *A Management Odyssey: The Royal Dockyards, 1714–1914* (1994), pp. 78–92.

36. John Scott Russell, "On the Education of Naval Architects in England and France," *TINA* 4 (1863), pp. 163–185; *Annual of the Royal School of Naval Architecture and Marine Engineering* (1871–1874); Physick, *The Victoria and Albert Museum* (1982), pp. 148–150.

37. Brown, *A Century of Naval Construction* (1983), pp. 39–40; Lambert, *The Foundations of Naval History* (1998), pp. 31–35; Leggett, *Shaping the Royal Navy* (2015), pp. 120–125. My posthumous thanks to Louis Rydill for our many discussions, both during and after my studies at UCL (1985–1986), where I was the first American naval architect since before World War II to graduate with a master of science under the British naval constructor's course.

38. Robertson, "Technical Education in the British Shipbuilding and Marine Engineering Industries, 1863–1914" (1974), pp. 230–232.

39. Cooley, *Scientific Blacksmith* (1947); Hattendorf, *Sailors and Scholars* (1984), p. 13. My posthumous thanks to Harry Benford, former professor and longtime friend and colleague, for additional information on the history of my alma mater, the University of Michigan.

40. Dunbaugh, *A Centennial History of Webb Institute of Naval Architecture* (1994); *XIII-A, Massachusetts Institute of Technology: One Hundred Years* (2001).

41. Fukasaku, "In-Firm Training at Mitsubishi Nagasaki Shipyard, 1884–1934" (1991); Matsumoto, *Technology Gatekeepers for War and Peace* (2006), pp. 74–78.

42. Ferreiro, *Ships and Science* (2007), pp. 51–62, 301–303.

43. John Farey Jr., "An Approximative Rule for Calculating the Velocity with Which a Steam Vessel Will Be Impelled through Still Water," *TICE/MPICE* 1 (1836), pp. 111–116.

44. Morgan and Creuze, *Papers on Naval Architecture* (1829–1832). My thanks to Louise Bloomfield of Lloyd's Register for additional information.

45. Sinclair, *Philadelphia's Philosopher Mechanics* (1975); Larrie D. Ferreiro, "A Biographical Sketch of John Willis Griffiths from Primary and Archival Sources," *NRJ* 52, no. 4 (2007), pp. 221–228.

46. Melville, *Proceedings of the International Engineering Congress* (1894); Bertin, *La Marine des États-Unis* (1896).

47. Barnaby, *The Institution of Naval Architects, 1860–1960* (1960); Newton, *The Royal Institution of Naval Architects, 1960–1980* (1981). I give additional thanks to Trevor Blakeley.

48. Clarke, *A Century of Service to Engineering and Shipbuilding, 1884–1984* (1984).

49. ATMA, *Association Technique Maritime et Aéronautique Centième Anniversaire* (1989); Lehmann, *100 Jahre Schiffbautechnische Gesellschaft* (1999); Alimento et al., *Associazione Italiana di Tecnica Navale 1947–2017* (2017).

50. Meader, *ASNE: The First 100 Years* (1988).

51. "Proceedings of the Preliminary Meeting of the Penn Institute of Engineers and Naval Architects as a Section of the Franklin Institute," *JFI* 136 (1893), p. 471; Thomas, *Speed on the Ship!* (1993).

52. Ujifusa, *A Man and His Ship* (2013).

53. Knight and MacNaughton, *The Encyclopedia of Yacht Designers* (2005), pp. 388–390; Bruce Rosenblatt, interviews with author, October 2007; Walker, *Ships and Shipbuilders* (2010), p. 231.

54. Nevesbu website; MTG Marintechnik website; BMT Group website, all accessed March 2018.

55. Walker, *Ships and Shipbuilders* (2010), pp. 191–194; Lessenich, "Schiffbau: Neue Schiffsformen," accessed May 2019.

56. Maria Beasley, "Life-Raft," U.S. Patent 258,191, filed 16 May 1882.

57. Lacy, *The History of the Female Shipwright* (2008); Doe, *Enterprising Women and Shipping in the Nineteenth Century* (2009); Walker, *Ships and Shipbuilders* (2010), pp. 131–132.

58. Bruce, *The Life of William Denny* (1888), pp. 86–89, 99, quotes p. 87; Holberton, Oral history interview (1983).

59. Layne, *Women in Engineering* (2009), p. 90; Wealleans, *Designing Liners* (2006), pp. 80–83, 142–143; Stanley, *From Cabin "Boys" to Captains* (2016), pp. 87–89.

60. Zappas, "Constance Tipper Cracks the Case of the Liberty Ships" (2015); Susan Caccavale (daughter of Elaine Kaplan), interview with author, October 2010; Arthur Welch (Rachel Welch's son), interview with author, November 2017.

61. Benford, "What Every Young Girl Should Know Before She Marries a Naval Architect" (1961), quote p. 9; Beaufoy and Beaufoy, *Nautical and Hydraulic Experiments* (1834), quote p. xxviii; Milka Duno, interviews with author, January 2011.

62. *Time* 50, no. 13 (28 September 1942).

## Chapter 6

1. Scaife, *From Galaxies to Turbines* (2000), pp. 262–265; Carlton, *Marine Propellers and Propulsion* (2007), pp. 209–210.

2. Alic, *Trillions for Military Technology* (2007), pp. 35–36.

3. De Beer, *The Sciences Were Never at War* (1960).

4. The Royal Arsenal Woolwich, established in the seventeenth century, carried out gunpowder and explosives experiments, but it was primarily a manufactory and not dedicated to research.

5. Richard W. L. Gawn, "The Admiralty Experimental Works, Haslar," *TINA* 97 (1955), pp. 1–35; Kenneth C. Barnaby and Anthony L. Dorey, "A Towing Tank Storm," *TINA* 107 (1965), pp. 265–272; Brown, *The Way of the Ship in the Midst of the Sea* (2006), pp. 220–234.

6. Johnston and Buxton, *The Battleship Builders* (2013), p. 108; Yarrow, *Alfred Yarrow, His Life and Work* (1923), pp. 206–209; Bailey, *Ships in the Making* (1995); George S. Baker, "Methodical Experiments with Mercantile Ship Forms," *TINA* 54 (1913), pp. 162–180.

7. Maurice E. Denny, "The British Shipbuilding Research Association: The First Six Years," *TINA* 93 (1951), pp. 40–55; *PNA2* (1967), p. 319; Meek, *There Go the Ships* (2003), pp. 59–85; Ian Varcoe, "Co-operative Research Associations in British Industry," *Minerva: A Review of Science, Learning and Policy* 19, no. 3 (1981), pp. 433–463.

8. Claudio Boccalatte, "La vasca navale della Spezia e la nascita della moderna architettura navale in Italia" (The La Spezia naval model basin and the birth of modern naval architecture in Italy), *Bollettino d'Archivio dell'Ufficio Storico della Marina Militare* (Archival bulletin of the Navy Historical Office) 27 (2013), pp. 37–94; R. W. Munro Ltd., *R.W. Munro Ltd. Centenary, 1864–1964* (1964), p. 25; INSEAN, *La Vasca Navale di Roma nel cinquantenario della sua istituzione* (1977); Cosentino and Stanglini, *Il Corpo del Genio Navale* (2006), pp. 37–38, 111.

9. Horst Nowacki, "Zur Entwicklung der Schiffshydrodynamik im 20. Jahrhundert" (On the development of ship hydrodynamics in the 20th century), in Lehmann, *100 Jahre Schiffbautechnische Gesellschaft*, vol. 1 (1999), pp. 219–259; Lehmann, *Schiffbautechnische Forschung in Deutschland*

(2003–2004), vol. 1; HSVA, *HSVA@100* (2013); Weir, *Building the Kaiser's Navy* (1992), pp. 81–84, 155; Vienna Model Basin website, accessed April 2018. Figure 6.2: Kempf and Sottorf, "The High-Speed Tank of the Hamburg Shipbuilding Company" (1931/1934).

10. Krylov, *Professor Krylov's Navy* (2014); Krylov State Research Center website, accessed January 2018; Paul Wlodkowski, "Taming Poseidon's Beast of Uncertainty: The Auspicious Debut of Marine Engineer Aleksej Nikolaevich Krylov, 1863–1945," *TNS/IJHET* 85, no. 1 (2015), pp. 140–158.

11. DGA/DCN, *1906–1988: quatre-vingt-deux ans d'histoire du Bassin d'essais des Carènes de Paris* (1988); DGA, *100 Years of History: The Bassin d'essais des carènes* (2006); Gaudard, "Le bassin des carènes et le service technique des constructions navales à Balard (2009)." The French nation calls itself "the Hexagon" after its shape on the map.

12. Groothedde, Meurs, and Veltmeijer, *Pride: 75 Years of Maritime Research* (2007); Davids and Schippers, "Innovations in Dutch Shipbuilding in the First Half of the Twentieth Century" (2008), pp. 213–214.

13. Ramírez Gabarrús, *La Construcción Naval Militar Española, 1730–1980* (1980), p. 134; CEHIPAR, *Canal de Experiencias hidrodinámicas de El Pardo* (2001); Ferreiro, "Shipbuilders to the World" (2010).

14. Wright, *History of the Washington Navy Yard* (1921), vol. 2, p. 244. The commandant's order allowing the EMB to be used as a swimming pool is surprising even for that era; as a former University of Michigan "tank rat" (a student who helps maintain the towing tank), I can attest that swimming in model basin waters is neither healthy nor family friendly.

15. Taylor, *The Speed and Power of Ships* (1910); Allison, Keppel, and Nowicke, *D.W. Taylor* (1988); Carlisle, *Where the Fleet Begins* (1998).

16. Cooley, *Scientific Blacksmith* (1947); Don Leggett, "Replication, Replacing and Naval Science in Comparative Context, c. 1868–1904," *BJHS* 46, no. 1 (2013), pp. 1–21; Miloh, *Mathematical Approaches in Hydrodynamics* (1991), pp. xv–xxi; Marshall Tulin, interviews with author, January 2006.

17. Taniguchi, "Historical Review of Research and Development in Ship Hydrodynamics" (1984); Matsumoto and Sinclair, "How Did Japan Adapt Itself to the Scientific and Technological Revolution at the Turn of the 20th Century? An Experiment in Transferring the Experimental Tank to Japan" (1994); Matsumoto, *Technology Gatekeepers for War and Peace* (2006), pp. 26–49, 118–144.

18. ITTC website, accessed May 2018.

19. Noalhat, *Navigation aérienne et navigation sous-marine: deux faces d'un même problème* (1910).

20. This section is largely derived from Ferreiro, "The Mutual Influence of Aircraft Aerodynamics and Ship Hydrodynamics in Theory and Experiment" (2014). See that paper for more comprehensive references. See also Hagler, *Modeling Ships and Space Craft* (2013).

21. Eckert, "Approaching Reality by Idealization: How Fluid Resistance Was Studied by Ideal Flow" (2017), pp. 199–200. While submarine designers contributed to the *R38* airship, its successor

the *R101* was instrumental in developing the teardrop-shaped hull of the USS *Albacore* experimental submarine in 1952, which paved the way for all modern submarine hull forms.

22. Endrikat, "Maritime Classification Societies' Role in the Development of Aviation Verification and Validation Processes" (2016).

23. Carlisle, *Where the Fleet Begins* (1998), pp. 114–115, 134.

24. Kuhn, *The Structure of Scientific Revolutions* (1962).

25. Blackburn, *The Theory and Science of Naval Architecture Familiarly Explained* (1836), pp. 39–43, quote p. 40.

26. Eckert, *The Dawn of Fluid Dynamics* (2006), p. 107–128.

27. Kempf and Foerster, *Hydromechanische Probleme des Schiffsantriebs* (1932).

28. Karl Schoenherr, "Resistance of Flat Surfaces Moving through Fluid," *TSNAME* 40 (1932), pp. 279–313; Carlisle, *Where the Fleet Begins* (1998), pp. 116–118.

29. Among the scientists examining circulation theory at the 1932 Hamburg conference was Melitta Schiller, later to become the sister-in-law of Claus von Stauffenberg, leader of the 1944 Valkyrie plot to assassinate Adolf Hitler. Although she was implicated in the plot, she survived retribution only to be shot down in combat.

30. Note that the naval architect Jean Dieudonné (1900–1972) should not be confused with the mathematician Jean Dieudonné (1906–1992).

31. Kenneth Davidson and Leonard Schiff, "Turning and Course-Keeping Qualities," *TSNAME* 54 (1946), pp. 152–200.

32. Bailey, *Ships in the Making* (1995), p. 43; Carlisle, *Where the Fleet Begins* (1998), p. 71; DGA/DCN, *1906–1988: quatre-vingt-deux ans d'histoire du Bassin d'essais des Carènes de Paris* (1988), p. 22.

33. Ferreiro, "The Mutual Influence of Aircraft Aerodynamics and Ship Hydrodynamics in Theory and Experiment" (2014), pp. 254–255; Carlisle, *Where the Fleet Begins* (1998), pp. 92, 119.

34. Westwood, *Russian Naval Construction, 1905–45* (1994), pp. 177–208.

35. Bailey, *Ships in the Making* (1995), pp. 119–154.

36. Carlisle, *Where the Fleet Begins* (1998), pp. 161–166, 179–184.

37. Harold E. Saunders, "General Notes on an Inspection Trip to Europe in October–December 1945," *JASNE/NEJ* 58, no. 4 (1946), pp. 529–559; Weir, *Forged in War* (1993), pp. 68–79; Carlisle, *Where the Fleet Begins* (1998), pp. 194–196; Niedermair, Niedermair Files, boxes 1, 16.

38. The history and all proceedings of ITTC are from the ITTC website, accessed June 2018.

39. ATTC website, accessed June 2018; Robert L. Townsin and Mohammed A. Mosaad, "The ITTC Line—Its Genesis and Correlation Allowance," *TNA*, September 1985, pp. E359–E362.

40. Ship Structure Committee website, accessed August 2017.

41. Bush, *Science, the Endless Frontier* (1945).

42. Bowen, *Ships, Machinery and Mossbacks* (1954), pp. 345–360; Sapolsky, *Science and the Navy: The History of the Office of Naval Research* (1990); Buderi, *Naval Innovation for the 21st Century* (2013); Wenk, *Making Waves* (1995), pp. 41–52.

43. Mutel, *Flowing through Time* (1998).

44. Symposium on Naval Hydrodynamics (1956–present); Sidney Reed, interviews with author, September 2005. Dr. Reed was a former ONR program manager who worked with Phillip Eisenberg.

45. Astute readers will recognize that the chapter title comes from the famous work by Latour and Woolgar, *Laboratory Life* (1979).

46. This section is largely derived from Murray et al., "Cutwaters before Rams" (2017); and Larrie D. Ferreiro, "The Social History of the Bulbous Bow," *T&C* 52, no. 2 (2011), pp. 335–359. See these papers for more comprehensive references.

47. Hüllen, *Leitfaden für den Unterricht im Schiffbau* (1908), plate 9.

48. Ferreiro, "The Social History of the Bulbous Bow" (2011), p. 348.

49. Ibid., p. 349.

50. Saunders, *Hydrodynamics in Ship Design* (1957–1965), vol. I, p. 369.

51. Most of the following section is based on my extensive interview with the late Takao Inui at the University of Tokyo in October 2003). I am indebted to him for providing me comprehensive notes and files to round out his extraordinary recollections. Additional thanks go to Hitoshi Narita for assisting in making the arrangements for the meeting.

52. Note that many antisubmarine destroyers and frigates since the late 1950s have been equipped with large sonar domes, protruding forward of and below the keel, that resemble bulbous bows. However, these bulbs are to ensure adequate sonar coverage around the ship and are not specifically intended for reducing wavemaking resistance.

## Chapter 7

1. Horst Nowacki, personal communication, June 2013; Nowacki, "Five Decades of Computer-Aided Ship Design" (2010), p. 957.

2. Booker, *A History of Engineering Drawing* (1963); Baynes and Pugh, *The Art of the Engineer* (1981); Hambly, *Drawing Instruments, 1580–1980* (1988). A synthetic history of computer-aided design (but not engineering) is given in Cardoso Llach, *Builders of the Vision: Software and the Imagination of Design* (2015). For Soviet computing, see Gerovitch, *From Newspeak to Cyberspeak* (2002).

3. The phrase "ghost in the machine" began as a philosophical concept in the 1940s, arising from the views of René Descartes on the role of the mind (ghost) in the body (machine). By the

1970s the phrase had become a shorthand for computer software and is today frequently used in popular culture to describe artificial intelligence.

4. Ferreiro, *Ships and Science* (2007), pp. 42–45; Hambly, *Drawing Instruments, 1580–1980* (1988), pp. 100–101; Rålamb, *Skeps Byggerij eller adelig öfnings tionde tom* (1691), plate A; Robinson, *The Marine Room of the Peabody Museum of Salem* (1921), p. 90; Richard Barker, "Two Architectures—a View of Sources and Issues," in Nowacki and Lefèvre, *Creating Shapes in Civil and Naval Architecture* (2001), vol. 1, pp. 41–133, at pp. 74–76.

5. Murray, *A Treatise on Ship-Building and Navigation* (1754), p. 40; Stalkartt, *Naval Architecture* (1781), pp. 158–159; Steel, *The Shipwright's Vade-Mecum* (1805), p. 218 (a vade-mecum, literally "comes with me," is a handbook); d'Etroyat, *Traité élémentaire d'architecture navale* (1846), p. 2; Griffiths, *A Treatise on Marine and Naval Architecture* (1850), p. 165; Monjo i Pons, *Curso métodica de arquitectura naval* (1856), plate 15; Hambly, *Drawing Instruments, 1580–1980* (1988), pp. 101–104; Nowacki, "Splines im Schiffbau" (2000); Ferreiro, *Ships and Science* (2007), pp. 40–41.

6. Hambly, *Drawing Instruments, 1580–1980* (1988), p. 101; Colin Tipping, "Technical Change and the Ship Draughtsman," *MM* 84, no. 4 (1998), pp. 458–469, quote p. 459.

7. d'Etroyat, *Traité élémentaire d'architecture navale* (1846), p. 2; Jobst Lessenich, "Draughting Curves Used in Ship Design," in Nowacki and Lefèvre, *Creating Shapes in Civil and Naval Architecture* (2009), pp. 425–434. Many of these large instrument-makers went out of business at the end of the twentieth century, but small start-up companies have noted both the need and the nostalgia for ships' curves and now fabricate them with—ironically—advanced computer-aided manufacturing systems.

8. James W. Queen & Co., *Priced and Illustrated Catalogue of Mathematical Instruments* (1883), p. 61; Howard I. Chapelle Papers, letter to Charles Lienert, 5 January 1970, Smithsonian Archives Record Unit 7228 (1969–1975), Washington, DC: Smithsonian Institution.

9. William H. White and William John, "On the Calculation of the Stability of Ships and Some Matters of Interest Connected Therewith," *TINA* 12 (1871), pp. 77–127, quote p. 78. For the different methods of determining stability, see Reed, *A Treatise on the Stability of Ships* (1885); and Pollard and Dudebout, *Architecture navale* (1891–1894), vol. 2.

10. Victor-Marcel-Michel Andrade, "Note relative au planimtere d'Amsler" (Note concerning Amsler's planimeter), *MGM*, 2nd series, no. 1 (1875), pp. 33–35; McGee, "The Amsler Integrator and the Burden of Calculation" (1998); Brown, *A Century of Naval Construction* (1983), p. 265; Brown, *Warrior to Dreadnought* (1997), p. 209. The Tchebycheff method is named for its inventor, the Russian mathematician Pafnuty Lvovich Chebyshev.

11. Brown, *A Century of Naval Construction* (1983), p. 163. The unnamed constructor was not punished and led a distinguished career afterward.

12. *Time* 55, no. 5 (23 January 1950), pp. 54–60, quote p. 57.

13. Edward Alvey Wright, "Naval Mathematics at the David Taylor Model Basin," *JASNE/NEJ* 69 (1957), pp. 205–230; Richstone, "The Applied Mathematics Laboratory of the David W. Taylor Model Basin" (1961).

14. Boslaugh, *When Computers Went to Sea* (1999); Van Atta, *Transformation and Transition* (2003), vol. 1, pp. 5, 61.

15. Even Mehlum and Paul F. Sørensen, "Example of an Existing System in the Ship-Building Industry: The Autokon System," *PRS* 321, no. 1545 (1971), pp. 219–233; *SDC3* (1980), pp. 610–614; Noble, *Forces of Production* (1984), pp. 107–143; Le Pavic, "Une histoire des techniques de l'arsenal de la Marine de Lorient" (2015), pp. 247–255.

16. Cardoso Llach, *Builders of the Vision* (2015), pp. 69–72; Nowacki, "Five Decades of Computer-Aided Ship Design" (2010), p. 957; Bournemouth University Oral History Research Unit, Vosper-Thornycroft shipyard interviews, accessed May 2018.

17. Manley St. Denis and Willard J. Pierson Jr., "On the Motion of Ships in Confused Seas," *TSNAME* 61 (1953), pp. 280–357; Saunders, *Hydrodynamics in Ship Design* (1957–1965), vol. 2, pp. 609–637; Korvin-Kroukovsky, *Theory of Seakeeping* (1961).

18. "Ship Structural Design," special issue, *JASNE/NEJ* 114, no. 2 (2002); Legaz Alamansa, "Computer Aided Ship Design: A Brief Overview" (2015); Ship Structure Committee website, accessed May 2018; John Rosborough, "SHCP-Early History." https://stability-shcp.com/2018/06/shcp-early-history, accessed June 2018.

19. Brown and Moore. *Rebuilding the Royal Navy* (2012), p. 17; Yuille, *I Fathered a GODDESS* (2012); Peter Gale, personal communication, June 2018.

20. Nowacki, "Five Decades of Computer-Aided Ship Design" (2010), p. 957.

21. Hess and Smith, "Calculation of Potential Flow about Arbitrary Bodies" (1962). Horst Nowacki independently and almost simultaneously developed a similar numerical solution to the potential flow problem; see *Potentialtheoretische Strömungs- und Sogberechnungen für schiffsähnliche Körper* (1963). Additional information thanks to Horst Nowacki.

22. Blazek, *Computational Fluid Dynamics* (2006), pp. 1–4; Zhang et al., "Application of CFD in Ship Engineering Design Practice and Ship Hydrodynamics" (2006).

23. John J. Nachstheim and Lawrence Dennis Ballou, "Present Status of Computer-Aided Design and Construction: Is That All There Is?," *JASNE/NEJ* 82, no. 1 (1970), pp. 33–43; Philip Sims and Jennifer Lin, "Common Studies Necessary to Support Early Stage Ship Alternative Evaluation," *JASNE/NEJ* 125, no. 2 (2013), pp. 134–143. Additional information courtesy of Phil Sims.

24. James L. Bates, "Discussion of Methods Used in Making Preliminary Approximations to Weights and Dimensions of Vessels," NARAUSA RG19, 448, box 43, file 6, 12 December 1913; Ernest Edwin Bustard, "Preliminary Calculations in Ship Design," *TNECIES* 57 (1941), pp. 179–206. For extensive discussions on tacit knowledge in ship design, see Thiesen, *Industrializing American Shipbuilding* (2006); and Leggett, *Shaping the Royal Navy* (2015).

25. John Harvey Evans, "Basic Design Concepts," *JASNE/NEJ* 71, no. 4 (1959), pp. 671–678; Nowacki, "A Farewell to the Design Spiral" (2016).

26. Murphy, Sabat, and Taylor, "Least Cost Ship Characteristics by Computer Techniques" (1963); Philip Anklowitz, "Computer Methods and Use in Ship Design," *JASNE/NEJ* 76, no. 6 (1964),

pp. 929–936 (quote p. 933); Schmidt, "Preliminary Design with the Aid of a Computer" (1964); Robert S. Johnson, "Automation in Pre-contract Definition Ship Design," *JASNE/NEJ* 81, no. 3 (1969), pp. 89–97.

27. Schmidt, *A Computer-Aided Feasibility Design Method for Destroyer-Type Ships* (1965); Cotton, *COnceptual DEsign of SHIPS Model CODESHIP* (1971); Johnson, "Computer-Aided Ship Design" (1976); Reed, "Ship Synthesis Model for Naval Surface Ships" (1976); Reeves, *CNA's Conceptual Design and Cost Models for High-Speed Surface Craft* (1984); Friedman, *U.S. Destroyers* (2004), pp. 369–377; Peter Gale, personal communication, June 2018. Additional information from Phil Sims, Tom Messenger, and Jim Mills.

28. Bureau of Ships, *Study of Bismarck* (1941).

29. Graham and Mills, *Comparative Naval Architecture* (1971); James W. Kehoe, "Warship Design—Ours and Theirs," and Hebert A. Meier, "Methodology for Analyzing Foreign Warships," both in McGwire and McDonnell, *Soviet Naval Influence* (1977); Rohwer and Monakov, *Stalin's Ocean-Going Fleet* (2001), pp. 62–63; William H. Buckley, "Improved Design for Seakeeping: Seaway Criteria and Related Developments," *MT* 46, no. 2 (2009), pp. 74–90.

30. *ASSET Advanced Ship and Submarine Evaluation Tool Training* (2012).

31. "Concept Design System (CONDES)" (1985), master of science degree course notes, University College London, author's collection; Colwell, *User's Manual for the SHOP5 System* (1988); David Andrews, "The Art and Science of Ship Design," *IJME* 85 (2007), pp. 9–26; Paramarine website, accessed January 2018.

32. Brown, *Nelson to Vanguard* (2012), p. 187. Baker's statement was first quoted in a 1954 journal article.

33. Ferreiro, *Ships and Science* (2007), pp. 303–304; Sánchez Carrión, "Los Ingenieros de Marina" (2009), vol. 1, pp. 166–172.

34. James L. Bates, "Discussion of Methods Used in Making Preliminary Approximations to Weights and Dimensions of Vessels," NARAUSA RG19, 448, box 43, file 6, 12 December 1913; Brown, *A Century of Naval Construction* (1983), pp. 72–73; Brown, *Warrior to Dreadnought* (1997), pp. 198–201.

35. *LX Amphibious Assault Ship Preliminary / Contract Design Management Plan*. Washington, DC: Naval Sea Systems Command, 1993. Author's Collection.

36. Bureau of Labor Statistics, "Occupational Outlook Handbook," https://www.bls.gov/ooh/, accessed July 2017.

## Epilogue

1. Kossiakoff and Sweet, *Systems Engineering: Principles and Practice* (2003); Nowacki, "A Farewell to the Design Spiral" (2016).

2. Papanikolaou, *Risk-Based Ship Design* (2009); Alberto Francescutto and Apostolos D. Papanikolaou, "Buoyancy, Stability, and Subdivision: From Archimedes to SOLAS 2009 and the Way Ahead," *PIMechE* 225, no. 1 (2011), pp. 17–32.

3. Ashe, "Naval Ship Design" (2016); Guy Gibbons, interviews with author, December 2011–January 2012. Guy Gibbons was knighted in 2003, ostensibly for his work in naval ship classification.

4. O'Rourke, *Navy Littoral Combat Ship (LCS) Program: Background, Issues, and Options for Congress* (2010), quote p. 59. This paragraph draws on my notes from the meeting on 18 October 1999 between NAVSEA, the American Bureau of Shipping, shipyard representatives, and design agents regarding naval vessel criteria.

5. Ferreiro, "Comparing Ship versus Aircraft Development Costs" (2018).

6. Stevens, *Engineering Mega-systems* (2010).

7. Levison, *The Box* (2008); Ferreiro, "The Warship since the End of the Cold War" (2016); DDG 1000 SAR Selected Acquisition Report (2010).

8. EMSHIP European Master's Course website, accessed May 2018.

9. "Que no existe en el mundo una carrera y una profesión más ingratas para los que tienen la desgracia de dedicarse a ellas, que las de la arquitectura naval, cuyas obras, aun las más suntuosas, las construidas con más cuidado, las más costosas a la nación, y las que más afanes y sinsabores han causado á los Jefes que las han combinado y dirigido, las ven estos desaparecer de su vista en un momento como el humo, ... sin que quede el más mínimo vestigio que pueda atestar ... después de sus desapariciones, el celo, el cuidado y los incesantes desvelos, tan poco apreciados, del desgraciado Jefe que las ha ejecutado." Sánchez Carrión, "Los Ingenieros de Marina" (2009), vol. 1, p. 293.

10. Olson, *Madame Fourcade's Secret War* (2019), pp. 184–192, 304–305, 357–359.

11. D'Oliveira, *A Corveta Portuguesa Dos Anos 70* (1999); Rogério D'Oliveira, interviews with author, July 2012. Additional thanks to Sandro Mendonça.

12. Andrew B. Summers, interviews with author, June 2018. Andy Summer's actual title in the DDG 51 and DDG 1000 projects was ship design manager.

# Selected Bibliography

For the sake of brevity, specific citations for the very large number papers appearing in the transactions and proceedings of the Royal Society, BAAS, RINA, SNAME, ATM and ATMA, ASNE, and others, as well as articles in key journals such as *Mariner's Mirror* and *Technology and Culture*, are given in the endnotes and not repeated here.

When books appear in multiple editions, I generally refer to the first edition.

### Abbreviations

| | |
|---|---|
| **ABS** | American Bureau of Shipping. New York 1862–1999, Houston, 2000–present. |
| **AMC** | *Annales Maritimes et Coloniales* (Maritime and colonial yearbook). Paris, 1816–1847. |
| **AN** | *American Neptune*. Salem, MA, 1941–2002. |
| **ANF** | Archives Nationales de France (French National Archives). Paris. |
| **ASNE** | American Society of Naval Engineers. Washington, DC, 1888–1983, Alexandria, VA, 1983–present. |
| **ATM/ATMA** | Association Technique Maritime (Maritime Technical Association). Paris, 1889–1923; Association Technique Maritime et Aéronautique (Maritime and Aeronautical Technical Association), 1924–present. |
| **ATTC** | American Towing Tank Conference. http://www.sname.org/attc/home. 1938–present. |
| **BAAS Report** | British Association for the Advancement of Science, Report. London, 1831–2009. |
| **BATM/BATMA** | *Bulletin de l'Association Technique Maritime*. Paris, 1889–1923; *Bulletin de l'Association Technique Maritime et Aéronautique*, 1924–present. |
| **BJHS** | *British Journal for the History of Science*. Cambridge, 1962–present. |
| **BV** | Bureau Veritas. Antwerp, 1828–1830; Paris, 1830–present. |
| **CTTS** | *Comparative Technology Transfer and Society*. Baltimore, 2003–2009. |
| **H&T** | *History and Technology*. London, 1983–present. |
| **ICE** | Institution of Civil Engineers. London, 1818–present. |
| **IJME** | *International Journal of Maritime Engineering*. London, 2003–present. |
| **IJNH** | *International Journal of Naval History*. Washington, DC, 2002–present. http://www.ijnhonline.org/. |

| | |
|---|---|
| IMCO | Inter-Governmental Maritime Consultative Organization. London, 1959–1982. Followed by IMO. |
| IMO | International Maritime Organization. London, 1982–present, www.imo.org. |
| IN | *Ingenieria Naval* (Naval engineering). Cartagena and Madrid, 1929–present. |
| INA | Institution of Naval Architects. London, 1860–1959. Followed by RINA. |
| ITTC | International Towing Tank Conference. 1933–present, www.ittc.org. |
| JASNE/NEJ | *Journal of the American Society of Naval Engineers*, 1883–1961, *Naval Engineers Journal*, 1961–present. ASNE, Washington, DC, 1888–1983, Alexandria, VA, 1983–present. |
| JFI | *Journal of the Franklin Institute*. Philadelphia, 1826–present. |
| JSR | *Journal of Ship Research* (SNAME). New York, 1957–1986; Jersey City, NJ 1987–2013; Alexandria, VA, 2014–present. |
| JSTG | *Jahrbuch der Schiffbautechnischen Gesellschaft* (Yearbook of the shipbuilding society). Berlin, 1900–present. |
| LR | Lloyd's Register. London, 1760–present. |
| MASIF | *Mémoires de l'Académie des Sciences de l'Institut de France* (Memoirs of the Academy of Sciences of the Institute of France). Paris: Gauthier-Villars, 1798–1943. |
| MGM | *Mémorial du Génie Maritime* (Memoirs of the Naval Engineering Service). Toulon and Paris, 1847–1914. |
| MM | *Mariner's Mirror*. London, 1911–present. |
| MT | *Marine Technology* (SNAME). New York, 1964–1986. |
| NARAUSA | National Archives and Records Administration of the United States of America. Washington, DC. |
| NAUK | National Archives of the United Kingdom. Richmond, UK. |
| NINH | *New Interpretations in Naval History: Selected papers from Naval History Symposium held at the United States Naval Academy*. Annapolis, MD: Naval Institute Press; and Newport, RI: Naval War College, 1981–present. |
| NMM | National Maritime Museum. Greenwich, UK. |
| NRJ | *Nautical Research Journal*. Beaufort, NC, 1948–present. |
| PIMechE | *Proceedings of the Institution of Mechanical Engineers*. Birmingham, UK, 1847–1876; London, 1877–present. |
| PNA1 | *Principles of Naval Architecture*. 2 vols., edited by Henry Rossell and Lawrence Chapman. New York: SNAME, 1939. |
| PNA2 | *Principles of Naval Architecture*. 2nd ed., edited by John P. Comstock. New York: SNAME, 1967. |
| PNA3 | *Principles of Naval Architecture*. 3rd ed., edited by Edward V. Lewis. 3 vols. Jersey City, NJ: SNAME, 1988. |
| PPSG | *Proceedings of the Philosophical Society of Glasgow*. Glasgow, 1841–1901. |
| PRS | *Proceedings of the Royal Society*. London, 1854–present. |
| PTRS | *Philosophical Transactions of the Royal Society*. London, 1665–present. |
| RHM | *Revue d'Histoire Maritime* (Maritime history review). Paris, 1997–present. |
| RHN | *Revista de Historia Naval* (Naval history review). Madrid, 1983–present. |
| RINA | Royal Institution of Naval Architects. London, 1960–present. Preceded by INA. |

## Selected Bibliography

| | |
|---|---|
| *RM* | *Rivista Marittima* (Maritime journal). Rome, 1868–present. |
| *RMC* | *Revue Maritime et Coloniale* (Maritime and colonial review). Paris, 1861–1896. |
| *SHDM* | Service Historique de la Défense Archives de La Marine (Defense Historical Service, Navy Archives). Vincennes, France. |
| *SDC3* | *Ship Design and Construction*. 3rd ed., edited by Robert Taggart. New York: SNAME, 1980. |
| *SDC4* | *Ship Design and Construction*. 4th ed., edited by Thomas Lamb. 2 vols. Jersey City, NJ: SNAME, 2003. |
| *SNAME* | Society of Naval Architects and Marine Engineers. New York, 1893–1986; Jersey City, NJ, 1987–2013; Alexandria, VA, 2014–present. |
| *T&C* | *Technology and Culture*. Norman, OK, 1959–present. |
| *TICE/MPICE* | *Transactions of the Institution of Civil Engineers / Minutes of the Proceedings of the Institution of Civil Engineers*. London, *TICE* 1836–1842; *MPICE* 1841–1935. Followed by *Journal of the Institution of Civil Engineers*, 1935–1951, then *Proceedings of the Institution of Civil Engineers*, 1952–1991, then *Proceedings of the Institution of Civil Engineers* (various parts), 1992–present. |
| *TIESS* | *Transactions of the Institution of Engineers and Shipbuilders in Scotland*. Glasgow, 1857–present. |
| *TINA/TRINA* | *Transactions of the Institution of Naval Architects*. London, 1860–1959; *Transactions of the Royal Institution of Naval Architects*, 1960–present. |
| *TIS* | *Technology in Society*. Amsterdam, 1979–present. |
| *TNA* | *The Naval Architect*. London: RINA, 1971–present. |
| *TNECIES* | *Transactions of the North-East Coast Institution of Engineers and Shipbuilders*, Newcastle-upon-Tyne, UK, 1885–1992. |
| *TNM/LMN* | *The Northern Mariner / Le marin du nord*. Ontario, 1991–present. |
| *TNS/IJHET* | *Transactions of the Newcomen Society / International Journal for the History of Engineering and Technology*. London, *TNS* 1920–2009; *IJHET* 2009–present. |
| *TRSE* | *Transactions of the Royal Society of Edinburgh*. Edinburgh, 1785–present. |
| *TSNAME* | *Transactions of the Society of Naval Architects and Marine Engineers*. New York, 1893–1986; Jersey City, NJ, 1987–2013; Alexandria, VA, 2014–present. |
| *WT* | *Warship Technology*. London: Royal Institution of Naval Architects, 1987–present. |

### Primary Sources

ABS (American Bureau of Shipping). *Rules for the Classification and Construction of Steel Ships*. New York: ABS, 1880–present.

Adams, John, Abigail Adams, and Charles F. Adams. *Familiar Letters of John Adams and His Wife Abigail Adams, during the Revolution*. New York: Hurd & Houghton, 1876.

American Marine Standards Committee, Department of Commerce. *Stability and Loading of Ships. Final Report of the Special Committee on Stability and Loading*. Washington, DC: Government Printing Office, 1929.

American Railway Engineering and Maintenance of Way Association. *General Specifications for Steel Railway Bridges*. Chicago: American Railway Engineering and Maintenance of Way Association, 1910.

*Annual of the Royal School of Naval Architecture and Marine Engineering*. 4 vols. London: Henry Sotheran, 1871–1874.

Aristophanes. *The Comedies of Aristophanes*, edited by Benjamin Rogers. 5 vols. London: Bell, 1907–1919.

Ashe, Glenn, ed. "Naval Ship Design." In *Proceedings of the 16th International Ship and Offshore Structures Congress*, edited by Paul A. Frieze and Ramanand Ajit Shenoi. 3 vols. Southampton, UK: University of Southampton Press, 2016, vol. 2, pp. 217–266.

*ASSET Advanced Ship and Submarine Evaluation Tool Training*. Carderock, MD: NSWC Carderock Division Code 20, 2012.

Atherton, Charles. *The Capability of Steam Ships*. Woolwich, UK: John Grant, 1853.

Attwood, Edward L. *Text-Book of Theoretical Naval Architecture*. London: Longmans, Green, 1899.

Beaufoy, Mark. "On the Stability of Vessels." *Annals of Philosophy* 7 (1816), pp. 184–204.

Beaufoy, Mark, and Henry Beaufoy. *Nautical and Hydraulic Experiments*. Lambert, UK: Henry Beaufoy, 1834.

Benford, Harry [Pamela Ritter, pseud.]. "What Every Young Girl Should Know Before She Marries a Naval Architect." Unpublished pamphlet. Ann Arbor: University of Michigan, 1961.

Bentham, Mary Sophia. *The Life of Brigadier General Sir Samuel Bentham*. London: Longman, 1862.

Bernoulli, Daniel. *Die Werke von Daniel Bernoulli* (The works of Daniel Bernoulli). 8 vols. Basel, Switzerland: Birkhäuser Verlag, 1982–2004.

Bertin, Louis-Émile. *La Marine des États-Unis* (The United States Navy). Paris: E. Bernard, 1896.

Blackburn, Isaac. *The Theory and Science of Naval Architecture Familiarly Explained*. Plymouth, UK: Bartlett, 1836.

BMT Group website. https://www.bmt.org/.

Bonjean, Antoine Nicolas François. *Nouvelles Echelles de Déplacement et de Centre de Gravité de Carène, pour des Vaisseaux de guerre* (New scales of displacement; and the center of gravity of the hull, for vessels of war). Lorient: Baudoin, 1810.

Bouguer, Pierre. *Traité du Navire, de sa Construction, et de ses Mouvemens* (Treatise of the ship, its construction and its movements). Paris: Charles-Antoine Jombert, 1746.

Bourgois, Siméon. *Études sur les Manœuvres des Combats sur Mer* (Studies on the maneuvers of combat at sea). Paris: Berger-Levrault et Cie, 1876.

Bourgois, Siméon. *Memoire sur la resistance de l'eau au mouvement des corps et particulierement des batiments de mer* (Memoir on the resistance of water to the movement of bodies and particularly those of sea-going ships). Paris: Arthus Bertrand, 1857.

# Selected Bibliography

Bourgois, Siméon. *Recherches Théoriques et Expérimentales sur les Propulseurs Helicoïdes* (Theoretical and experimental research on helicoidal propulsors). Paris: Arthus Bertrand, 1845.

Bourne, John. *A Catechism of the Steam Engine*. London: John Williams, 1847.

Bourne, John. *A Treatise on the Screw Propeller: With Various Suggestions of Improvement*. London: Longman, 1852.

Bournemouth University Oral History Research Unit, Vosper-Thornycroft shipyard interviews, https://histru.bournemouth.ac.uk/Oral_History/Talking_About_Technology/shipbuilding/shipbuilding_2.htm.

Bowen, Harold G. *Ships, Machinery and Mossbacks: The Autobiography of a Naval Engineer*. Princeton, NJ: Princeton University Press, 1954.

Bravais, August. *Sur l'Equilibre des corps flottants* (On the equilibrium of floating bodies). Paris: Arthus Bertrand, 1840.

Brunel Collection, University of Bristol Library Special Collections, Bristol, UK.

Bureau of Construction and Repair. *Instructions for [Standard Ship] Displacement and Stability Calculations under the Bureau of Construction and Repair*. Washington, DC: Government Printing Office, 1898, 1923.

Bureau of Ships. *Study of Bismarck*. Washington DC: Preliminary Design Branch, Bureau of Ships, December 1941.

Bush, Vannevar. *Science, the Endless Frontier: A Report to the President*. Washington, DC: Government Printing Office, 1945.

Calkoen, Jan Frederick van Beeck. *Wiskundige scheeps-bouw en bestuur* (Mathematical shipbuilding and management). Amsterdam: Allard, 1805.

Cathérineau, Jean. *Construction navale, traité élémentaire du système Cathérineau* (Ship construction: Elementary treatise of the Cathérineau system). Bordeaux: P. Chaumas, 1854.

CEHIPAR. *Canal de Experiencias hidrodinámicas de El Pardo* (El Pardo Model Basin). Madrid: Imprenta Ministerio de Defensa, 2001.

[Chatfield, Henry]. *An Apology for English Ship-Builders; Showing That It Is Not Necessary the Country Should Look to the Navy for Naval Architects*. London: Effingham Wilson, 1833.

Chatfield, Henry. *On the Advantages of Observing a Ship's Inclination at Sea*. London: Sherwood, Gilbert, & Piper, 1831.

Cochrane, Edward L., and John C. Niedermair. *Subdivision, Stability, and Construction of Merchant Ships*. Bureau of Construction and Repair Bulletin No. 8. Washington, DC: Government Printing Office, 1935.

Colwell, James L. *User's Manual for the SHOP5 System: A Concept Exploration Model for Monohull Frigates and Destroyers*. Technical Communication 88/302. Halifax: Defence Research & Development Atlantic, 1988.

Committee on Safety of Construction, International Conference on Safety of Life at Sea. *Report and Minutes of Proceedings*. London: Foreign Office, 1914.

Cooley, Mortimer. *Scientific Blacksmith*. Ann Arbor: University of Michigan Press, 1947.

Cotton, James L. *COnceptual DEsign of SHIPS Model CODESHIP*. 4 vols. CNA Systems Evaluation Group Research Contribution 159, December 1971. Arlington, VA: Center for Naval Analyses archives.

Damásio, Bruno, and Sandro Mendonça. "Modeling Insurgent-Incumbent Dynamics: Vector Autoregressions, Multivariate Markov Chains, and the Nature of Technological Competition." Working Paper 044-2018, July 2018. Lisbon: Research in Economics & Mathematics.

*DDG 1000 SAR Selected Acquisition Report*. 31 December 2010.

d'Etroyat, Adrien. *Traité élémentaire d'architecture navale* (Elementary treatise on naval architecture). Lorient, France: Gousset, 1846.

d'Eyncourt, Eustace Henry William Tennyson. *A Shipbuilder's Yarn: The Record of a Naval Constructor*. London: Hutchinson, 1948.

D'Oliveira, Rogério. *A Corveta Portuguesa Dos Anos 70* (The Portuguese corvette of the 1970s). Lisbon: Comissao Cultural de Marinha, 1999.

Dupin, Charles. *Applications de géométrie et de méchanique: à la marine aux ponts et chaussées, etc., pour faire suite aux Développements de géométrie* (Applications of geometry and mechanics: From the navy to bridges and roads, etc. following [the book] Developments in geometry). Paris: Bachelier, 1822.

Dupuy de Lôme, Stanislas Charles Henri Laurent. *Atlas joint au Rapport sur les Bâtiments en Fer* (Atlas accompanying the Report on Iron Ships). Toulon, France: A. Vincent, 1843.

Dupuy de Lôme, Stanislas Charles Henri Laurent. *Mémoire sur la Construction des Bâtiments en Fer* (Memoir on the construction of iron ships). Paris: Arthus Bertrand, 1844.

Dupuy de Lôme, Stanislas Charles Henri Laurent. *Notice sur les travaux scientifiques de M. Dupuy de Lôme* (Notice on the scientific works of M. Dupuy de Lôme). Paris: Gauthier-Villars, 1866.

Eardley-Wilmot, John. *Reminisces of the Late Thomas Assheton Smith, Esq*. London: Everett, 1902.

EMSHIP European Master's Course—Advanced Design in Ship and Offshore Structures website. http://www.emship.eu/.

Euler, Leonhard. *Leonhardi Euleri Opera Omnia* (Leonhard Euler's complete works). 72 vols. Basel, Switzerland: Birkhäuser, 1911–2009.

Fairbairn, William. *The Life of Sir William Fairbairn, Bart*, edited by William Pole. London: Longman, 1877.

Fairbairn, William. *Treatise on Iron Ship Building: Its History and Progress*. London: Longman, 1865.

Ferreiro, Larrie D. "Comparing Ship versus Aircraft Development Costs." *Naval Postgraduate School Acquisition Research Symposium 2018*. Monterey, CA: Naval Postgraduate School, 2018.

*Final Report of a Board of Investigation to Inquire into the Design and Methods of Construction of Welded Steel Merchant Vessels.* Washington, DC: Government Printing Office, 1946.

Flachat, Eugène. *Navigation à vapeur transocéanienne* (Transocean steam navigation). 3 vols. Paris: J. Baudry, 1866.

Franklin, Benjamin. "Letter to Alphonsus Le Roy Containing Sundry Maritime Observations." *Transactions of the American Philosophical Society* 2 (1785), pp. 294–329.

Froude, William. *The Papers of William Froude, 1810–1879.* London: INA, 1955.

*General Specifications for Building Vessels of the United States Navy.* Various editions. Washington, DC: Government Printing Office, 1908–1999.

George Steers Papers, 1851–1856. Ann Arbor: William Clements Library, University of Michigan.

Gore, Charles. *Result of Two Series of Experiments towards Ascertaining the Respective Velocity of Floating Bodies, Varying in Form; and towards Determining the Form Best Adapted to Stability.* London: Black, 1799.

Gower, Richard Hall. *Original Observations regarding the Inability of Ships to Perform Their Duty with Promptitude and Safety, with Suggestions for Their Improvement as Practised on Board the Transit etc.* London: Hullmandel, 1833.

Graham, Clark, and James L. Mills Jr. *Comparative Naval Architecture: Surface Combatants. NAVSHIPS Presentation to the Chief of Naval Architecture.* June 1971.

Griffiths, John W. *A Treatise on Marine and Naval Architecture, or Theory and Practice Blended in Shipbuilding.* New York: Pudny & Russell, 1850.

Groothedde, Gerrit Jan, Frank Meurs, and Hans Veltmeijer. *Pride: 75 Years of Maritime Research.* Wageningen, Netherlands: MARIN (Maritime Research Institute Netherlands), 2007.

Hess, Paul, and Apollo Milton Olin (A.M.O.) Smith. "Calculation of potential flow about arbitrary bodies." Douglas Aircraft Corporation Report No. E.S. 40622, carried out under Bureau of Ships Fundamental Hydromechanics Research Program NS 715–102, Contract No. Nonr 2722(00), administered by the David Taylor Model Basin. Long Beach, CA: Douglas Aircraft, 1962.

Holberton, Frances E. OH 50. Oral history interview by James Baker Ross, Potomac, Maryland. Minneapolis: Charles Babbage Institute, University of Minnesota, 14 April 1983.

Holmen dockyard (Denmark) database. http://www.orlogsbasen.dk.

Hovgaard, William. *Structural Design of Warships.* London: E. & F. N. Spon, 1915.

HSVA (Hamburgische Schiffbau-Versuchsanstalt). *HSVA@100, 1913–2013: A Century of Pivotal Research, Innovation and Progress for the Maritime Industry.* Hamburg, Germany: Hansa, 2013.

Hugo, Victor. *Les Misérables.* 5 vols. Paris: Pagnerre, 1862.

Hüllen, Johann Theodor Adolf van. *Leitfaden für den Unterricht im Schiffbau* (Guidelines for instruction in shipbuilding). Berlin: Ernst Siegfried Mittler & Sohn, 1908.

Hutchison, William. *A Treatise on Naval Architecture*. Liverpool: Billinge, 1794.

Independence Seaport Museum, John Lenthall Collection. Philadelphia.

INSEAN (Istituto Nazionale per Studi ed Esperienze di Architettura Navale). *La Vasca Navale di Roma nel cinquantenario della sua istituzione* (The Rome Model Basin at the 50th anniversary of the institution). Rome: INSEAN, 1977.

Isherwood, Benjamin F. *Experimental Researches in Steam Engineering*. 2 vols. Philadelphia: William Hamilton, 1863–1865.

James W. Queen & Co. *Priced and Illustrated Catalogue of Mathematical Instruments*. Philadelphia: James W. Queen & Co., 1883.

Johnson, Robert S. "Computer-Aided Ship Design." In *Computers in the Navy*, edited by Jan Prokop, pp. 174–207. Annapolis, MD: Naval Institute Press, 1976.

Juan y Santacilia, Jorge. *Examen Marítimo, Theórico Práctico* (Maritime examination, theoretical and practical). Madrid: Manuel de Mena, 1771.

Kanigel, Robert. *The One Best Way: Frederick Winslow Taylor and the Enigma of Efficiency*. Cambridge, MA: MIT Press, 1997.

Kemp, Dixon. *Yacht Architecture: A Treatise on the Laws Which Govern the Resistance of Bodies Moving in Water, Propulsion by Steam and Sail; Yacht Designing; and Yacht Building*. London: Horace Cox, 1897.

Kempf, Günther, and Ernst Foerster, eds. *Hydromechanische Probleme des Schiffsantriebs* (Hydromechanical problems of ship propulsion). Hamburg, Germany: HSV, 1932.

Kempf, Günther, and Walter Sottorf. "The High-Speed Tank of the Hamburg Shipbuilding Company." NACA Technical Memorandum 735. Washington, DC: NACA, 1934. Translation of "Der neue Schleppkanal fur hohe Geschwindigkeiten der Hamburgischen Schiffbau-Versuchsanstalt." *Werft-Reederei-Hafen* 12 (1931), pp. 175–180.

Kipling, Rudyard. *The Day's Work*. London: Macmillan, 1898.

Korvin-Kroukovsky, Boris V. *Theory of Seakeeping*. New York: SNAME, 1961.

Kossiakoff, Alexander, and William N. Sweet. *Systems Engineering: Principles and Practice*. New York: Wiley, 2003.

Krylov, Aleksey Nikolaevich. *Professor Krylov's Navy: Memoir of a Naval Architect*, translated by Laura Meyerovich. London: Magnet, 2014. Translation of Krylov, *Moi Vospominaniya* (My memories). Leningrad: Izd-vo Akademii nauk SSSR, 1942.

Krylov State Research Center website. http://krylov-center.ru/eng/.

Labrousse, Henri. *Des Propulseurs Sous-Marins* (On submarine propulsors). Paris: Bureau de la Revue Générale de l'Architecture et des Travaux Public, 1845.

Lacy, Mary. *The History of the Female Shipwright*. Greenwich, UK: National Maritime Museum, 2008.

Lagerhjelm, Pehr, Johan Henrik af Forselles, and Georg Samuel Kallstenius. *Hydrauliska försök, anställda vid Fahlu Grufva, åren 1811–1815* (Hydraulic tests, undertaken at the Fahlu Mine, years 1811–1815). 2 vols. Stockholm: Cederborgska, 1818–1822.

Lammeren, Wilhelmus Petrus Antonius van, *Resistance, Propulsion and Steering of Ships*. Haarlem, Netherlands: H. Stam, 1948.

Latour, Bruno, and Steve Woolgar. *Laboratory Life: The Construction of Scientific Facts*. Beverly Hills, CA: Sage, 1979.

*Literary Gazette and journal of the belles lettres, arts, sciences*, etc. London, 1817–1862.

Longfellow, Henry Wadsworth. *The Seaside and the Fireside*. Boston: Ticknor, Reed, & Fields, 1849.

Loon, Folkert Nicolaas van. *Beschouwing van den Nederlandschen scheepsbouw met betrekking tot deszelfs zeilaadje* (Reflections on Dutch shipbuilding related to better sailing). Haarlem, Netherlands: Loosjes, 1820; facsimile, De Boer Maritime, 1980.

Loon, Folkert Nicolaas van. *Handleiding tot den Burgerlijken Scheepsbouw* (Manual for civil shipbuilding). Workum Netherlands: H. Brandenburgh, 1838.

Marestier, Jean-Baptiste. *Mémoire sur les bateaux à vapeur des Etats-Unis d'Amérique* (Memoir on the steamboats of the United States of America). Paris: Imprimerie Royale, 1824. *Memoir on steamboats of the United States of America*, translated by Sidney Withington. Mystic, CT: Marine Historical Association, 1957.

Masefield, John. *Salt-Water Poems and Ballads*. New York: Macmillan, 1916.

McGwire, Michael, and John McDonnell, eds. *Soviet Naval Influence: Domestic and Foreign Dimensions*. New York: Praeger, 1977.

Melville, George W. *Proceedings of the International Engineering Congress, Division of Marine and Naval Engineering and Naval Architecture*. 2 vols. New York: Wiley, 1894.

Monjo i Pons, Joan. *Curso métodica de arquitectura naval aplicada a la construcción de los buques mercantes* (Methodical course of naval architecture applied to the construction of merchant ships). Barcelona, Spain: Imprenta de Jose Tauló, 1856.

Montagu, Robert. *Naval Architecture: A Treatise on Ship-Building and the Rig of Clippers*. London: Colburn, 1852.

Montgéry, Jacques-Philippe Mérigon de. *Mémoire sur les navires en fer* (Memoir on iron ships). Paris: Bachelier, 1824.

Moreau, Philippe-Jacques. *Principes fondamentaux de l'équilibre et du mouvement des corps flottans dans deux milieux résistans* (Fundamental principles of equilibrium and movement of floating bodies in two resistant media). Brest, France: Lefournier, 1830.

Morgan, William, and Augustin Francis Bullock Creuze. *Papers on Naval Architecture and Other Subjects Connected with Naval Science*. 4 vols. London: G. B. Whittaker, 1829–1832.

MTG Marintechnik website. https://www.mtg-marinetechnik.de.

Murphy, Robert D., Donald J. Sabat, and Robert J Taylor. "Least Cost Ship Characteristics by Computer Techniques." SNAME Chesapeake Section paper, October 1963.

Murray, Mungo. *A Treatise on Ship-Building and Navigation.* London: D. Henry & R. Cave, 1754.

Murray, William M., Larrie D. Ferreiro, John Vardalas, and Jeffery G. Royal. "Cutwaters before Rams: An Experimental Investigation into the Origins and Development of the Waterline Ram." *International Journal of Nautical Archaeology* 46 (2017), pp. 72–82.

NAVSEA. *Warship Appearance.* NAVSEA Report T907A-81-RPT-040. Washington, DC: Department of the Navy, 1981.

Nevesbu website. http://www.nevesbu.com.

Niedermair, John C. Niedermair Files, 17 boxes. Bethesda, MD: Office of the Curator of Models, Naval Surface Warfare Center Carderock Division.

Niedermair, John C. *Reminiscences of John C. Niedermair (Naval Architect—Bureau of Ships).* Annapolis, MD: U.S. Naval Institute, 1978.

Noalhat, Henri. *Navigation aérienne et navigation sous-marine: deux faces d'un même problème* (Aerial navigation and submarine navigation: Two faces of the same problem). Paris: L. Geisler, 1910.

Nowacki, Horst. "Potentialtheoretische Strömungs- und Sogberechnungen für schiffsähnliche Körper" (Potential-theoretical flow and suction calculations for ship-like bodies). PhD diss., University of Berlin, 1963. Published in *Jahrbuch de Schiffbautechnischen Gesellschaft* 57 (1963).

O'Rourke, Ronald. *Navy Littoral Combat Ship (LCS) Program: Background, Issues, and Options for Congress.* Washington, DC: Congressional Research Service, 29 November 2010.

Papanikolaou, Apostolos. *Risk-Based Ship Design: Methods, Tools and Applications.* Berlin: Springer, 2009.

Paramarine website. http://paramarine.qinetiq.com/.

Plimsoll, Samuel. *Our Seaman: An Appeal.* London: Virtue, 1873.

Poisson, Siméon Denis. *Traité de Mecanique* (Treatise of mechanics). 2 vols. Paris: Bachelier, 1833.

Rahola, Jaakko. "The Judging of the Stability of Ships and the Determination of the Minimum Amount of Stability, Especially Considering the Vessels Navigating Finnish Waters." PhD diss., Technical University of Finland, 1939.

Rålamb, Åke Classon, *Skeps Byggerij eller adelig öfnings tionde tom* (Ship building from exercises for young noblemen). Vol. 10, Niclas Wankijfs Tryckerij, Stockholm, 1691. Facsimile, A. B. Malmö Ljustrycksanstalt, Malmö, 1943.

Rankine, William John Macquorn. *A Manual of the Steam Engine and Other Prime Movers.* London: Richard Griffin, 1859.

# Selected Bibliography

Rankine, William John Macquorn, ed. *Shipbuilding, Theoretical and Practical*. London: Mackenzie, 1866.

Rawson, Kenneth J., and Eric C. Tupper. *Basic Ship Theory*. 2 vols. London: Longman, 1983.

Rawson, Robert. *The Screw Propeller; An Investigation of Its Geometrical and Physical Properties, and Its Application to the Propulsion of Vessels*. London: Whittaker, 1851.

Reech, Frédéric. *Cours de mécanique d'après la nature généralement flexible et élastique des corps* (Course of mechanics according to the generally flexible and elastic nature of bodies). Paris: Carilian-Goeury et Dalmont, 1852.

Reech, Frédéric. *Mémoire sur les machines à vapeur et leur application à la navigation* (Memoir on steam engines and their application to navigation). Paris: Arthus Bertrand, 1844.

Reed, Edward. *A Treatise on the Stability of Ships*. London: Charles Griffin, 1885.

Reed, Michael Robert. "Ship Synthesis Model for Naval Surface Ships." Master's thesis, MIT, 1976.

Reeves, John M. L. "CNA's Conceptual Design and Cost Models for High-Speed Surface Craft." CNA Professional Paper 381, April. Arlington, VA: Center for Naval Analyses archives, 1983.

*Report of the Board of Trade Inquiry into the Loss of the Vestris and the Findings of the Court*. Liverpool, UK: Charles Birchall, 1929.

Richstone, Morris. "The Applied Mathematics Laboratory of the David W. Taylor Model Basin." *Communications of the ACM* 4 (1961), pp. 372–375.

Robinson, John. *The Marine Room of the Peabody Museum of Salem*. Salem, MA: Peabody Museum, 1921.

Ruskin, John. *The Works of John Ruskin*, edited by E. T. Cook and Alexander Wedderburn. 39 vols. London: George Allen, 1903–1912.

Russell, John Scott. *The Fleet of the Future: Iron or Wood?* London: Longman, 1861.

Russell, John Scott. John Scott Russell Notebooks, Ms. 516, books 1–7 (circa 1859). London: Science Museum Library.

Russell, John Scott. *The Modern System of Naval Architecture*. 3 vols. London: Day & Son, 1865.

R. W. Munro Ltd. *R. W. Munro Ltd. Centenary, 1864–1964: A Century of Family Enterprise in Instrument Making and Precision Engineering*. London: R. W. Munro, 1964.

Saldanha, Luiz Philippe. *Memoir on the Novel Formation of the Bottom of Ships and Vessels, Proposed by the Brazilian Naval Architect Trajano Augusto de Carvalho*. Philadelphia: R. Magee & Son, 1876.

Saunders, Harold E. *Hydrodynamics in Ship Design*. 3 vols. New York: SNAME, 1957–1965.

Schmidt, John W. *A Computer-Aided Feasibility Design Method for Destroyer-Type Ships*. CNA Research Contribution 52, August. Arlington, VA: Center for Naval Analyses archives, 1965.

Schmidt, John W. "Preliminary Design with the Aid of a Computer." SNAME New England Section paper, May 1964.

Schumacher, Theodore L. *Stability and Compartmentation of Ships*. Bureau of Construction and Repair Bulletin No. 14. Washington, DC: Government Printing Office, 1938.

Scribanti, Angelo. *La statica della nave: esposta in base al principio del minimo lavoro di assestamento* (Ship statics: Exposition on the basis of the minimum work of righting [arm]). Milan: Hoepli, 1928.

Seaton, Albert E. *A Manual of Marine Engineering*. London: Charles Griffin, 1907.

Seguin, Marc (*ainé*, the elder). *Mémoire sur la Navigation à Vapeur* (Memoir on steam navigation). Paris: Bachelier, 1828.

Sewell, John, ed. *A Collection of Papers on Naval Architecture*. London: Sewell, 1791, 1792, 1795, 1800.

"Ship Design Division Stability Criteria." Manuscript. Washington, DC: Bureau of Ships, 1957.

Ship Structure Committee website. http://www.shipstructure.org/.

Smith, Francis Pettit. *On the Introduction and Progress of the Screw Propeller: With Statistics of the Comparative Economy of Screw Ships and Paddle Vessels for Her Majesty's Service*. London: Longman, 1856.

*Stabilité des Bâtiments de Surface* (Surface ship stability). Instruction Technique no. 11-03 DCN Index B. Paris: Direction des Constructions Navales, 1998.

*Stability of Ships*. Admiralty Book of Reference (BR) 298(42). Bath, UK: Directorate of Naval Construction (DNC) Department, March 1942.

Stalkartt, Marmaduke. *Naval Architecture, Or, the Rudiments and Rules of Ship Building*. London: Boydell, 1781.

Steel, David. *The Elements and Practice of Naval Architecture*. London: P. Steel, 1805.

Steel, David. *The Shipwright's Vade-Mecum*. London: P. Steel Navigation Warehouse, 1805.

Stevens, Renee. *Engineering Mega-Systems: The Challenge of Systems Engineering in the Information Age*. Boca Raton, FL: CRC Press, 2010.

Stibolt, Ernst Vilhelm. *Afhandling om Skibes Kiølbrækkelighed* (Treatise on keelbreaking of ships). Copenhagen: Stein, 1784.

Symposium on Naval Hydrodynamics. Washington, DC: Office of Naval Research, 1956–present.

Taniguchi, Kaname. "Historical Review of Research and Development in Ship Hydrodynamics." Mitsubishi Technical Bulletin No. 164. Tokyo: Mitsubishi Heavy Industries, 1984.

Taylor, David W. *Resistance of Ships and Screw Propulsion*. New York: Macmillan, 1893.

Taylor, David W. *The Speed and Power of Ships; A Manual of Marine Propulsion*. New York: Wiley, 1910.

Thibault, Louis-Adrien. *Recherches expérimentales sur la résistance de l'air et particulièrement sur l'impulsion du vent considéré comme force motrice, sur la voilure des vaisseaux* (Experimental research on the resistance of air and in particular on the impulsion of wind considered as a motive force on the sails of vessels). Brest, France: Lefournier & Deperiers, 1826.

Tideman, Bruno J. *Memoriaal van de marine* (Memoir of the navy). Amsterdam: van Heteren, 1876–1880.

Tredgold, Thomas. *A Practical Essay on the Strength of Cast Iron*. London: J. Weale, 1842.

Tupinier, Jean-Marguerite. *Mémoires du Baron Tupinier, directeur des ports et arsenaux (1779–1850)* (Memoirs of Baron Tupinier, director of ports and arsenals). Edited by Bernard Lutun. Paris: Desjonquères, 1994.

UNESCO Convention for the Safeguarding of Intangible Cultural Heritage. "The Watertight-Bulkhead Technology of Chinese Junks." Nomination File 00321, Fifth session, Kenya, 2010.

U.S. Navy Bureau of Construction and Repair. *The Stability of Ships and Damage Control*. Washington, DC: Government Printing Office, 1931.

Verne, Jules. *Twenty Thousand Leagues Under the Sea*, translated by William Butcher. London: Oxford University Press, 1998.

Vienna Model Basin / Schiffbautechnische Versuchsanstalt In Wien website. http://www.sva.at/.

White, William. *A Manual of Naval Architecture*. London: J. Murray, 1877.

Wiard, Norman. *Iron vs Wood for Great Ships*. New York: Crichton, 1867.

Wickenden, William E. *A Comparative Study of Engineering Education in the United States and in Europe*. Lancaster, PA: Lancaster Press, 1929.

Yuille, Ian. *I Fathered a GODDESS: Autobiography of a Naval Scientist*. Bedfordshire, UK: Authors Online Limited, 2012.

Zhang Zhi-rong, Liu Hui, Zhu Song-ping, and Zhao Feng. "Application of CFD in Ship Engineering Design Practice and Ship Hydrodynamics." *Journal of Hydrodynamics*, ser. B, 18, no. 3 (2006), pp. 315–322.

## Secondary Sources

Abell, Wescott. *The Shipwright's Trade*. Cambridge: Cambridge University Press, 1948.

Achútegui Rodríguez, Juan José, ed. *I Simposio de historia de las técnicas, la construcción naval y la navegación* (First symposium of the history of technology, naval construction and navigation). Santander, Spain: Universidad de Cantabria, 1996.

Addis, William, ed. *Structural and Civil Engineering Design*. Aldershot, UK: Ashgate, 1999.

Addis, William. *Structural Engineering: The Nature of Theory and Design*. New York: Ellis Horwood, 1990.

Ahlström, Göran. *Engineers and Industrial Growth: Higher Technical Education and the Engineering Profession during the Nineteenth and Early Twentieth Centuries: France, Germany, Sweden, and England*. London: Croom Helm, 1982.

Alberi Auber, Paolo. "Historical Developments in Naval Propulsion (1829–1830). New Insight into Feathering Paddle Wheels (Morgan Wheel) and the Screwpropeller, Invented in 1829." *European Transport/Trasporti Europei* 23 (2003), pp. 58–61.

Albion, Robert G. *Forests and Sea Power; the Timber Problem of the Royal Navy, 1652–1862*. Cambridge, MA: Harvard University Press, 1926; reprint, Annapolis, MD: Naval Institute Press, 1999.

Albion, Robert G. *The Rise of New York Port, 1815–1860*. New York: Charles Scribner's Sons, 1939; reprint, Boston: Northeastern University Press, 1984.

Albion, Robert G. *Square Riggers on Schedule: The New York Sailing Packets to England, France, and the Cotton Ports*. Hamden, CT: Archon Books, 1965.

Alic, John A. *Trillions for Military Technology: How the Pentagon Innovates and Why It Costs So Much*. New York: Palgrave Macmillan, 2007.

Alimento, Mario, Claudio Boccalatte, Gianfranco Damilano, and Bruno Della Loggia. *Associazione Italiana di Tecnica Navale 1947–2017* (Italian Association of Marine Technology, 1947–2017). Pisa: Edizioni ETS, 2017.

Allen, Robert C. *The British Industrial Revolution in Global Perspective*. Cambridge: Cambridge University Press, 2009.

Allington, Peter, and Basil Greenhill. *The First Atlantic Liners: Seamanship in the Age of Paddle Wheel, Sail and Screw*. London: Conway Maritime Press, 1997.

Allison, David K., Ben G. Keppel, and Carol Elizabeth Nowicke. *D.W. Taylor*. Washington, DC: Government Printing Office, 1988.

American Bureau of Shipping. *The History of the American Bureau of Shipping, 150th Anniversary*. Houston, TX: ABS, 2013.

Andersen, Håkon W., and John P. Collett. *Anchor and Balance: Det norske Veritas 1864–1989*. Oslo: Cappelens, 1989.

Arjava, Jouni. *Raholan Kriteeri: Professori Jaakko Raholan elämä ja työ* (Rahola criterion: Professor Jaakko Rahola's life and work). Helsinki: Oy Merkur, 2002. Abridged translation *The Rahola Criterion—The Life and Work of Professor Jaakko Rahola*. Edited by Risto Jalonen. Translated by Susan Sinisalo. Espoo: Aalto University, 2015. https://aaltodoc.aalto.fi/handle/123456789/15447.

ATMA. *Association Technique Maritime et Aéronautique Centième Anniversaire* (Centennial of the Maritime and Aeronautical Technical Association). Paris: ATMA, 1989.

Augustin Normand, Paul. *La Genèse de l'Hélice Propulsive* (The genesis of the screw propulsor). Paris: Académie de Marine, 1962.

Bailey, David. *Ships in the Making: A History of Ship Model Testing at Teddington and Feltham, 1910–1994*. London: Lloyd's of London Press, 1995.

Bamford, Paul. *Forests and French Sea Power, 1660–1789*. Toronto: University of Toronto Press, 1956.

# Selected Bibliography

Bargoni, Franco, Franco Gay, and Valerio Gay. *Navi a Vela e navi Miste Italiane, 1861–1887* (Italian sailing ships and mixed-propulsion ships, 1861–1887). Rome: Ufficio storico della marina militare, 2001.

Barker, Richard. "Iron Ships in Green Fields, 1777–1833: Wilkinson's Legacy." *Journal of the Broseley Local History Society* 30 (2008), pp. 20–33.

Barnaby, Kenneth C. *The Institution of Naval Architects, 1860–1960*. London: RINA, 1960.

Battesti, Michèle. *La Marine de Napoléon III* (The navy of Napoleon III). 2 vols. Vincennes, France: Service historique de la marine, 1997.

Baynes, Ken, and Francis Pugh. *The Art of the Engineer*. Woodstock, NY: Overlook Press, 1981.

Beauchesne, Geneviève. *Historique de la construction navale à Lorient de 1666 à 1770* (History of naval construction at Lorient from 1666 to 1770). Vincennes, France: Service Historique de la Marine, 1980.

Benford, Harry, and John C Mathes. *Your Future in Naval Architecture: With Information on Marine Engineering*. New York: Richards Rosen Press, 1968.

Bennett, Stuart. *A History of Control Engineering, 1800–1930*. London: Peregrinus, 1979.

Berthiau, Jean André. "Des maîtres entretenus aux ingénieurs, 1819–1971" (From on-call masters to engineers, 1819–1971). *Techniques and Culture* 45, 2005. http://tc.revues.org/1401.

Betts, Charles V. ed. *Vanguard to Dreadnought: 30 years of Naval Construction 1983–2013*. London: RINA, 2018.

Biran, Adrian, and Rubén López-Pulido. *Ship Hydrostatics and Stability*. Amsterdam: Elsevier, 2003.

Blazek, Jiri. *Computational Fluid Dynamics: Principles and Applications*. Amsterdam: Elsevier, 2006.

Blind, Knut. *The Economics of Standards: Theory, Evidence, Policy*. Cheltenham, UK: Edward Elgar, 2004.

Boisson, Philippe. *Safety at Sea: Policies, Regulations and International Law*. Paris: Edition Bureau Veritas, 1999.

Booker, Peter Jeffrey. *A History of Engineering Drawing*. London: Chatto & Windus, 1963.

Bopp-Vigne, Catherine. "Emigrés français de Constantinople en Russie pendant la Révolution" (French émigrés from Constantinople in Russia during the Revolution). In *L' influence française en Russie au XVIIIe* (French influence in Russia in the 18th century), edited by Jean-Pierre Poussou, Anne Mézin, and Yves Perrret-Gentil, pp. 411–427. Paris: Presses de l'Université de Paris-Sorbonne, 2004.

Borde, Christian. "The Madness of Frédéric Sauvage of Boulogne (1786–1857) and the Birth of the French Propeller." In *Science and the French and British Navies, 1700–1850*, edited by Pieter van der Merwe, pp. 131–146. Greenwich, UK: National Maritime Museum, 2003.

Boslaugh, David L. *When Computers Went to Sea: The Digitization of the United States Navy*. Los Alamitos, CA: IEEE Computer Society Press, 1999.

Boudriot, Jean, and Gérard Delacroix. *Le Fleuron: Vaisseau de 64 canons 1729* (Fleuron: 64-gun vessel, 1729). Nice: Editions Omega, 1995.

Bovis, Alain. *Hydrodynamique navale: théorie et modèles* (Naval hydrodynamics: theory and models). Paris: Les Presses de l'ENSTA, 2009.

Bradley, Margaret. *Charles Dupin (1784–1873) and His Influence on France: The Contributions of a Mathematician, Educator, Engineer, and Statesman*. London: Cambria Press, 2011.

Breemer, Jan S. "The Great Race: Innovation and Counter-Innovation at Sea, 1840–1890." Corbett Centre for Maritime Policy Studies, Paper 2, January 2011.

Brindle, Steven. *Brunel: The Man Who Built the World*. London: Weidenfield & Nicolson, 2005.

Briot, Claude, and Jacqueline Briot. *Les Clippers Français* (French clippers). Douarnenez, France: Chasse-Marée, 1993.

Brisou, Dominique. *Accueil, Introduction et Développement de l'Énergie Vapeur dans la Marine Militaire Française au XIX$^e$ Siècle* (Acceptance, introduction and development of steam energy in the French navy in the 19th century). 2 vols. Vincennes, France: Service Historique de la Marine, 2001.

Brown, David K. *Before the Ironclad: Development of Ship Design, Propulsion and Armament in the Royal Navy, 1815–60*. London: Conway Maritime Press, 1990.

Brown, David K. *A Century of Naval Construction: The History of the Royal Corps of Naval Constructors*. London: Conway Maritime Press, 1983.

Brown, David K. "The Development of Subdivision in Merchant Ships." In *International Conference on Watertight Integrity and Ship Survivability*. London: RINA, 1996.

Brown, David K. *The Grand Fleet: Warship Design and Development, 1906–1922*. Barnsley, UK: Seaforth, 1997.

Brown, David K. *Nelson to Vanguard: Warship Design and Development 1923–1945*. Barnsley, UK: Seaforth, 2012.

Brown, David K. *Warrior to Dreadnought: Warship Design and Development 1860–1905*. London: Chatham, 1997.

Brown, David K. *The Way of the Ship in the Midst of the Sea: The Life and Work of William Froude*. Penzance, UK: Periscope, 2006.

Brown, David K., and George Moore. *Rebuilding the Royal Navy: Warship Design since 1945*. Barnsley, UK: Seaforth, 2012.

Bruce, Alexander. *The Life of William Denny, Ship-Builder, Dumbarton*. London: Hodder & Stoughton, 1888.

Brunel, Isambard. *The Life of Isambard Kingdom Brunel, Civil Engineer*. London: Longmans, Green, 1870.

Buchanan, R. Angus. *Brunel: The Life and Times of Isambard Kingdom Brunel*. London: Hambledon, 2002.

Buderi, Robert. *Naval Innovation for the 21st Century: The Office of Naval Research since the End of the Cold War*. Annapolis, MD: Naval Institute Press, 2013.

Bulley, Anne. *The Bombay Country Ships, 1790–1833*. Richmond, UK: Curzon, 2000.

*Bureau Veritas 1828/1978: A Record of 150 Years*. Neuilly-sur-Seine: PEMA 2B, 1978.

Burgess, Douglas R., Jr. *Engines of Empire: Steamships and the Victorian Imagination*. Stanford, CA: Stanford University Press, 2016.

Busch, John L. *Steam Coffin: Captain Moses Rogers and the Steamship Savannah Break the Barrier*. New Canaan, CT: Hodos Historia, 2010.

Butler, John A. *Atlantic Kingdom: America's Contest with Cunard in the Age of Sail and Steam*. Washington, DC: Brassey's, 2001.

Cardoso Llach, Daniel. *Builders of the Vision: Software and the Imagination of Design*. New York: Routledge, 2015.

Cardwell, Donald S. L. *From Watt to Clausius: The Rise of Thermodynamics in the Early Industrial Age*. Ithaca, NY: Cornell University Press, 1971.

Carlisle, Rodney P. *Where the Fleet Begins: A History of the David Taylor Research Center, 1898–1998*. Washington, DC: Naval Historical Center, 1998.

Carlton, John S. *Marine Propellers and Propulsion*. Oxford: Elsevier, 2007.

Carr-Saunders, Alexander Morris, and Paul Alexander Wilson. *The Professions*. Oxford: Clarendon Press, 1933.

Casanueva González, José Francisco, and Antonio Jose Fraidias Becerra. "El *Real Fernando*: El Primer Vapor Español." *Revista de Historia Naval* (Naval history journal) 8, no. 28 (1990), pp. 49–59.

Cerulus, Frans A. "Daniel Bernoulli and Leonhard Euler on the Jetski." In *Two Cultures: Essays in Honour of David Speiser*, edited by Kim Williams, pp. 73–96. Basel, Switzerland: Birkhäuser Verlag, 2006.

Chandler, Alfred D., Jr. *The Visible Hand: The Managerial Revolution in American Business*. Cambridge, MA: Harvard University Press, 1977.

Chapelle, Howard I. *The History of the American Sailing Navy: The Ships and Their Development*. New York: Bonanza Books, 1949.

Chapelle, Howard I. *The History of American Sailing Ships*. New York: Norton, 1935.

Chapelle, Howard I. *The Search for Speed under Sail, 1700–1855*. New York: Bonanza Books, 1967.

Chatenet, Madeleine du. "L'amiral Jean-Bapstiste de Traversay (1754–1831)." In *L' influence française en Russie au XVIIIe* (French influence in Russia in the 18th century), edited by Jean-Pierre

Poussou, Anne Mézin, and Yves Perrret-Gentil, pp. 487–499. Paris: Presses de l'Université de Paris-Sorbonne, 2004.

Christen, Carole, and François Vatin. *Charles Dupin (1784–1873): Ingénieur, savant, économiste, pédagogue et parlementaire du Premier au Second Empire* (Engineer, scholar, economist, teacher and parliamentarian of the First and Second Empires). Rennes, France: Presses Universitaires de Rennes, 2009.

Christensen, Dan Ch., ed. *European Historiography of Technology.* Odense, Denmark: Odense University Press, 1993.

Clark, Basil E. *Steamboat Evolution: A Short History.* Self-published, Lulu / Fogdog, 2006.

Clarke, David. "The Foundations of Steering and Maneuvering." In *Maneuvering and Control of Marine Craft 2003,* edited by Joan Batlle and Mogens Blanke, pp. 2–16. Oxford: Elsevier, 2004.

Clarke, Joseph F. *A Century of Service to Engineering and Shipbuilding, 1884–1984.* Newcastle upon Tyne, UK: North East Coast Institution of Engineers & Shipbuilders, 1984.

Clarke, Joseph F. *The Changeover from Wood to Iron Shipbuilding: Occasional Papers in the History of Science and Technology,* no. 3. Newcastle upon Tyne: Newcastle upon Tyne Polytechnic, 1986.

Clarke, Joseph F., and Frank Storr. *The Introduction of the Use of Mild Steel into the Shipbuilding and Marine Engine Industries.* Newcastle upon Tyne, UK: Newcastle upon Tyne Polytechnic, 1983.

Coles, Howard E. "The Contribution of British Defence Departments to Technical Education and Instruction from about 1700." PhD diss., University of Manchester, 1984.

Cooling, Benjamin Franklin. *Gray Steel and Blue Water Navy: The Formative Years of America's Military-Industrial Complex, 1881–1917.* Mamden, CT: Archon Books, 1979.

Corlett, Ewan. *The Iron Ship: The Story of Brunel's SS Great Britain.* London: Conway Maritime Press, 1975.

Cosentino, Michele, and Ruggero Stanglini. *Il Corpo del Genio Navale* (The corps of naval engineers). Rome: Selex sistemi integrati, 2006.

Cotte, Michel. *De l'espionnage industriel à la veille technologique* (From industrial espionage to technological surveillance). Belfort, France: Presses universitaires de Franche-Comté, 2005.

Crothers, William. *The American Clipper-Built Ship, 1850–1856.* Camden, ME: International Marine, 1997.

Dahl, Eric J. "Naval Innovation: From Coal to Oil." *Joint Forces Quarterly* 50 (2000–2001), pp. 5–56.

Darrigol, Olivier. *Worlds of Flow: A History of Hydrodynamics from the Bernoullis to Prandtl.* Oxford: Oxford University Press, 2005.

Davids, Mila, and Hans Schippers. "Innovations in Dutch Shipbuilding in the First Half of the Twentieth Century." *Business History* 50, no. 2 (2008), pp. 205–225.

Davies, Michael. *Belief in the Sea: State Encouragement of British Merchant Shipping and Shipbuilding*. London: Lloyd's of London Press, 1992.

Davin, Emmanuel. "Un toulonnais, Jean-Louis Barrallier, ingénieur de la marine, constructeur d'un port et de vaisseaux anglais, 1751–1834" (A Toulonnais, Jean-Louis Barrallier, naval engineer, constructor of ports and English vessels, 1751–1834). *Provence historique* 3, no. 12 (1953), pp. 140–148.

De Beer, Gavin. *The Sciences Were Never at War*. London: Nelson, 1960.

Dedet, Christian. *Les Fleurs d'acier du Mikado* (The steel flowers of the Mikado). Paris: Flammarion, 1993.

Delis, Apostolos. *Mediterranean Wooden Shipbuilding: Economy, Technology and Institutions in Syros in the Nineteenth Century*. Leiden, Netherlands: Brill, 2016.

Depeyre, Michel. "Éperon et Bélier, entre Histoire et Techniques" (Spur and ram, between history and technology). In *L'Evolution de la pensée navale* (The evolution of naval thought), edited by Hervé Coutau-Bégarie. Vol. 7, pp. 25–38. Paris: Economica, 1999.

Deybach, Frédéric. "Intact Stability Criteria for Naval Ships." Master's thesis, MIT, 1997.

DGA. *100 Years of History: The Bassin d'essais des carènes*. Paris: DGA Délégation générale pour l'armement, internal publication, 2006.

DGA/DCN. *1906–1988: quatre-vingt-deux ans d'histoire du Bassin d'essais des Carènes de Paris* (1906–1988: 88 years of history of the Paris Model basin). Paris: DGA Délégation générale pour l'armement / DCN Direction des constructions navales, internal publication, 1988.

Dickinson, Henry Winram. *Robert Fulton, Engineer and Artist*. London: John Lane, 1913.

Dirkzwager, Jan M. "De voor geschiedenis van de opleiding tot scheepsbouwkundig ingenieur aan de Technische Universiteit Delft" (The history of the naval engineering degree at the Technical University of Delft). *Roering / Vereniging van Ingenieurs van Defensie* 1 (1994), pp. 1–14.

Dirkzwager, Jan M. *Dr. B.J. Tideman, 1834–1883: Grondlegger van de Moderne Scheepsbouw in Nederland* (Dr. B. J. Tideman, 1834–1883: Founder of modern shipbuilding in the Netherlands). Leiden, Netherlands: Brill, 1970.

Doe, Helen. *Enterprising Women and Shipping in the Nineteenth Century*. Woodbridge, UK: Boydell Press, 2009.

Dunbaugh, Edwin L. *A Centennial History of Webb Institute of Naval Architecture*. Glen Cove, NY: Webb Institute of Naval Architecture, 1994.

Eckert, Michael. "Approaching Reality by Idealization: How Fluid Resistance Was Studied by Ideal Flow." In *Mathematics as a Tool: Tracing New Roles of Mathematics in the Sciences*, edited by Johannes Lenhard and Martin Carrier, pp. 197–214. Cham, Switzerland: Springer, 2017.

Eckert, Michael. *The Dawn of Fluid Dynamics*. Weinheim, Germany: Wiley-VCH, 2006.

Eddy, Richard. *Joshua Humphreys, His Life and Times*. New York: Vantage Press, 2005.

Emmerson, George S. *John Scott Russell: A Great Victorian Engineer and Naval Architect*. London: Murray, 1977.

Endrikat, Jonathan D. "Maritime Classification Societies' Role in the Development of Aviation Verification and Validation Processes." Master's thesis, Stevens Institute of Technology, 2016.

Eskew, Garnett E. *Our Navy's Ships and Their Builders, 1775–1961 [i.e. 1883]: The Epic Story of the Evolution, Design and Construction of the U.S. Fleet*. Manuscript. Washington, DC: Bureau of Ships, 1962.

Fassett, Frederick Gardiner, Jr., ed. *The Shipbuilding Business in the United States of America*. 2 vols. New York: SNAME, 1948.

Fernández González, Francisco. *Aportación de Cataluña a la arquitectura naval: Valor tecnológico de los veleros del siglo XIX* (Contribution of Catalonia to naval architecture: Technological value of sailing ships of the 19th century). Barcelona, Spain: Museu Marítim de Barcelona, 2009.

Fernández González, Francisco, Larrie D. Ferreiro, and Horst Nowacki, eds. *Technology of the Ships of Trafalgar*. Madrid: ETSIN-UPM, 2006.

Ferreiro, Larrie D. "The Aristotelian Heritage in Early Naval Architecture: From the Venice Arsenal to the French Navy, 1500–1700." *Theoria* 25, no. 2 (2010), pp. 227–241.

Ferreiro, Larrie D. *Brothers at Arms: American Independence and the Men of France and Spain Who Saved It*. New York: Knopf, 2016.

Ferreiro, Larrie D. "Développements et avantages tactiques du doublage en cuivre des coques des navires français, britanniques et espagnols" (Developments and tactical advantages of coppering the hulls of French, British and Spanish warships)." In *Les Marines de la Guerre d'Indépendance Américaine 1763–1783* (The navies of the War of American Independence 1763–1783), edited by Olivier Chaline, Philippe Bonnichon, and Charles-Philippe de Vergennes. 2 vols., vol. 2, pp. 37–65. Paris: Presses de l'Université Paris-Sorbonne, 2013–2018.

Ferreiro, Larrie D. "The Mutual Influence of Aircraft Aerodynamics and Ship Hydrodynamics in Theory and Experiment." *Archive for History of Exact Sciences* 68, no. 2 (2014), pp. 241–263.

Ferreiro, Larrie D. "Organizational Trust in Naval Ship Design Bureaus: France, Great Britain, and the United States." *Acquisition Review Quarterly* 5, no. 3 (1998), pp. 285–296.

Ferreiro, Larrie D. "Shipbuilders to the World: Evolution and Revolution in Spanish and Chilean Shipbuilding from the Cold War to the 21st Century: A Study in International Technology Transfer in the Naval Industries." *International Journal of Naval History* 8, no. 3 (2010). http://www.ijnhonline.org/issues/volume-8-2009/dec-2009-vol-8-issue-3/.

Ferreiro, Larrie D. *Ships and Science: The Birth of Naval Architecture in the Scientific Revolution, 1600–1800*. Cambridge, MA: MIT Press, 2007.

Ferreiro, Larrie D. "The Warship since the End of the Cold War." In *Maritime Strategy and Global Order: Markets, Resources, Security*, edited by Daniel Moran and James A. Russell, pp. 209–238. Washington, DC: Georgetown University Press, 2016.

Firebaugh, Millard S., ed. *Naval Engineering and American Sea Power*. 2nd ed. Dubuque, IA: Kendall/Hunt, 2000.

Fishbourne, Edmund Gardiner. *Lectures on Naval Architecture*. London: John Russell Smith, 1846.

Flexner, James T. *Steamboats Come True: American Inventors in Action*. New York: Viking Press, 1944.

Foster, Kevin J. "The Search for Speed under Steam: The Design of Blockade Running Steamships, 1861–1865." Master's thesis, East Carolina University, 1991.

Fowler, William J., Jr. *Steam Titans: Cunard, Collins and the Epic Battle for Commerce on the North Atlantic*. London: Bloomsbury, 2017.

Fox, Stephen. *Transatlantic: Samuel Cunard, Isambard Brunel, and the Great Atlantic Steamships*. New York: HarperCollins, 2003.

Fragiacomo, Paolo. *L'industria come continuazione della politica. La cantieristica italiana 1861–2011* (Industry as a continuation of politics. Italian shipbuilding 1861–2011). Milan: FrancoAngeli, 2012.

Frängsmyr, Tore, J. L. Heilbron, and Robin E. Rider. *The Quantifying Spirit in the 18th Century*. Berkeley: University of California Press, 1990.

Fratta, Arturo, ed. *La fabbrica delle navi: Storia della cantieristica nel Mezzogiorno d'Italia* (The fabrication of ships: The history of shipbuilding in southern Italy). Naples: Electa, 1990.

Friedman, Norman. *U.S. Destroyers: An Illustrated Design History*. Annapolis, MD: Naval Institute Press, 2004.

Frieze, Paul A., and R. Ajit Shenoi, eds. *Proceedings of the 16th International Ship and Offshore Structures Congress*. 2 vols. Southampton, UK: University of Southampton Press, 2006.

Fukasaku, Yukiko. "In-Firm Training at Mitsubishi Nagasaki Shipyard, 1884–1934." In *Industrial Training and Technological Innovation: A Comparative and Historical Study*, edited by Howard F. Gospel, pp. 120–139. London: Routledge, 1991.

Gabriele, Mariano. *Benedetto Brin*. Rome: Ufficio storico della marina militare, 1998.

García-Torralba Pérez, Enrique. *Navíos de la Real Armada 1700–1860* (Ships of the Royal Navy 1700–1860). Madrid: Fondo Editorial de Ingeniería Naval, 2016.

Gardiner, Robert, ed. *The Advent of Steam: The Merchant Steamship before 1900*. London: Conway Maritime Press, 1993.

Gardiner, Robert. *Frigates of the Napoleonic Wars*. London: Chatham, 2000.

Gardiner, Robert, and Andrew Lambert, eds. *Steam, Steel and Shellfire: The Steam Warship 1815–1905*. London: Conway Maritime Press, 1992.

Garzke, William H., and John B Woodward. *Titanic Ships, Titanic Disasters: An Analysis of Early Cunard and White Star Superliners*. Jersey City, NJ: SNAME, 2002.

Gatti, Luciana. *Un raggio di convenienza: Navi mercantili, costruttori e proprietari in Liguri nella prima metà dell'Ottocento* (A range of convenience: Ligurian merchant ships, constructors and owners in the first half of the 19th century). Genoa: Società Ligure di Storia Patria, 2008.

Gaudard, Valérie. "Le bassin des carènes et le service technique des constructions navales à Balard : un exemple de cité scientifique à Paris" (The model basin and the Naval Construction Technical Service at Balard: An example of a scientific city in Paris). *In Situ: Revue de Patrimoines* 10 (2009). https://journals.openedition.org/insitu/3806.

Gerovitch, Slava. *From Newspeak to Cyberspeak: A History of Soviet Cybernetics*. Cambridge, MA: MIT Press, 2002.

Glete, Jan. *Navies and Nations: Warships, Navies and State Building in Europe and America 1500–1860*. 2 vols. Stockholm: Academitryck AB Edsbruk, 1993.

Goodwin, Peter. *The Construction and Fitting of the English Man of War, 1650–1850*. Annapolis, MD: Naval Institute Press, 1987.

Greene, Jack, and Alessandro Massignani. *Ironclads at War: The Origin and Development of the Armored Warship, 1854–1891*. Conshohocken, PA: Combined, 1998.

Griffiths, Denis. *Steam at Sea: Two Centuries of Steam-Powered Ships*. London: Conway Maritime Press, 1997.

Griffiths, Denis, Andrew Lambert, and Fred M. Walker. *Brunel's Ships*. London: Chatham, 1999.

Guedj, Youri. "L'ingénieur, le Génie Maritime et l'idée de Progrès Théoriques de l'Architecture Navale à l'aube du XIX$^e$ Siècle" (The engineer, the Naval Constructors Corps, and the idea of theoretical progress at the dawn of the nineteenth century). Master's thesis, Université Paris Diderot, 2016.

Haas, James M. *A Management Odyssey: The Royal Dockyards, 1714–1914*. Lanham, MD: University Press of America, 1994.

Hagler, Gina. *Modeling Ships and Space Craft: The Science and Art of Mastering the Oceans and Sky*. New York: Springer, 2013.

Hambly, Maya. *Drawing Instruments, 1580–1980*. London: Sotheby's, 1988.

Hamilton, Charles I. *Anglo-French Naval Rivalry, 1840–1870*. Oxford: Clarendon Press, 1993.

Hamilton, Charles I. *The Making of the Modern Admiralty: British Naval Policy-Making 1805–1927*. Cambridge: Cambridge University Press, 2011.

Harland, John. *Seamanship in the Age of Sail*. London: Conway Maritime Press, 1984.

Harley, Charles K. "The Shift from Sailing Ships to Steamships, 1850–1890: A Study in Technological Change and Diffusion." In *Essays on a Mature Economy: Britain after 1840*, edited by Donald N. McCloskey, pp. 215–237. Princeton, NJ: Princeton University Press, 1971.

Hattendorf, John B. *Sailors and Scholars: The Centennial History of the U.S. Naval War College*. Newport, RI: Naval War College Press, 1984.

# Selected Bibliography

Hawkey, Arthur. *Black Night off Finisterre: The Tragic Tale of an Early British Ironclad*. Shrewsbury, UK: Airlife, 1999.

Headrick, Daniel R. *The Tentacles of Progress: Technology Transfer in the Age of Imperialism, 1850–1940*. Oxford: Oxford University Press, 1988.

Hepburn, Richard D. *History of American Naval Dry Docks*. Arlington, VA: Noesis, 2003.

Herd, Robert J. "Rahola—40 Years On." Sydney: RINA Australian Branch—Technical Library. 1979. https://www.rina.org.uk/download5485.html.

Herreshoff, L. Francis. *Capt. Nat Herreshoff, the Wizard of Bristol*. Dobbs Ferry, NY: Sheridan House, 1953.

Hills, Richard L. *Life and Inventions of Richard Roberts, 1789–1864*. Ashbourne, UK: Landmark, 2002.

Hilton, George W. *Eastland: Legacy of the Titanic*. Stanford, CA: Stanford University Press, 1995.

Hocker, Fred, and Cheryl Ward. *The Philosophy of Shipbuilding: Conceptual Approaches to the Study of Wooden Ships*. College Station: Texas A&M University Press, 2004.

Howells, John. "The Response of Old Technology Incumbents to Technological Competition: Does the Sailing Ship Effect Exist?" *Journal of Management Studies* 39, no. 7 (2002), pp. 887–907.

Hudson, Pat. *The Industrial Revolution*. London: Edward Arnold, 1992.

Hughes, Thomas P. *Networks of Power: Electrification in Western Society, 1880–1930*. Baltimore: Johns Hopkins University Press, 1983.

Hutchins, John B. G. *The American Maritime Industries and Public Policy, 1789–1914*. Cambridge, MA: Harvard University Press, 1941.

Hyman, Anthony. *Charles Babbage: Pioneer of the Computer*. Princeton, NJ: Princeton University Press, 1982.

Jackson, Gordon, and David M. Williams, eds. *Shipping, Technology, and Imperialism: Papers Presented to the Third British-Dutch Maritime History Conference*. Aldershot: Scholar Press, 1996.

Jeremy, David I. "Damming the Flood: British Government Efforts to Check the Outflow of Technicians and Machinery, 1780–1843." *Business History Review* 51, no. 1 (Spring 1977), pp. 1–34.

Johnman, Lewis, and Hugh Murphy. *British Shipbuilding and the State since 1918*. Exeter, UK: University of Exeter Press, 2002.

Johnman, Lewis, and Hugh Murphy. "Welding and the British Shipbuilding Industry." In *The Royal Navy, 1930–2000: Innovation and Defence*, edited by Richard Harding, pp. 89–116. London: Frank Cass, 2005.

Johnston, Ian, and Ian Buxton. *The Battleship Builders: Constructing and Arming British Capital Ships*. Barnsley, UK: Seaforth, 2013.

Jones, Nicolette. *The Plimsoll Sensation: The Great Campaign to Save Lives at Sea*. London: Little, Brown, 2006.

Jouin, Yves, ed. *600 ans de constructions navales* (600 years of naval construction). Revue Historique des Armées, no. 1. Paris: Ministère des Armées, 1974.

Kamp, Adolph Frederik, ed. *De Technische Hogeschool te Delft, 1905–1955* (Technical University of Delft, 1905–1955). 's-Gravenhage: Staatsdrukkerij- en Uitgeverijbedrijf, 1955.

Kaukiainen, Yrjö. "Shrinking the World: Improvements in the Speed of Information Transmission, c. 1820–1870." *European Review of Economic History* 5 (2001), pp. 1–28.

Kelly, Andrew, and Melanie Kelly, eds. *Brunel: In Love with the Impossible*. Bristol, UK: Bristol Cultural Development Partnership, 2006.

Kipp, Jacob W. "The Russian Navy and the Problem of Technological Transfer: Technological Backwardness and Military-Industrial Development, 1853–1876." In *Russia's Great Reforms, 1855–1881*, edited by Ben Eklof et al., pp. 115–138. Bloomington: Indiana University Press, 1994.

Knight, Lucia Del Sol, and Daniel Bruce MacNaughton. *The Encyclopedia of Yacht Designers*. New York: Norton, 2005.

Kobylinski, Lech K., Sigismund Kastner, Vadim L. Belenky, and Nikita B. Sevastianov. *Stability and Safety of Ships*. 2 vols. Vol. 1, *Regulations and Operations*; vol. 2, *Risk of Capsizing*. Amsterdam: Elsevier, 2003.

Kranakis, Eda. *Constructing a Bridge: An Exploration of Engineering Culture, Design, and Research in Nineteenth-Century France and Spain*. Cambridge, MA: MIT Press, 1997.

Krisman, Kevin J., and Arthur B. Cohn. *When Horses Walked on Water: Horse-Powered Ferries in Nineteenth-Century America*. Washington, DC: Smithsonian Institution Press, 1998.

Kuhn, Thomas. *The Structure of Scientific Revolutions*. Chicago: University of Chicago Press, 1962.

Labaree, Benjamin, et al. *America and the Sea: A Maritime History*. Mystic, CT: Mystic Seaport, 1998.

Lambert, Andrew. *The Foundations of Naval History: John Knox Laughton, the Royal Navy and the Historical Profession*. London: Chatham, 1998.

Lambert, Andrew. *The Last Sailing Battlefleet: Maintaining Naval Mastery 1815–1850*. London: Conway Maritime Press, 1991.

Lambert, Andrew. "Science and Seapower: The Navy Board, the Royal Society and the Structural Reforms of Sir Robert Seppings." *Transactions of the Naval Dockyards Society* 1 (2006), pp. 9–19.

Lambert, Andrew. *Trincomalee: The Last of Nelson's Frigates*. Annapolis, MD: Naval Institute Press, 2002.

Lambert, Andrew. *Warrior: Restoring the World's First Ironclad*. London: Conway Maritime Press, 1987.

Layne, Margaret E. *Women in Engineering: Pioneers and Trailblazers*. Reston, VA: ASCE Press, 2009.

Layton, Edwin T., Jr. *The Revolt of the Engineers: Social Responsibility and the American Engineering Professions*. Baltimore: Johns Hopkins University Press, 1986.

Legaz Alamansa, María José. "Computer Aided Ship Design: A Brief Overview." *SeMA Journal Boletin de la Sociedad Española de Matemática Aplicada* 72, no. 1 (2015), pp. 47–59.

Leggett, Don. *Shaping the Royal Navy: Technology, Authority and Naval Architecture, c. 1830–1906*. Manchester, UK: Manchester University Press, 2015.

Leggett, Don, and Richard Dunn. *Re-inventing the Ship: Science, Technology and the Maritime World, 1800–1918*. Farnham, UK: Ashgate, 2012.

Lehmann, Eike. "The Historical Development of the Strength of Ships." In *The History of Theoretical, Material and Computational Mechanics—Mathematics Meets Mechanics and Engineering*, edited by Erwin Stein, pp. 267–295. Berlin: Springer-Verlag, 2014.

Lehmann, Eike. *100 Jahre Schiffbautechnische Gesellschaft* (100 years of the Society of Shipbuilders). 3 vols. Berlin: Springer, 1999.

Lehmann, Eike. *Schiffbautechnische Ausbildung in Deutschland* (Shipbuilding training in Germany). 2 vols. Hamburg, Germany: Seehafen-Verlag, 2001–2002.

Lehmann, Eike. *Schiffbautechnische Forschung in Deutschland* (Shipbuilding research in Germany). 2 vols. Hamburg, Germany: Seehafen-Verlag, 2003–2004.

Lemmers, Alan. *Techniek op schaal: Modellen en het technologiebeleid van de Marine 1725–1885* (Technology to scale: Models and technology policy of the navy, 1725–1885). Amsterdam: De Bataafsche Leeuw, 1996.

Lemmers, Alan. *Van werf tot facilitair complex: 350 jaar marinegeschiedenis op Kattenburg* (From shipyard to complex facility: 350 years of maritime history at Kattenburg). The Hague: Nederlands Instituut voor Militaire Historie, 2011.

Le Pavic, Fabrice. "Une histoire des techniques de l'arsenal de la Marine de Lorient dans la seconde moitié du XXe siècle: les Constructions neuves" (A history of the technologies of the Lorient Naval Arsenal in the second half of the 20th century: New construction). PhD diss., Université de Nantes, 2015.

Lessenich, Jobst. "Schiffbau: Neue Schiffsformen" (Shipbuilding: New ship forms). Deutschen Schiffahrtsmuseum. https://web.archive.org/web/20140312062452/http://www.dsm.museum/medien/17/4863/lessenich_neuform.pdf.

Levison, Marc. *The Box: How the Shipping Container Made the World Smaller and the World Economy Bigger*. Princeton, NJ: Princeton University Press, 2008.

Lindsay, William S. *History of Merchant Shipping and Ancient Commerce*. 4 vols. London: Sampson Low, Marston, Low, and Searle, 1874–1876.

Lutun, Bernard. *1814–1817 ou l'Epuration dans la Marine Française* (1814–1817 or the purification of the French navy). Paris: Harmattan, 2005.

Lutun, Bernard. *Liste générale des élèves et des ingénieurs du corps des ingénieurs-constructeurs de la marine puis des ingénieurs du génie maritime, 1765–1967* (General list of students and engineers of the Corps of Engineer-Constructors of the Navy and the Naval Engineering Corps, 1765–1967). Paris: Harmattan, 2013.

Lux, Jonathan. *Classification Societies*. London: Lloyd's of London Press, 1993.

MacDougall, Philip, ed. *Chatham Dockyard, 1815–1865: The Industrial Transformation*. Farnham, UK: Ashgate, 2009.

MacGregor, David R. *British and American Clippers: A Comparison of Their Design, Construction and Performance in the 1850s*. London: Conway Maritime Press, 1993.

MacGregor, David R. *Fast Sailing Ships: Their Design and Construction, 1775–1875*. Annapolis, MD: Naval Institute Press, 1988.

Maindron, Ernest. *Les Fondations de prix à l'Académie des Sciences: les lauréats de l'Académie, 1714–1880* (The foundations of the Academy of Sciences prize: Academy winners, 1714–1880). Paris: Gauthier-Villars, 1881.

Manning, Frederic. *The Life of Sir William White*. London: John Murray, 1923.

Manstan, Roy, and Frederic Frese. *Turtle: David Bushnell's Revolutionary Vessel*. Yardley, PA: Westholme, 2010.

*Marine et Technique au XIX$^e$ Siècle* (Navy and technology in the 19th century). Vincennes, France: Service Historique de la Marine, 1988.

Marsden, Ben. "The Administration of the 'Engineering Science' of Naval Architecture at the British Association for the Advancement of Science." In *Technological Development between Economy and Administration in Great Britain and Germany*, pp. 67–94. Vol. 20 of *Jarbuch für Europäische Verwaltungsgeschichte* (Yearbook of European administrative history), edited by Erk Volkmar Heyen. Baden-Baden, Germany: Nomos, 2008.

Marzari, Mario. *Progetti per l'Imperatore: Andrea Salvini, ingegnere a l'Arsenal 1802–1817* (Projects for the emperor: Andrea Salvini, engineer at the Arsenal). Trieste: B&M Fachin, 1990.

Massie, Robert K. *Dreadnought: Britain, Germany and the Coming of the Great War*. New York: Random House, 1991.

Matsumoto, Miwao. *Technology Gatekeepers for War and Peace: The British Ship Revolution and Japanese Industrialization*. New York: Palgrave Macmillan, 2006.

Matsumoto, Miwao, and Bruce Sinclair. "How Did Japan Adapt Itself to the Scientific and Technological Revolution at the Turn of the 20th Century? An Experiment in Transferring the Experimental Tank to Japan." *Japan Journal for Science, Technology and Society* 3 (1994), pp. 133–155.

McBride, William M. *Technological Change and the United States Navy, 1865–1945*. Baltimore: Johns Hopkins University Press, 2000.

McCarthy, Michael. *Ships' Fastenings: From Sewn Boat to Steamship.* College Station: Texas A&M University Press, 2005.

McGee, David B. "The Amsler Integrator and the Burden of Calculation." *Material Culture Review* 48 (1998), pp. 57–74.

McGee, David B. "Floating Bodies, Naval Science: Science, Design and the Captain Controversy, 1860–1870." PhD diss., University of Toronto, 1994.

McGlaughlin, Stephen. *Russian and Soviet Battleships.* Annapolis, MD: Naval Institute Press, 2003.

McKay, Richard C. *Some Famous Sailing Ships and Their Builder Donald McKay.* New York: Putnam's Sons, 1928.

Meader, Bruce. *ASNE: The First 100 Years.* Alexandria, VA: American Society of Naval Engineers, 1988.

Meek, Marshall. *There Go the Ships.* Spennymoor, UK: Memoir Club, 2003.

Meiksens, Peter, and Chris Smith, eds. *Engineering Labour: Technical Workers in Comparative Perspective.* London: Verso, 1996.

Mendonça, Sandro. "The Sailing Ship Effect: Reassessing History as a Source of Insight on Technical Change." *Research Policy* 42 (2013), pp. 1724–1738.

Millerson, Geoffrey. *The Qualifying Associations: A Study in Professionalization.* London: Routledge, 1964.

Miloh, Touvia. *Mathematical Approaches in Hydrodynamics.* Philadelphia: SIAM, 1991.

Modelski, George, and William R. Thompson. *Seapower in Global Politics, 1494–1993.* London: Macmillan, 1988.

Molina, Ricardo Hernández. *El ingenio de Blasco de Garay, 1539–1543* (The genius of Blasco de Garay). Cádiz, Spain: Servicio de Publicaciones de la Universidad de Cádiz, 1996.

Mollat, Michel, ed. *Les Origines de la Navigation à Vapeur* (Origins of steam navigation). Paris: Presses Universitaires de France, 1970.

Morriss, Roger. *Cockburn and the British Navy in Transition.* Exeter, UK: University of Exeter Press, 1997.

Musson, Albert E., and Eric Robinson. *Science and Technology in the Industrial Revolution.* Manchester UK: Manchester University Press, 1968.

Mutel, Cornelia Fleischer. *Flowing through Time: A History of the Iowa Institute of Hydraulic Research.* Iowa City: Iowa Institute of Hydraulic Research, 1998.

Nakaoka, Tetsuro. "From Shipbuilding to Automobile Manufacturing." In *The Introduction of Modern Science and Technology to Turkey and Japan*, edited by Feza Günergun and Kuriyama Shigehisa, pp. 37–54. Kyoto: International Research Centre for Japanese Studies, 1998.

Needham, Joseph. *Science and Civilisation in China.* 7 vols., 27 parts. Cambridge: Cambridge University Press, 1954–2020.

Newton, Robert N. *The Royal Institution of Naval Architects, 1960–1980.* London: RINA, 1981.

Noble, David F. *Forces of Production: A Social History of Industrial Automation.* New York: Alfred A. Knopf, 1984.

North, Douglass C. "Sources of Productivity Change in Ocean Shipping, 1600–1850." *Journal of Political Economy* 76, no. 5 (1968), pp. 953–970.

Nowacki, Horst. "A Farewell to the Design Spiral." Invited Note Presented at the Mini-Symposium on Ship Design, Ship Hydrodynamics and Maritime Safety, Athens, National Technical University of Athens Ship Design Laboratory, 30 September 2016.

Nowacki, Horst. "Five Decades of Computer-Aided Ship Design." In "Computer-Aided Ship Design: Some Recent Results and Steps Ahead in Theory, Methodology and Practice, Dedicated to Professor Horst Nowacki on the Occasion of His 75th birthday." Special issue, *Computer Aided Design* 42, no. 11 (November 2010), pp. 956–959.

Nowacki, Horst. "Splines im Schiffbau" (Splines in shipbuilding). In *Das Schiff aus dem Computer* (The ship out of the computer), *Proceedings of the 21st Duisburg Colloquium Ship and Ocean Technology*. Duisburg, Germany: Institute of Ship Technology, 2000, pp. 27–53.

Nowacki, Horst, and Wolfgang Lefèvre, eds. *Creating Shapes in Civil and Naval Architecture: A Cross-Disciplinary Comparison.* Berlin: Max Planck Institute for the History of Science, preprint 338. 2 vols., 2001; reprint, Leiden, Netherlands: Brill, 2009.

O'Brien, Phillips P. *Technology and Naval Combat in the Twentieth Century and Beyond.* London: Frank Cass, 2001.

O'Har, George Michael. "Shipbuilding, Markets and Technological Change in East Boston." PhD. diss., MIT, 1995.

Olivier, David H. *German Naval Strategy, 1856–1888: Forerunners to Tirpitz.* London: Frank Cass, 2004.

Olson, Lynne. *Madame Fourcade's Secret War: The Daring Young Woman Who Led France's Largest Spy Network against Hitler.* New York: Random House, 2019.

Olsson, Lars O. *Technology Carriers: The Role of Engineers in the Expanding Swedish Shipbuilding System.* Gothenburg, Sweden: Chalmers University of Technology, 2000.

Ostenc, Michel, ed. *La Marine Italienne de l'Unité à Nos Jours* (The Italian navy from unification to today). Paris: Economica, 2005.

Paine, Lincoln. *The Sea and Civilization: A Maritime History of the World.* New York: Alfred A. Knopf, 2013.

Parkes, Oscar. *British Battleships 1860–1950: A History of Design, Construction and Armament.* London: Seeley, Service, 1966.

Parkinson, Roger. *The Late Victorian Navy: The Pre-dreadnought Era and the Origins of the First World War.* Woodbridge, UK: Boydell Press, 2008.

Pastore, Christopher. *Temple to the Wind: The Story of America's Greatest Naval Architect and His Masterpiece, Reliance*. Guildford, CT: Lyons Press, 2005.

Pedisich, Paul E. *Congress Buys a Navy: Politics, Economics and the Rise of American Naval Power, 1881–1921*. Annapolis, MD: Naval Institute Press, 2016.

Peter, Jean. *Le port et l'arsenal de Toulon sous Louis XIV* (The port and arsenal of Toulon under Louis XIV). Paris: Economica, 1995.

Petersen, Peter B. "Fighting for a Better Navy: An Attempt at Scientific Management (1905–1912)." *Journal of Management* 16, no. 1 (1990), pp. 151–166.

Physick, John. *The Victoria and Albert Museum: The History of Its Building*. Oxford: Phaidon, 1982.

Pollard, Jules Ambroise, and Auguste Dudebout. *Architecture navale; Théorie du Navire* (Naval architecture; ship theory). 4 vols. Paris: Gauthier-Villars, 1890–1894.

Pollard, Sidney, and Paul Robertson. *The British Shipbuilding Industry, 1870–1914*. Cambridge, MA: Harvard University Press, 1979.

Porter, Theodore M. *Trust in Numbers: The Pursuit of Objectivity in Science and Public Life*. Princeton, NJ: Princeton University Press, 1996.

Prinsep, George A. *An Account of Steam Vessels and of Proceedings Connected with Steam Navigation in British India*. Calcutta: Government Gazette Press by G. H. Huttmann, 1830.

Pugh, Philip. *The Cost of Seapower: The Influence of Money on Naval Affairs from 1815 to the Present Day*. London: Conway Maritime Press, 1986.

Pugsley, Alfred. *The Works of Isambard Kingdom Brunel: An Engineering Appreciation*. Cambridge: Cambridge University Press, 1976.

Quartermaine, Peter. *Building on the Sea: Form and Meaning in Modern Ship Architecture*. London: Academy Editions, 1996.

Ramírez Gabarrús, Manuel. *La Construcción Naval Militar Española, 1730–1980* (Spanish naval construction, 1730–1980). Madrid: Bazán, 1980.

Rank, Ludwig. *Die Theorie des Segelns in ihrer Entwicklung: Geschichte eines Problems der nautischen Mechanik* (The evolution of the theory of sailing: The history of a nautical mechanics problem). Berlin: Dietrich Reimer Verlag, 1984.

Rasmussen, Frank A. "Statslig eller Privat? Relationerne mellem Orkogsvaerrftet og Burmeister & Wain 1843–1882" (State-owned or private? The relations between the Danish Naval Dockyard and Burmeister & Wain, 1843–1882). In *Handels-og Søfartsmuseet på Kronborg, 1993* (Trade and shipping museum of Kronborg Yearbook), pp. 93–120, 1993.

Renn, Jurgen, and Matteo Valleriani. *Galileo and the Challenge of the Arsenal*. Berlin: Max Planck Institute for the History of Science, preprint 179, 2001; reprinted in *Nuncius* 16, no. 2 (2001), pp. 481–503.

Reynolds, Terry. *Stronger than a Hundred Men: A History of the Vertical Water Wheel.* Baltimore: Johns Hopkins University Press, 1983.

Rippon, Peter M. *Evolution of Engineering in the Royal Navy.* 2 vols. Tunbridge Wells, UK: Spellmount, 1988.

Roberts, Steven. "The Introduction of Steam Technology in the French Navy, 1818–1852." PhD diss., University of Chicago, 1976.

Roberts, William P. *History of the Construction Corps of the United States Navy.* Washington, DC: Government Printing Office, 1937.

Robertson, Paul L. "Technical Education in the British Shipbuilding and Marine Engineering Industries, 1863–1914." *Economic History Review* 27, no. 2 (1974), pp. 222–235.

Rodríguez González, Agustín Ramón. *Isaac Peral, Historia de una frustración* (Isaac Peral, a story of frustration). Madrid: Ediciones Grafite, 2007.

Rohwer, Jürgen, and Mikhail S. Monakov. *Stalin's Ocean-Going Fleet: Soviet Naval Strategy and Shipbuilding Programmes 1935–1953.* London: Frank Cass, 2001.

Røksund, Arne. *The Jeune École: The Strategy of the Weak.* Leiden, Netherlands: Brill, 2007.

Roland, Alex, W. Jeffrey Bolster, and Alexander Keyssar. *The Way of the Ship: America's Maritime History Reenvisioned, 1600–2000.* Hoboken, NJ: Wiley, 2008.

Rolt, Lionel Thomas Caswell. *Isambard Kingdom Brunel: A Biography.* London: Longmans, Green, 1957.

Ropp, Theodore. *The Development of a Modern Navy: French Naval Policy 1871–1904.* Annapolis, MD: Naval Institute Press, 1987.

Rosenberg, Nathan, and Walter G. Vincenti, *The Britannia Bridge: The Generation and Diffusion of Technological Knowledge.* Cambridge, MA: MIT Press, 1978.

Rostovsev, Eugeny A. *Stolichnyy universitet Rossiyskoy imperii: uchenoe soslovie, obshchestvo i vlast', vtoraya polovina XIX—nachalo XX v.* (Metropolitan University of the Russian Empire: An academic class, society and power, second half of the 19th–beginning of the 20th century). Moscow: ROSSPEN, 2017.

Rousmaniere, John. *The Low Black Schooner: Yacht America, 1851–1945.* Mystic, CT: Mystic Seaport Museum, 1986.

Sánchez Carrión, José María. "Los Ingenieros de Marina: Motores de la Renovación y Tecnificación de la Construcción Naval Española (1770–1827)—Su Organización, Academia y Realizaciones" (The naval engineers: Engines of the renewal and modernization of Spanish naval construction [1770–1827]—their organization, academy and achievements). 3 vols. PhD diss., Escuela Técnica Superior de Ingenieros Navales, Universidad Politécnica de Madrid, 2009.

Sapolsky, Harvey M. *Science and the Navy: The History of the Office of Naval Research.* Princeton, NJ: Princeton University Press, 1990.

Scaife, William Garrett. *From Galaxies to Turbines: Science, Technology and the Parsons Family.* Bristol, UK: Institute of Physics, 2000.

Schaffer, Simon. "'The Charter'd Thames': Naval Architecture and Experimental Spaces in Georgian Britain." In *The Mindful Hand: Inquiry and Invention from the Late Renaissance to Early Industrialization*, edited by Lissa Roberts, Simon Schaffer, and Peter Dear, pp. 279–305. Amsterdam: Koninkliijke Nederlandse Akademie van Wetenschappen, 2007.

Schaffer, Simon. "Fish and Ships: Models in the Age of Reason." In *Models: The Third Dimension of Science*, edited by Soraya de Chadarevian and Nick Hopwood, pp. 71–105. Stanford, CA: Stanford University Press, 2004.

Schaub, Jeremy P. "U.S. Navy Shipboard Damage Control: Innovation and Implementation during the Interwar Period." Master's thesis, U.S. Army Command & General Staff College, 2014.

Schuster, Leslie A. *A Workforce Divided: Community, Labor and the State in Saint-Nazaire's Shipbuilding Industry, 1880–1910*. Westport, CT: Greenwood Press, 2002.

Seaton, Albert E. *The Screw Propeller: And Other Competing Instruments for Marine Propulsion*. London: C. Griffin, 1909.

Sechrest, Larry. "American Shipbuilders in the Heyday of Sail: Their Rise and Decline," Mises Institute Working Paper, 1998. http://mises.org/journals/scholar/sechrest2.PDF.

*II centenario de las enseñanzas de ingenieria naval, 1772–1972–3* (Second centennial of teaching naval engineering, 1772–1972/1973). Madrid: RAEC, 1975.

Sheridan, Jonathan Andrew. "Synthesis of Aesthetics for Ship Design." Master's thesis, University of Southampton, 2013.

Shinn, Terry. *L'école polytechnique, 1794–1914: savoir scientifique & pouvoir social* (École Polytechnique, 1794–1914: Scientific knowledge and social power). Paris: Presses de la Fondation Nationale des Sciences Politiques, 1980.

Sicard, Daniel. *Saint-Nazaire et la construction navale* (Saint-Nazaire and naval construction). Saint-Nazaire, France: Ecomusée de Saint-Nazaire, 1991.

Silva Suárez, Manuel, ed. *Técnica e ingeniería en España*. (Technology and engineering in Spain). 8 vols. Zaragoza, Spain: Prensas de la Universidad de Zaragoza, 2013.

Sinclair, Bruce. *Early Research at the Franklin Institute: The Investigation into the Causes of Steam Boiler Explosions 1830–1837*. Philadelphia: Franklin Institute, 1966.

Sinclair, Bruce. *Philadelphia's Philosopher Mechanics: A History of the Franklin Institute 1824–1865*. Baltimore: Johns Hopkins University Press, 1975.

Smith, Crosbie. *Coal, Steam and Ships: Engineering, Enterprise and Empire on the Nineteenth-Century Seas*. Cambridge: Cambridge University Press, 2018.

Smith, Crosbie. "The 'Crinoline' of Our Steam Engineers: Reinventing the Marine Compound Engine, 1850–1885." In *Geographies of Nineteenth-Century Science*, edited by David N. Livingstone and Charles W. J. Withers, pp. 229–254. Chicago: University of Chicago Press, 2011.

Smith, Crosbie, and Ben Marsden. *Engineering Empires: A Cultural History of Technology in Nineteenth-Century Britain*. New York: Palgrave Macmillan, 2005.

Smith, Edgar C. *A Short History of Naval and Marine Engineering*. Cambridge: Cambridge University Press, 1938.

Sobel, Dava. *Longitude*. New York: Walker, 1995.

Solar, Peter M. "Opening to the East: Shipping between Europe and Asia, 1770–1830." *Journal of Economic History* 73, no. 3 (2013), pp 625–661.

Sondhaus, Lawrence. *The Naval Policy of Austria-Hungary, 1867–1918: Navalism, Industrial Development, and the Politics of Dualism*. West Lafayette, IN: Purdue University Press, 1994.

SPEI (Service de propagande, édition, information), ed. *Bi-centenaire du Génie Maritime 1765–1965* (Bicentennial of maritime engineering 1765–1965). Imprimèrie de Montligeon, La Chapelle-Montligeon, 1965.

Sprout, Harold, and Margaret Sprout. *The Rise of American Naval Power 1776–1918*. Princeton, NJ: Princeton University Press, 1939.

*Stability and Buoyancy of U.S. Naval Surface Ships*. Design Data Sheet (DDS) 079–1. Hyattsville, MD: Naval Ship Engineering Center, 1975.

Staccioli, Valerio. *Mare Scienza e Tecnica: L'industrializzazione delle attività marittime nella Trieste dell'Ottocento* (Ocean science and technology: Industrialization in maritime activities in Trieste in the 19th century). Trieste, Italy: Civico Museo del Mare, 1990.

Stanley, Jo. *From Cabin 'Boys' to Captains: 250 Years of Women at Sea*. Stroud, UK: History Press, 2016.

Steffy, J. Richard. *Wooden Ship Building and the Interpretation of Shipwrecks*. College Station: Texas A&M University Press, 1994.

Stevens, Francis B. "The First Steam Screw Propeller Boats to Navigate the Waters of Any Country." *Stevens Indicator* 10, no. 2 (April 1893), pp. 102–130.

Stewart, Matthew. *Monturiol's Dream: The Extraordinary Story of the Submarine Inventor Who Wanted to Save the World*. New York: Pantheon Books, 2003.

Sutcliffe, Andrea. *Steam: The Untold Story of America's First Great Invention*. New York: Palgrave Macmillan, 2004.

Sutherland, Robert James Mackay, ed. *Structural Iron, 1750–1850*. Aldershot, UK: Ashgate, 1997.

Sutton, Jean. *Lords of the East: The East India Company and Its Ships, 1600–1874*. London: Conway Maritime Press, 2000.

Thiesen, William H. *Industrializing American Shipbuilding: The Transformation of Ship Design and Construction, 1820–1920*. Gainesville: University Press of Florida, 2006.

*XIII-A, Massachusetts Institute of Technology: One Hundred Years*. Cambridge, MA: MIT Press, 2001.

Thomas, William duBarry. *Speed on the Ship!* Jersey City, NJ: SNAME, 1993.

Timoshenko, Stephen. *Engineering Education in Russia*. New York: McGraw-Hill, 1959.

Timoshenko, Stephen. *History of Strength of Materials, with a Brief Account of the History of Theory of Elasticity and Theory of Structures*. New York: McGraw-Hill, 1953.

Todd, Daniel, and Michael Lindberg. *Navies and Shipbuilding Industries: The Strained Symbiosis*. Westport, CT: Praeger, 1996.

Todhunter, Isaac. *A History of the Theory of Elasticity and of the Strength of Materials, from Galilei to the Present Time*. 2 vols. Cambridge: Cambridge University Press, 1886.

Toll, Ian W. *Six Frigates: The Epic History of the Founding of the U.S. Navy*. New York: Norton, 2006.

Tomblin, Barbara. "From Sail to Steam: The Development of Steam Technology in the United States Navy 1838–1865." PhD diss., Rutgers University, 1988.

Tredrea, John, and Eduard Sozaev. *Russian Warships in the Age of Sail, 1696–1860*. Barnsley, UK: Seaforth, 2010.

True, Frederick W. *History of the First Half-Century of the National Academy of Sciences 1863–1913*. Washington, DC: National Academy of Sciences, 1913.

Tyler, David B. *The American Clyde: A History of Iron and Steel Shipbuilding on the Delaware from 1840 to World War I*. Newark: University of Delaware Press, 1958.

Ujifusa, Steven. *Barons of the Sea and Their Race to Build the World's Fastest Clipper Ship*. New York: Simon & Schuster, 2018.

Ujifusa, Steven. *A Man and His Ship: America's Greatest Naval Architect and His Quest to Build the S.S. United States*. New York: Simon & Schuster, 2013.

Unger, Richard W. "The Technology and Teaching of Shipbuilding 1300–1800." In *Technology, Skills and the Pre-Modern Economy in the East and the West*, edited by Maarten Prak and Jan Luiten van Zanden, pp. 161–204. Leiden, Netherlands: Brill, 2013.

Vallée-Poussin, Paloma de la. "Froude parle français: étude des recherches croisées entre France et Angleterre sur le roulis des navires" (Froude speaks French: Study of research exchanges between Franc and England on ship rolling). Nantes, France: Colloque international, Université de Nantes, *La construction navale et ses objets: nouvelles approches, nouveaux outils* (Naval construction and its objectives: new approaches, new tools), 20–21 September 2012.

Vallée-Poussin, Paloma de la. "Histoire des théories de stabilité des corps flottants 1727–1879" (History of the theories of the stability of floating bodies, 1727–1879). PhD diss., Louvain, Belgium, Université Catholique de Louvain, 2015.

Van Atta, Richard, ed. *Transformation and Transition: DARPA's Role in Fostering an Emerging Revolution in Military Affairs*. Paper P-3698. 2 vols. Alexandria, VA: Institute for Defense Analyses, 2003.

Vandersmissen, Hans. "Folkert van Loon, schakel tussen wetenschap en ambacht" (Folkert van Loon, the link between science and craft). *Spiegel der Zeilvaart* 2 (2004), pp. 43–47.

Varende, Jean de la. *Les Augustin-Normand: Sept générations de Constructeurs de Navires* (The Augustin-Normands: Seven generations of shipbuilders). Mayenne, France: Floch, 1960.

Vaughan, Adrian. *Isambard Kingdom Brunel: Engineering Knight-Errant*. London: Murray, 1991

Villain-Gandossi, Christiane, ed. *Deux siècles de constructions et chantiers navals, milieu XVIIe—milieu XIXe siècle* (Two centuries of naval shipyards and construction, mid-17th to mid-19th centuries). Paris: Éditions du CTHS, 2002.

Vincenti, Walter G. *What Engineers Know and How They Know It: Analytical Studies from Aeronautical Engineering*. Baltimore: Johns Hopkins University Press, 1990.

Vos, Ron de. *Nederlandse Clippers* (Dutch clippers). Haarlem, Netherlands: Van Wijnen, 2003

Wadia, Rittonjee Ardeshir. *The Bombay Dockyard and the Wadia Master Builders*. Bombay: R. A. Wadia, 1957.

Walker, Fred M. "The Capsize of the *Daphne* in 1883: The World's Worst Launching Tragedy and Its Consequences." In *Learning from Marine Incidents II International Conference*. London: RINA, 2002.

Walker, Fred M. *Ships and Shipbuilders: Pioneers of Design and Construction*. Barnsley, UK: Seaforth, 2010.

Walker, Fred M. *Song of the Clyde: A History of Clyde Shipbuilding*. Cambridge: P. Stephens, 1984.

Watson, Nigel. *Lloyd's Register: 250 Years of Service*. London: Lloyd's Register, 2010.

Watson, Peter. *The German Genius*. New York: Harper, 2010.

Wealleans, Anne. *Designing Liners: A History of Interior Design Afloat*. London: Routledge, 2006.

Weir, Gary. *Building the Kaiser's Navy: The Imperial Navy Office and German Industry in the von Tirpitz Era, 1890–1919*. Annapolis, MD: Naval Institute Press, 1992.

Weir, Gary. *Forged in War: The Naval-industrial Complex and American Submarine Construction, 1940–1961*. Washington, DC: Naval Historical Center, Department of the Navy, 1993.

Wend, Henry B. *Recovery and Restoration: US Foreign Policy and the Politics of Reconstruction of West Germany's Shipbuilding Industry, 1945–1955*. Westport, CT: Praeger, 2001.

Wenk, Edward, Jr. *Making Waves: Engineering, Politics, and the Social Management of Technology*. Urbana: University of Illinois Press, 1995.

Westwood, John N. *Russian Naval Construction, 1905–45*. London: Macmillan, 1994.

Wilensky, Harold L. "The Professionalization of Everyone?" *American Journal of Sociology* 70, no. 2 (1964), pp. 137–158.

Witthöft, Hans Jürgen. *Meyer Werft: Innovative Shipbuilding from Papenburg*. Hamburg, Germany: Koehler, 2005.

Witthöft, Hans Jürgen. *Tradition und Fortschritt: 125 Jahr Blohm + Voss* (Tradition and progress: 125 years Blohm and Voss). Hamburg, Germany: Koehler, 2002.

Wright, Carroll Q. "History of the Washington Navy Yard 1799–1921." Manuscript. 2 vols. Washington, DC: Naval History & Heritage Command library, 1921.

Wright, Thomas. "Ship Hydrodynamics 1710–1880." PhD diss., University of Manchester, 1983.

Yarrow, Lady Eleanor Barnes. *Alfred Yarrow, His Life and Work*. London: Edward Arnold, 1923.

Yener, Emir. *From the Sail to the Steam: Naval Modernization in the Ottoman, Russian, Chinese and Japanese Empires, 1830–1905*. Saarbrücken, Germany: Lambert Academic, 2010.

Zappas, Kelly. "Constance Tipper Cracks the Case of the Liberty Ships." *Journal of the Minerals, Metals and Materials Society* 67, no. 12 (2015), pp. 2774–2776.

Zorlu, Tuncay. *Innovation and Empire in Turkey: Sultan Selim III and the Modernisation of the Ottoman Navy*. London: Tauris Academic Studies, 2008.

# Index

Adams, John, 191–193
Admiralty coefficient, 110, 112
Aircraft aerodynamics and ship hydrodynamics, 128, 133, 136, 137, 188, 222, 247–255, 291
Airy, George Biddell, 83
American Bureau of Shipping, 149, 151, 153, 159, 174, 176, 248, 260, 305
American Towing Tank Conference (ATTC), 260
Amidships symbol, 189
Amsler integrators, 283, 285–286, 299
Andrews, Thomas, 215, 226, 232
*Archimedes*, 76, 129, 134
Archimedes screw, 71, 71–78, 132
*Arleigh Burke* (DDG 51), 296, 310
Association Technique Maritime et Aéronautique (ATMA), 222
Atherton, Charles, 109–110, 114
Atwood, George, 26, 98, 103, 218

Babbage, Charles, 4, 46, 97
Baker, George Stephen, 227, 236
Baker, Rowland, 297
Barillon, Émile-Georges, 207, 241
Barnaby, Nathaniel, 92, 103, 118, 122, 145, 192, 200, 202, 213
Barnes, Frederick Kynaston, 103
Barrallier, Jean-Louis, 28
Beaufoy, Margaretta, 230
Beaufoy, Mark, 13, 30–32, 43, 48–49, 111, 113, 114, 117, 119, 218, 230, 238

Bell, Henry, 64–65, 69
Bell, William, 98–99, 102
Benford, Harry Bell, ix, xviii, 230
Bentham, Samuel, 2, 41, 152
Bernoulli, Daniel, 61–62, 69, 71, 97
Bertin, Louis-Émile, 105, 153, 194, 195, 207, 217, 221, 240, 245
Biles, John Harvard, 145, 166, 179
Bomanjee Wadia, Jamsetjee, 34
Bonjean, Antoine Nicholas François, 26, 27, 153
Bonjean curves, 26, 143, 153
Bouguer, Pierre, 26, 30, 41, 43, 97, 111, 128–129, 138–139, 141, 182, 303
Boulton and Watt, 2, 63, 65, 85, 108, 109
Boundary layer theory, 210, 250–252
Bourgois, Siméon, 113–114, 132, 135
Bouyón, Honorato, 308–310
Brard, Roger, 207, 241
Brin, Benedetto, 197, 210, 237
Britain as world economic and military leader, x–xi, 2, 24, 105, 163
Britannia tubular railroad bridge, 5, 140
  and *Great Eastern*, 14–16, 84–85, 140, 145
British Association for the Advancement of Science (BAAS), 46–47, 50–52, 83, 102, 107, 110–111, 114, 123, 136, 140, 218, 233
British Merchant Shipping Act of 1854, 151
Brown, David Keith, xvii, 89, 117, 277, 285, 297
Brunel, Henry, 116, 118, 123

Brunel, Isambard Kingdom, xi, 1–22, 46, 67, 70, 97, 140, 144, 174, 218, 230
 and *Great Eastern*, 10–22, 24, 84, 96, 99–100
 and screw propeller testing, 9–10, 76–77, 129–131,136
 biography of, 2–7, 18–20
Brunel, Marc, 2–3, 27, 85
Bubnov, Ivan Grigoryevich, 209, 240
Bulbous bow, 264–276
Bulbous forefoot, 268–272
Bulkheads, watertight, 41, 87, 151–153, 156, 158, 175, 186
Bureau Veritas, 149–151, 153, 173, 248, 305
Bürkner, Hans, 184
Burrill, Leonard, 180–181, 253

Calkoen, Jan Frederick van Beeck, 42
*Captain* (HMS), 102–105, 116, 188, 179, 199–200, 202, 282
Cavé, François, 70, 78
Chapman, Fredrik Henrik af, 42, 189, 211, 280
*Charlotte Dundas*, 64–65, 69
Chatfield, Henry, 98, 212–213
Chicago World's Fair Columbian Exposition of 1893, 220–222
Circulation theory, 133, 252–253
Classification societies, 148–160, 161, 173, 175–178, 298
 and aircraft, 248
 and navies, 304–306
Claxton, Christopher, 7–9
*Clermont (North River Steamboat)*, 65
Clipper ships, xi, 51–52, 55, 150, 192, 220, 230
Cochrane, Edward Lull, 180
Cole, Henry, 13, 214
Coles, Cowper Phipps, 102–103, 199–200
Collins Line, 67
Computational fluid dynamics (CFD), 291–292
Computer-aided ship design and analysis, xii, 145, 178, 186, 211, 260, 277–278, 286–301
Conference on Hydromechanical Problems of Ship Propulsion, Hamburg 1932, 251–253

Construction Corps (US Navy), 202–205, 216
Containers (ISO shipping), 307
Cooley, Mortimer, 216, 244
Copper sheathing, 23, 41, 43–44, 46, 55, 81–82, 121, 218
Corpo del Genio Navale (Italy), 197–198, 210
Corps du Génie Maritime (France), 193–194, 196, 197, 200, 219, 309
Costs
 aircraft versus ship development, 306–307
 hull systems versus combat systems, 308
 of warships (*Warrior*, *Dreadnought*, modern destroyer), 90, 198, 306
Cousteau, Jacques, ix
Cramp shipyard, 80, 163, 167–170, 172, 175
Crimean War, 15, 80, 82, 83, 89, 90, 102, 197, 199
Cuerpo de Ingenieros de Marina (Spain), 193–196, 208, 308–309
Cunard Line, 9, 15, 67, 156
Cuniberti, Vittorio, 69, 198

Davidson, Kenneth Seymour Moorhead, 255, 260
Denny, William, 65, 171, 179, 226, 236, 286
Denny shipyard and model basin, 127, 128, 163–166, 170, 175, 227, 237, 246
Design spiral, xii, 293–299, 303
d'Eyncourt, Eustace Tennyson, 202
Diagonal framing, in wooden ships, 32–40
Dieudonné, Jean, 207, 254, 309
D'Oliveira, Rogério Silva Duarte Geral, xviii, 309–310
*Dreadnought*, 198–199, 267
Dupin, Pierre Charles François, 26–27, 40, 98, 194
Dupuy de Lôme, Stanislas-Charles-Henri-Laurent, 87–90, 112–114, 134, 194, 207, 230, 247
Durand, William Frederick, 247, 249

*Eastland*, 157
Edye, John, 45

Ericsson, John (Johan), 75–78, 102, 203, 211, 230, 233
Euler, Leonhard, 26, 41–43, 59–61, 69, 71, 97, 114, 128, 139
Evans, John Harvey, 293–295

Fairbairn, William, xi, 5–6, 12, 218
  and ship structural theory, 22, 85, 140–145, 150, 174
Fausto, Vettor, 59
Ferris, Theodore Ernest, 224
*Fighting Temeraire, The* (Turner painting), 66
Fincham, John, 98, 212–213
Fishbourne, Edmund Gardiner, 50, 122
Fisher, John Arbuthnot "Jacky," 198
Fitch, John, 57, 63, 108
Flachat, Eugène, 105–108, 113
Fox, Josiah, 37, 280, 281, 283
French Academy of Sciences, 2, 25, 61, 69, 71, 97, 107, 128, 193, 234
Friction line, turbulent, 250–252, 260–261
Froude, Robert Edmund, 123–127, 132–134, 196, 198–199, 234–235, 237, 267–268, 298
Froude, William, xi, 22, 132–134, 165–166, 170, 197, 199, 200, 217, 218, 219, 233, 236, 237, 238, 246, 256
  biography of, 18, 96–97, 123–125
  and *Great Eastern*, 18–20, 97–98
  and railroads, 5, 97
  and resistance and powering theory, 105–106, 114–123, 249–250, 268, 270
  and rolling, 97–105
  and screw propeller theory, 132–133, 252
Froude number, 96, 125–126
Fulton, Robert, 57, 64–67, 69, 71, 107

Galileo Galilei, 59, 113, 137–139
Garay, Blasco de, 60
Gerstner, František Josef, 101
Gibbons, Guy, 305
Gibbs, William Francis, 180, 223–224, 228, 232, 243, 270
Gilbert and Sullivan, 13, 122–123

Glavimans, Cornelis Jan, 40, 209
*Gloire*, 81, 90, 213
Gobert, Alexandre, 35–36, 78, 141
Goodall, Stanley, 175, 185, 202
Great Britain (nation). *See* Britain as world economic and military leader
*Great Britain* (ship), 9–10, 15, 21, 67, 76, 129–131, 136, 140, 230
*Great Eastern*, xi, 1, 3, 13–22, 24, 50, 67, 84, 85, 96–98, 102, 140, 145
Great Exhibition of 1851 (Crystal Palace), 13, 20, 53, 214, 219, 220, 258
*Great Western*, 7–10, 21, 67, 70
Great Western Railway, 4, 7, 8, 67, 97, 162
*Greyhound*, 96, 104, 118–123, 237
Griffiths, John Willis, 51–53, 192, 220
Grotius, Hugo, 157
Guppy, Thomas Richard, 7, 9

Havelock, Thomas Henry, 271
Herreshoff, Nathaniel, 54, 230
Hohlenberg, Henrik, 37–38
Hovgaard, William, 173, 175, 216
Hugo, Victor, 23–24, 55, 57, 79, 92
Humphreys, Joshua, 37, 40, 202
Humphreys, Samuel, 40, 202–203

Institution of Civil Engineers, 12, 218, 221
International Maritime Organization (IMO), 158
International Towing Tank Conference (ITTC), xiii, 246, 258, 260, 263, 273, 292
Inui, Takao, xvii, 271–276
*Iris* model, 235
Isherwood, Benjamin, 107, 203

*Jeune École*, 135, 197
*João Coutinho*, 309
Juan y Santacilia, Jorge, 26, 41, 42, 97, 138–139

Kármán, Theodore von, 251–252, 256
Kelvin, William Thomson, Lord, 84, 96, 104

Kempf, Günther, 211, 238, 251–252
Kipling, Rudyard, 92–93, 321n41
Korpus Korabel'nyh Inzhenerov (Russia), 197, 209
Krylov, Aleksey Nikolaevich, 197, 238, 240

Labrousse, Nicolas-Hippolyte, 134
Laird shipyard, 77, 80, 87, 103, 108
Lanchester, Frederick William, 133, 252
Lardner, Dionysius, 7–8, 13
Le Brun de Sainte Catherine, Jacques Balthazar, 27, 197
Lenthall, John, 203
Lerbs, Hermann, 211, 238, 253, 257, 289
Lloyd's Register, 149–156, 160, 219, 225, 248, 305
Longfellow, Henry Wadsworth, 50
*Lucy Ashton* trials, 236–237

Manby, Aaron, 68, 79, 84, 87
Maneuvering and control theory, 96, 127–129, 134–137, 219, 236, 237, 240, 241, 242, 244, 249, 253–255, 264
Marestier, Jean-Baptiste, 68
Masefield, John, 55
Maudslay engineering company, 2, 8, 68, 70, 130
May, William, 37–39
McKay, Donald, 52, 81, 230
Measured mile, 63, 108, 130
Mechanics' Institutes, 7, 212, 219
Meo, Giovanni (Jan, John) de, 258–259
Mérigon de Montgéry, Jacques Philippe, 85, 87
Merrifield, Charles, 113–116
Michell, John Henry, 271, 273
Model basins, xii, xiv, 96, 116–128, 134, 200, 204–205, 216, 223, 226, 228–229, 234–246, 247, 251–263, 264, 270, 273, 276, 287–288, 291, 294–296, 299, 301
Moseley, Henry, 5, 98–99, 102–103, 213

Napier, Charles, 79, 82
Napier, David, 5, 15, 64–65, 67

Napier, James Robert, 50, 110–111, 140, 165–166, 170
Napier, Robert, 15, 67, 163, 165–166, 170, 175
Naval architect, evolution of term, 192
Naval architecture
  consultancies and design firms, 223–225, 245
  definition of, ix–x
  journals and publications of, 24, 28–29, 50, 67, 103, 109, 135, 174, 186, 198, 210, 218–223, 233, 251, 345
  in popular culture, 191–192, 230–232
  professional societies of, 28–29, 32, 186, 218–223, 251, 274
  schools of, 27, 87, 92, 98, 111, 196, 203–204, 206–218, 219, 245
  in vertical integration of shipyards, x, 165–170
  women in, 160, 225–230
Naval constructors corps, 175, 192–205, 208, 211, 216, 219
Navier-Stokes equations, 119–121, 139, 291
Newcomen, Thomas, 62–63, 96
Niedermair, John Charles, 180, 185, 204
Nondimensional notation, 125–127, 196, 251, 254
*Normandie* (passenger liner), 197, 240, 271–272
Nowacki, Horst, xvii, 211, 277, 292
Nystrom, John William (Johan Vilhem), 211

Office of Naval Research (ONR), 263–264

Paddle wheels, 7, 9, 14, 15, 58, 60–66, 69–74, 76, 78, 79, 89, 112, 129, 131–132, 166–167, 220, 237
Parsons, Charles, 69, 233, 246
Patterson, William, 7–9
Plimsoll, Samuel, 152–155
Pollard and Dudebout, *Architecture navale*, 207
Popov, Andrei Alexandrovich, 197, 238
Prandtl, Ludwig, 210, 250, 252
Propellers. *See* Screw propellers
Purvis, Frank Prior, 123, 127, 165, 217, 246

# Index

Rahola Criteria, 182–184, 186, 304
Rahola Jr., Jaakko Juhani, 181–182, 188
Ram bows, 156, 264–269
Rankine, William John Macquorn, 104, 107, 110–111, 114, 124, 132–133, 140–141, 150, 165, 250, 252
Rankine streamlines, 110–112, 247–248
*Rattler* and *Alecto*, 76, 130–132
*Raven* and *Swan*, 115–116, 268, 270
Reech, Frédéric, 111–115, 207
Reed, Edward James, 103, 116, 118, 141, 145, 192, 199–200, 213, 219
Resistance and powering theory, 8, 12–13, 22, 25, 28–32, 41–54, 87, 95, 105–127, 166–167, 218, 229, 234–239, 243, 247–253, 259–261, 264, 267–276, 295, 298
Ressel, Josef, 73–77, 129
Rolling and dynamic stability, 18, 22, 95–106, 114, 118, 122, 127–128, 166, 178–188, 213, 218
Rotating-arm basin, 240, 254–255
Royal Corps of Naval Constructors (RCNC), 198–202, 215
Royal Institution of Naval Architects (RINA), 20, 22, 101, 102, 133, 192, 213, 221–223, 251
Royal Society of London, 13, 29, 40, 44, 46, 98, 111, 218, 233, 234, 236
Rudder theory. *See* Maneuvering and control theory
Ruskin, John, 23–24, 55, 57, 92
Russell, John Scott, xi, 10, 22, 102, 213–214, 221
  biography of, 11–13, 20
  and *Great Eastern*, 13–20, 85, 99, 145
  and wave-line theory, 12–13, 45–54, 101, 107, 108, 111, 113–115, 117, 218, 219, 271

Sadler, Herbert Charles, 216, 244
Safety of Life at Sea (SOLAS), 158, 179–180, 184, 304
Sailing ship effect, 24, 54–55, 81
Sané, Jacques-Noël, 25, 194, 196, 197
Sarchin and Goldberg Criteria, 186–188
Saunders, Harold Edward, 243–244
Sauvage, Pierre Louis Frédéric, 77
*Savannah*, 7, 66
Schank, John, 41, 43
Schoenherr, Karl Ernest, 251–252
Screw propellers, 1, 9, 14, 61–62, 65, 66, 69, 71–78, 95, 97, 129–132, 242, 257, 288, 290
  theory of, 95, 132–134, 252–253, 289, 292
Seppings, Robert, 38–40, 78, 218
Sewell, John, 28–32
Ship design process in the years 1800, 1900 and 2000 compared, 297–301
Ships' curves, 278–284, 298–299
Ship Structure Committee, 159–160, 260
Ship synthesis models, 278, 292–297, 299, 301
*Shipwright's Trade, The*, 80
Slade, Thomas, 199
Smith, Francis Pettit, 9, 74–75, 129
Smith correction, 143
Snodgrass, Gabriel, 34, 39, 78
Society for the Improvement of Naval Architecture, 28–32, 136, 218
Society of Naval Architects and Marine Engineers (SNAME), 186, 189, 222, 251, 274
Soetermeer, Cornelis, 27, 40
Splines, 278–282, 289, 298, 299
Spring styles (drawings), 298–299
Stability theory
  damaged, 152–153, 173, 178, 184–188, 290, 303–304
  intact, 26–30, 42, 97–98, 103–104, 148, 152, 164–170, 173, 178–188, 199, 218, 282–285, 289, 290, 299, 303–304
Standardization, benefits of, xii, 147–148, 162, 188–189
Standards, commercial versus naval shipbuilding, 173–188, 304–306
Stauffenberg, Claus von, 339n29
Steam engine development, 62–69
Steam engines
  compound, 68–69, 107, 147, 169, 200
  turbine, 69, 198, 217, 233
Steers, George, 53, 223

Stephenson, Robert, 4–5, 20, 53, 85
Stibolt, Ernst Vilhelm, 36, 37
Stosskopf, Jacques, 207, 309
Structural theory of ships
  balance on a wave, 140–145
  civil engineering heritage of, 9–17, 22, 84–85, 91, 140–145, 174–178, 218, 221, 278–280, 306
Structures
  composite wood/iron, 80–81, 150
  iron: early development of, 78–91; and comparison with wooden ship structures, 80–82; and reverse salients, 81–84
  steel: early development of, 91–92; and welding, 158–160
  wood, 32–41
Summers, Andrew Bradley, 310
Symington, William, 64, 69
Symonds, William, 45, 75, 212–213
Systems engineering, xii, 303–304, 306

Taurines, Auguste, 109, 131–132, 219
Taylor, David Watson, 204–205, 215, 216, 223, 224, 243, 247, 268–270, 295
Taylor, Frederick Winslow, 172
Tchebycheff method for numerical integration, 285–286
Testing tanks. *See* Model basins
Tideman, Bruno Joannes, 124–127, 196, 197, 209, 242, 249
Tipper, Constance, 160, 228–229
*Titanic*, 156–158, 179–180, 215, 226, 232
Tonello, Gaspare, 27, 171
Torquay model basin, 96, 116–127, 200
Towing tanks. *See* American Towing Tank Conference; International Towing Tank Conference; Model basins
Tredgold, Thomas, 85, 139
Triremes, 59, 90, 134, 173, 264–268
Troost, Laurens, 209, 242, 258–259

Verne, Jules, 20–21, 50
Vernon, John, 140–142

Vertical integration of shipyards, x, xii, 148, 161–172, 191–192, 206, 215
Vial du Clairbois, Honoré-Sébastien, 26, 98, 103, 218

*Warrior*, 84, 90, 145, 200, 213, 277
Waterjets, 61–63, 71
Watt, James, 2, 63, 96, 108
Watts, Isaac, 90, 103, 192, 199, 212
Watts, Phillip, 123, 198
Wave-line theory, 1, 12–13, 45–54, 101, 107, 108, 111, 113–115, 117, 218, 219, 271
Wave-line theory, and clipper ships, 50–52
Wave-line theory, and yacht *America*, 53–54
Wave theory, 48, 101–102
Waymouth, Bernard, 150
Webb, William Henry, 52, 216
Weinblum, Georg, 257, 271
Whirling arms, 43, 254–255
White, William Henry, 127, 141, 179, 200–202, 213, 215, 282
Wigley, Cyril, 271
*Wolf* structural trials, 145
Women in naval architecture, xii, xiv, 160, 225–230, 286
Woolley, Joseph, 213–214, 267–268
World War I, xi, 158–159, 171, 175, 184, 204, 209, 223, 224, 227, 251–252, 255–256, 309
World War II, xi, 158–159, 161, 175, 182, 184–185, 188, 197, 211, 217, 222, 224, 225, 227–229, 232, 233, 236, 237, 238, 240, 242–244, 246, 247, 254–258, 260, 261, 263, 271, 275, 286, 293, 295, 296, 309

*Yamashiro Maru*, 274–275
*Yamato*, 273
Young, Thomas, 40, 139
Yourkevitch, Vladimir Ivanovich, 197, 270–271

Zédé, Gustave, 247